Durum Wheat Products - Recent Advances

Durum Wheat Products - Recent Advances

Editor

Mike Sissons

MDPI • Basel • Beijing • Wuhan • Barcelona • Belgrade • Manchester • Tokyo • Cluj • Tianjin

Editor
Mike Sissons
Tamworth Agricultural
Institute
NSW Department of
Primary Industries
Calala
Australia

Editorial Office
MDPI
St. Alban-Anlage 66
4052 Basel, Switzerland

This is a reprint of articles from the Special Issue published online in the open access journal *Foods* (ISSN 2304-8158) (available at: www.mdpi.com/journal/foods/special_issues/Durum_Wheat_Products).

For citation purposes, cite each article independently as indicated on the article page online and as indicated below:

LastName, A.A.; LastName, B.B.; LastName, C.C. Article Title. *Journal Name* **Year**, *Volume Number*, Page Range.

ISBN 978-3-0365-9819-2 (Hbk)
ISBN 978-3-0365-9820-8 (PDF)

Contents

About the Editor

Mike Sissons

Mike Sissons is a Principal Research Scientist with the NSW Department of Primary Industries and is an Adjunct Professor with the University of New England. His research career spans cereal science covering barley malting and brewing, and durum wheat breeding and quality. His specialty is durum wheat science and technology where he has led many projects covering biotechnology, breeding, food science, and agronomic research resulting in the publication of 98 refereed papers and book chapters and was the editor in chief of the AACCI book "Durum wheat chemistry and technology, 2nd edition 2012". His research interests continue to broaden with a keen interest in functional foods, bran, and waste valorisation.

Preface to "Durum Wheat Products - Recent Advances"

Durum wheat is used around the world for a variety of food products, such as pasta, couscous, bulgur, bread, and cookies, to name a few. Developments in processing, breeding, new functional pasta, testing methods, and other technologies have been responsible for significant changes in durum-wheat-derived food products. The objective of this Special Issue is to summarize recent developments using a cross-disciplinary approach. A selection of peer reviewed papers covering this broad topic have been assembled. These will be of interest to anyone in the durum wheat supply chain, researchers, students, and the farming community.

Mike Sissons
Editor

foods

Editorial

Durum Wheat Products—Recent Advances

Mike Sissons

NSW Department of Primary Industries, Tamworth Agricultural Institute, 4 Marsden Park Road, Calala, NSW 2340, Australia; mike.sissons@dpi.nsw.gov.au

Citation: Sissons, M. Durum Wheat Products—Recent Advances. *Foods* **2022**, *11*, 3660. https://doi.org/10.3390/foods11223660

Received: 26 October 2022
Accepted: 8 November 2022
Published: 16 November 2022

Durum wheat is widely used in various products, including long and short dried pasta, fresh and sheeted pasta, couscous, bulgur and baked bread. The quality of the raw material, durum wheat, is critical for ensuring that these products meet their specifications. While the genotype chosen and environmental conditions during crop development are important for determining the final grain quality, its processing by milling, sizing or mixing and the end-product manufacture are important processes affecting the end product's quality. The key features of grains that strongly influence the yield are a good kernel size, high test weight, and high percentage of hard vitreous kernels, with minimal contaminants and no surface discoloration. While large grains provide more semolina, kernel-size uniformity is very important in the milling process. Wang et al. [1] evaluated the influence of kernel size and its potential interaction with genotype on key quality traits of durum wheat. Genotype was shown to have a strong impact on the test weight of small kernels and the milling yields of medium and large kernels. Millers are now striving to increase productivity, ensure food safety, and enhance sustainability efforts; for example, three studies on the better use of lower-value milling wastes (bran and germ) demonstrate the use of durum oil in focaccia to improve its resistance to oxidation [2], its use in biscuit making to decrease oxidative phenomena and increase bioactives [3], and its use to improve the stabilization of germ, to exploit the nutrients and bioactive compounds within wheat germ [4]. The key indicators of milling quality are the yields (total and semolina), ash content, and speck counts in the finished granular product. Recent milling developments in processing equipment and digital applications to improve quality and sustainability are reviewed by Sarkar and Fu [5]. Other durum wheat products are also considered, such as couscous, a product prepared from durum wheat semolina that agglomerates upon the addition of water and undergoes physical and thermal treatment. Its history, its manufacture on traditional and industrial scales, and its consumption in traditional and modern ways are reviewed by Hammami et al. to supplement the lack of scientific and technical descriptions for couscous [6]. Pasta makers require high-quality semolina meeting their industrial requirements, so methods to ascertain semolina quality are critical. The main factors determining the technological quality of semolina and approaches used for evaluating gluten quality are reviewed by Cecchini et al. [7]. There is a lack of standard tests across the industry; the tests have historically been adapted from methods for evaluating bread wheat and largely based on laboratory rather than industrial evaluation, so a better test to determine the behaviour of durum wheat semolina and the cooking quality of its corresponding pasta needs to be developed. Producing pasta while mostly using a few ingredients such as semolina, water and, perhaps, eggs is a complex process with many process variables. A change in a single variable, such as the type of raw material (refined vs. wholegrain semolina), can affect the entire process and product quality. Understanding the relationship between the processing variables and pasta quality is essential in "redesigning" the process when alternative raw materials (i.e., ingredients other than durum wheat semolina) are used [8]. An example is optimising the drying process for the manufacture of bulgur from grains of different quality to optimise the phytonutrient content [9].

There is a new trend in food consumption, especially in well-developed economies. Many consumers are increasingly interested in food that provides a benefit in preventing or reducing nutrition-related diseases, so-called "lifestyle or civilisation" diseases such as cancer, diabetes, cardiovascular disease, and obesity. These diseases afflict a large percentage of the population of Westernised countries, with the trend continuing to worsen in developing nations, and are the main non-communicable causes of death. Manufacturers are responding by trying to improve nutritional value or create presumed health benefits. Over the past decade or two, many studies have largely focused on a partial substitution of semolina or flour to create pasta products with improved nutritional value, such as higher fibre, better protein quality, and enrichment in essential fatty acids. However, few of these novel pasta products have shown clinical benefits. Sissons provides an update on durum-derived pasta products with proposed health benefits [10], while Di Pede et al. [11] provide an overview of the glycaemic indices of different pasta formulations. A new focus on algae as a food ingredient has been growing in Western diets. Macroalgae or seaweeds are low-calorie ingredients that are high in protein and iodine, low in fat, and a source of hydrocolloids, minerals, vitamins, and bioactive compounds (antioxidants). *Fucus vesiculosus*, a brown macroalga, was added to a pasta recipe, but, as is often the case, even relatively low incorporation impaired the quality of the pasta [12]. The issue with limited incorporation, which can limit protein and fibre enrichment, was investigated by Martín-Esparza et al. [13] in fresh tiger-nut-based tagliatelle using hydrocolloids as a structural agent to improve the textural characteristics, cooking properties, and fibre loss. Another approach to issues regarding substitution limits is described in a report on pasta enriched with encapsulated carrot waste; this waste is rich in carotenoids, and the encapsulation protects the pigments from thermal degradation. This resulted in pasta with significantly improved protein, fat, and ash contents [14]. An alternative approach to substituting semolina to create novel health-promoting pasta is the genetic manipulation of the starch and protein composition and modification of the hardness of durum wheat in order to improve its nutritional and technological value and expand its utilization for application to a wider range of end products. Increasing the amylose content, moving glutenin genes from bread wheat into durum to improve its bread-making performance, and making the kernel softer for use in biscuits, pizza, and other foods are the approaches reviewed by Lafiandra et al. [15].

Funding: This research received no external funding.

Conflicts of Interest: The author declares no conflict of interest.

References

1. Wang, K.; Taylor, D.; Chen, Y.; Suchy, J.; Fu, B.X. Effect of Kernel Size and Its Potential Interaction with Genotype on Key Quality Traits of Durum Wheat. *Foods* **2021**, *10*, 2992. [CrossRef] [PubMed]
2. Vurro, F.; Summo, C.; Squeo, G.; Caponio, F.; Pasqualone, A. The Use of Durum Wheat Oil in the Preparation of Focaccia: Effects on the Oxidative Stability and Physical and Sensorial Properties. *Foods* **2022**, *11*, 2679. [CrossRef] [PubMed]
3. Vurro, F.; Greco Miani, M.; Summo, C.; Caponio, F.; Pasqualone, A. Effect of Durum Wheat Oil on the Physico-Chemical and Sensory Features of Biscuits. *Foods* **2022**, *11*, 1282. [CrossRef] [PubMed]
4. Marzocchi, S.; Caboni, M.F.; Greco Miani, M.; Pasini, F. Wheat Germ and Lipid Oxidation: An Open Issue. *Foods* **2022**, *11*, 1032. [CrossRef] [PubMed]
5. Sarkar, A.; Fu, B.X. Impact of Quality Improvement and Milling Innovations on Durum Wheat and End Products. *Foods* **2022**, *11*, 1796. [CrossRef] [PubMed]
6. Hammami, R.; Barbar, R.; Laurent, M.; Cuq, B. Durum Wheat Couscous Grains: An Ethnic Mediterranean Food at the Interface of Traditional Domestic Preparation and Industrial Manufacturing. *Foods* **2022**, *11*, 902. [CrossRef] [PubMed]
7. Cecchini, C.; Bresciani, A.; Menesatti, P.; Pagani, M.A.; Marti, A. Assessing the Rheological Properties of Durum Wheat Semolina: A Review. *Foods* **2021**, *10*, 2947. [CrossRef] [PubMed]
8. Bresciani, A.; Pagani, M.A.; Marti, A. Pasta-Making Process: A Narrative Review on the Relation between Process Variables and Pasta Quality. *Foods* **2022**, *11*, 256. [CrossRef] [PubMed]
9. Dorra, S.T.; Farah, D.; Nesrine, H.; Wafa, A.; Youkabed, Z. Drying Behavior of Bulgur and Its Effect on Phytochemical Content. *Foods* **2022**, *11*, 1062. [CrossRef] [PubMed]
10. Sissons, M. Development of Novel Pasta Products with Evidence Based Impacts on Health—A Review. *Foods* **2022**, *11*, 123. [CrossRef] [PubMed]

11. Di Pede, G.; Dodi, R.; Scarpa, C.; Brighenti, F.; Dall'Asta, M.; Scazzina, F. Glycemic Index Values of Pasta Products: An Overview. *Foods* **2021**, *10*, 2541. [CrossRef] [PubMed]
12. Ribeiro, A.R.; Madeira, T.; Botelho, G.; Martins, D.; Ferreira, R.M.; Silva, A.M.S.; Cardoso, S.M.; Costa, R. Brown Algae Fucus vesiculosus in Pasta: Effects on Textural Quality, Cooking Properties, and Sensorial Traits. *Foods* **2022**, *11*, 1561. [CrossRef] [PubMed]
13. Martín-Esparza, M.E.; Raigón, M.D.; García-Martínez, M.D.; Albors, A. Role of Hydrocolloids in the Structure, Cooking, and Nutritional Properties of Fiber-Enriched, Fresh Egg Pasta Based on Tiger Nut Flour and Durum Wheat Semolina. *Foods* **2021**, *10*, 2510. [CrossRef] [PubMed]
14. Šeregelj, V.; Škrobot, D.; Kojić, J.; Pezo, L.; Šovljanski, O.; Tumbas Šaponjac, V.; Vulić, J.; Hidalgo, A.; Brandolini, A.; Čanadanović-Brunet, J.; et al. Quality and Sensory Profile of Durum Wheat Pasta Enriched with Carrot Waste Encapsulates. *Foods* **2022**, *11*, 1130. [CrossRef]
15. Lafiandra, D.; Sestili, F.; Sissons, M.; Kiszonas, A.; Morris, C.F. Increasing the Versatility of Durum Wheat through Modifications of Protein and Starch Composition and Grain Hardness. *Foods* **2022**, *11*, 1532. [CrossRef]

foods

MDPI

Article

Effect of Kernel Size and Its Potential Interaction with Genotype on Key Quality Traits of Durum Wheat

Kun Wang, Dale Taylor, Yuming Chen, Jerry Suchy and Bin Xiao Fu *

Grain Research Laboratory, Canadian Grain Commission, 303 Main Street, Winnipeg, MB R3C 3G8, Canada; kun.wang@grainscanada.gc.ca (K.W.); dale.taylor@grainscanada.gc.ca (D.T.); yuming.chen@grainscanada.gc.ca (Y.C.); jerry.suchy@grainscanada.gc.ca (J.S.)

* Correspondence: binxiao.fu@grainscanada.gc.ca; Tel.: +1-204-984-5605

Abstract: This study was conducted to evaluate the influence of kernel size and its potential interaction with genotype on durum wheat quality with emphases on kernel physical characteristics, milling performance, and color-related quality parameters. Wheat samples of seven genotypes, selected from the 2018 Canadian durum variety registration trial, were segregated into large (LK), medium (MK), and small-sized kernels (SK). In general, the kernel size greatly affected the durum wheat milling performance. Within a given size fraction, a strong impact of genotype was shown on the test weight of SK and the milling yields of MK and LK. Particularly, the MK fraction, segregated from the genotypes with superior milling quality, had a higher semolina yield than LK from the genotypes of inferior milling quality, inferring the importance of intrinsic physicochemical properties of durum kernels in affecting milling quality. SK exhibited inferior milling quality regardless of the genotypes selected. A strong impact of genotype was shown for the total yellow pigment (TYP) content and yellowness of semolina, while the kernel size had a significant impact on the brightness and redness of the semolina and pasta. Despite SK possessing much higher TYP, the semolina and pasta prepared from SK were lower in brightness and yellowness but with elevated redness.

Keywords: durum wheat; kernel size; genotype; milling quality; semolina quality; pasta color

Citation: Wang, K.; Taylor, D.; Chen, Y.; Suchy, J.; Fu, B.X. Effect of Kernel Size and Its Potential Interaction with Genotype on Key Quality Traits of Durum Wheat. *Foods* **2021**, *10*, 2992. https://doi.org/10.3390/foods10122992

Academic Editor: Mike Sissons

Received: 21 September 2021
Accepted: 2 December 2021
Published: 4 December 2021

1. Introduction

Durum wheat physical properties are very important in determining its commercial value. Strong associations have been reported between kernel physical characteristics and durum wheat milling performance, semolina composition, and pasta processing quality [1–6]. Emphasis has been on unveiling the relationship between test weight (TWT) to durum wheat milling potential by evaluating samples with a wide range of TWT, protein content, and kernel size distribution (KSD) [3–5]. Recent study in our laboratory has shown that kernel size is more effective than TWT in predicting the milling performance of durum wheat by assessing Canadian durum samples with a wide range of TWT and KSD [5].

In general, with the decrease of kernel size from large to medium, the semolina and total milling yields of durum wheat reduced gradually. A drastic decrease in milling quality was observed for small kernels passing through the no. 6 slotted sieve (2.38 mm aperture) [4,5] with much reduced milling yields coupled with elevated ash content. Baasandorj, Ohm, Manthey, and Simsek (2015) studied the impact of kernel size and mill type on the milling and baking quality of hard red spring wheat [7]. Compared with large-sized kernels, the small-sized kernels had a much lower flour yield because of the lower proportion of starchy endosperm to bran.

The kernel size of durum wheat can significantly affect not only the milling performance but also the semolina and pasta quality [3,5]. Semolina milled from SK exhibited higher protein content, finer granulation, and was higher in TYP but less bright in color with elevated ash [3,5]. Cooked pasta made from durum samples with a high proportion of SK had higher firmness but was duller in color.

While the impact of kernel size on semolina and pasta quality is well-documented, limited information is available on the response of genotype to the general relationships between kernel size and the key durum wheat quality parameters. Due to the variation in intrinsic quality, the degree of impact of kernel size on quality could be genotype dependent. Using milling performance as an example, it is not clear if the genotypes with superior milling quality would be less susceptible to kernel size variation than those of inferior milling quality, or vice versa. Genotypes with different intrinsic quality could respond differently to variations in kernel size.

On the other hand, differences in quality among genotypes could be affected by the variation in kernel size. Although TKW was shown to be highly correlated with semolina yield across four different durum varieties (R^2 = 0.92) evaluated by Wang and Fu (2020), greater variation in semolina yield was seen for larger kernels than for smaller ones [5]. The fact that the genotypic variation in durum milling performance was related to kernel size suggests a potentially greater role of genotype in the milling quality of large kernels than that of the small ones.

With the prevalence of hot and dry growing conditions on Canadian prairies in the last few years, some durum samples, although graded as No.1 or No. 2 Canada Western Amber Durum (CWAD), showed relatively wide range of KSD and milling quality [5]. To optimize the commercial value of durum wheat of different KSD and understand how quality parameters respond to kernel size variations, a thorough investigation is required to further elucidate the combined effect of kernel size and genotype on key durum wheat quality parameters.

Therefore, the objective of this study was to evaluate the influence of kernel size and its potential interaction with genotype on key durum wheat quality traits with emphases on the wheat physical properties, milling performance, and color-related quality attributes.

2. Materials and Methods

2.1. Wheat Samples

Seven genotypes were selected from the 2018 Canadian durum wheat variety registration trial based on their intrinsic differences in milling and color-related quality parameters. A composite of each genotype was prepared from wheat samples grown at nine locations across western Canada. Based on availability and grading information of wheat samples from the nine locations, a recipe was developed for the preparation of the wheat composites. All composites were graded as No.1 CWAD. Each of these variety composites was segregated into three size fractions using a Carter dockage tester (Simon-Day Ltd., Winnipeg, MB, USA) equipped with no. 6 (2.38 mm × 19.05 mm) and no. 7 (2.78 mm × 19.05 mm) slotted sieves. The segregated kernel size fractions were categorized as small-sized kernels (SK, through no.6 slotted sieve), medium-sized kernels (MK, passing no.7 but remained above no.6 slotted sieve), and large-sized kernels (LK, remained above no.7 slotted sieve).

2.2. Wheat Physical Properties

To accommodate the small sample size, the test weight (TWT) was measured using a 0.5 L container equipped with a cox funnel following the standard procedure described by the Canadian Grain Commission [8]. The value in gram per half liter was converted to kg per hectoliter using the test weight conversion chart for amber durum wheat. TKW was determined with an electronic seed counter (Model 750, The Old Mill Company, Savage, Maryland) using a 20 g sample of wheat of which all broken kernels were manually removed. KSD was determined on a series of slotted sieves (i.e., no. 6, 7, and 8). One hundred grams of wheat was subsampled and manually shaken for 30 s, after which the four fractions separated by the sieves were collected and weighted individually. All wheat physical tests were conducted in duplicate.

2.3. Standard Durum Milling Procedure

Following the mill flow previously described by Dexter et al. (1990) [9], original unsorted wheat samples were milled into semolina in duplicates of 2.3 kg lots with a four stand Allis-Chalmers laboratory mill (West Allis, WI, USA) in conjunction with a laboratory purifier. The mill room was controlled at 21 °C and 60% relative humidity. Semolina is defined as having less than 3% pass through a 149 μm sieve. The total milling yield is the combination of semolina and flour. Both the total and semolina yields are reported as a percentage of the cleaned wheat on a constant moisture basis. Semolina granules were prepared by adding the most refined flour stream(s) to semolina until 70% extraction was reached for quality analysis.

2.4. Micro-Milling and Purification Protocol

Wheat samples of various size fractions were milled to predict semolina and total milling yields following the micro-milling procedure previously developed by Wang et al. (2019) [10]. After tempering to a moisture content of 16% overnight, 200 g of wheat sample was ground with a Quadruma Junior (QJ)-II-G mill-semolina version (C.W. Brabender Instruments, Inc., South Hackensack, NJ, USA) with the original sifter removed. The resulting wholemeal was sifted through a universal laboratory sifter (Bühler MLUA GM sieve, Bühler AG) equipped with a bottom screen of 180 μm to remove the flour and a top screen of 630 μm to retain the bran-rich fraction. The unpurified semolina fraction (SY1) between the two screens was collected. Based on the prediction models developed by Wang et al. (2019) [10], semolina yield and total milling yield were calculated according to the amount of SY1 and bran-rich fraction. Formulas (1) and (2) are as follows:

$$\text{Semolina Yield (\%)} = 1.02 \times \text{Bran-rich fraction} + 1.80 \times \text{SY1} - 73.17. \quad (1)$$

$$\text{Total Milling Yield (\%)} = 0.62 \times \text{SY1} + 39.42 \quad (2)$$

To prepare refined semolina for analysis and pasta processing, the original purification steps described by Dexter et al. [9] were modified to accommodate the small semolina sample size with three purification and two sizing passages. A detailed description of the micro-milling and purification steps is illustrated in Figure 1. In a typical experiment, SY1 obtained from QJ semolina mill was passed over a laboratory purifier (Namad, Rome, Italy) equipped with four different sizing sieves (335, 425, 570, and 670μm). After the first purification (P1), large semolina granules collected in tray 4 and 5 were reduced with the first sizing roll (S1). The reduced semolina was sifted through a box sifter equipped with a 180 μm sieve for 30 s to remove the flour. The resulting fraction retained above the 180 μm sieve together with the semolina collected in tray 3 at P1 were subject to a second purification (P2). After P2, the semolina granules which remained in tray 4 and 5 were subject to a second sizing step (S2). The reduced fraction was sifted with a box sifter for 30 s to remove bran/shorts (>425 μm) and flour (<180 μm). The semolina fraction between 180 and 425 μm was combined with the semolina collected in tray 3 at P2 and transferred to the third purification (P3). Refined semolina was collected as tray 1 and 2 in P1, tray 1 and 2 in P2 and tray 1, 2, and 3 in P3. Tray 4 and 5 in P3 were defined as Feeds.

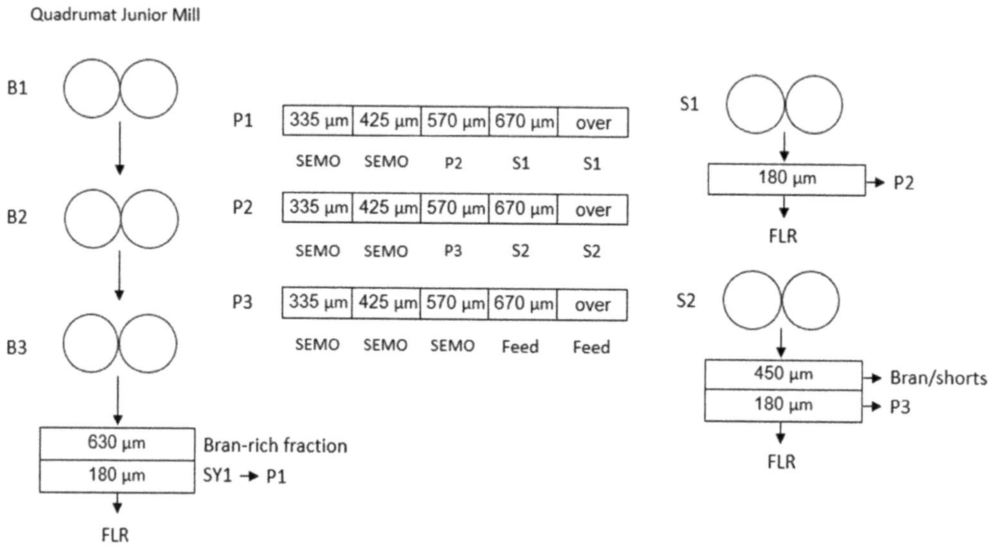

Figure 1. Durum micro-milling flow and purification procedure. B = break passage, FLR = flour, P = purifier, S = sizing passage, SEMO = semolina, SY1: unpurified semolina.

2.5. Semolina Quality Testing

The protein content of the whole wheat and semolina were measured following the method previously described by Williams et al. [11] with a LECO Truspec N CNA (combustion nitrogen analysis) analyzer (Saint Joseph, MI). Ground wheat meal was prepared using a Retsch ZM 200 mill (Retsch GmbH, Haan, Germany) equipped with a 0.5 mm screen (Trapezoid holes) at a speed of 14,000 rpm. Ash content, wet gluten, and gluten index were determined using AACC International approved methods 76-31.01 and 38-12.02, respectively [12]. Semolina color was measured with a Minolta colorimeter CR-410 (Konica Minolta Sensing, Inc., Tokyo, Japan) with a D65 illuminant. Color readings are expressed on the CIELAB color space system with L*, a* and b* parameters representing brightness, redness, and yellowness values, respectively. A micro scale rapid extraction procedure as described by Fu et al. [13] was used for the determination of the total yellow pigment (TYP) content of the semolina.

2.6. Spaghetti Processing and Color Measurement

Spaghetti were produced from semolina using a customized micro-extruder (Randcastle Extrusion Systems Inc., Cedar Grove, NJ, USA) following the method of Fu et al. [6]. Semolina was first mixed with water in a high-speed asymmetric centrifugal mixer (DAC 400 FVZ SpeedMixer, FlackTec, Landum, SC, USA) at water absorption of 31–32% to maintain a constant extrusion pressure of about 100 psi. Vacuum was applied to eliminate introduction of air bubbles and minimize oxidative degradation of the yellow pigment, after which the dough crumbs were extruded through a four-hole Teflon coated spaghetti die (1.8 mm). The fresh pasta was subsequently dried in a pilot pasta dryer (Bühler, Uzwil, Switzerland) coupled with a 325 min drying cycle and a maximum temperature of 85 °C. To measure spaghetti color, 6.5 cm bands of spaghetti strands were mounted on a white mat board with minimum interspace. Spaghetti color was determined using a Minolta colorimeter (CR-410) as described above.

2.7. Statistical Analysis

All data were analyzed with Microsoft Excel and SAS 9.4 Software (SAS Institute Inc., Gary, NC, USA). A 3 × 7 factorial experiment was applied to evaluate the impact of kernel size and genotype on key durum wheat quality characteristics by including 3 levels of kernel size (small, medium, and large) and 7 different genotypes (A to G) representing the major source of variations. Each segregated kernel size fraction from a selected genotype was treated as an independent sample. Significance of each factor as indicated by F values and percentage of variability assignable to each factor as measured by the ratio of sum of square to the total sum of squares was calculated. Tukey's test, which followed the analysis of variance, indicated significant differences with a level of $p < 0.05$.

3. Results and Discussion

3.1. Influence of Kernel Size and Genotype on Physical Properties of Durum Wheat

To understand the impact of kernel size, genotype, and their interactions on major durum wheat quality parameters, seven durum genotypes with variation in milling and color related quality attributes were segregated into three kernel size fractions using a Carter dockage tester. The wheat and semolina quality parameters of the unsorted samples are summarized in Table 1. The selected genotypes differed greatly in semolina and total milling yields, TYP, and gluten index, but with less variation in wheat physical properties (i.e., HVK, TWT, TKW, KSD), wheat protein, and ash contents. The semolina and total milling yields from the micro-milling procedure were comparable to those of standard laboratory milling except genotype D which showed higher semolina and total milling yields in the micro-milling process.

The significance of kernel size, genotype, and their interactions on major durum wheat quality parameters, as measured by the F value and percentage of variability assignable to each factor and their interactions, are summarized in Table 2. Significant impact was found for kernel size, genotype, and their interactions on all wheat quality parameters examined ($p < 0.001$). In terms of wheat physical properties, kernel size accounted for more than 80% of the variability in TWT and TKW with minor influences shown for genotypes and their interactions. Table 3 summarizes the impact of genotype on key quality parameters in relation to kernel size. TKW reduced drastically from 51.0 ± 1.8 g of LK to 36.1 ± 0.9 g of MK, but was only accompanied by a small decrease of TWT from 83.7 ± 0.7 kg/hL to 82.2 ± 0.6 kg/hL. Further decrease of kernel size from MK to SK led to a much greater reduction in average TWT from 82.2 kg/hL to 77.6 kg/hL, suggesting SK (TKW of 23.9 ± 0.4 g) was much less dense than the corresponding larger ones. A similar decrease in TKW and TWT was reported when a bulk CWAD cargo aggregate was fractioned into five different kernel sizes [5]. Wang and Fu reported that TWT is less effective than TKW in distinguishing the difference in kernel size [5].

Interestingly, the impact of genotype on TWT was greater for SK than for both MK and LK (Table 3). Although there was no significant difference in TKW of the SK fractions, SK possessed much greater variability in TWT, ranging from 74.5 to 80.6 kg/hL (F value = 465.6, $p < 0.001$) as compared to MK (81.1–82.6 kg/hL, F value = 73.3, $p < 0.001$) and LK (82.3 to 84.5 kg/hL, F value = 130, $p < 0.001$). On the other hand, greater variation in TKW among genotypes was shown for LK (48.3 to 52.8 g, F value = 23.81, $p < 0.001$) in comparison to MK (34.8–37.0 g, F value = 4.5, $p < 0.05$) and SK (23.2 to 24.4 g, F value = 1.9, ns).

Table 1. Wheat and semolina quality parameters of selected genotypes.

Line	Wheat Properties									Standard Milling		Micro-Milling		Semolina Properties-Standard Milling			
	Grade	HVK (%)	TWT (kg/hL)	TKW (g)	KSD (%)			Protein (%)	Ash (%)	SY (%)	TMY (%)	SY (%)	TMY (%)	Protein (%)	Ash (%)	TYP (ppm)	Gluten index(%)
					Small	Medium	Large										
A	1	94	83.6	48.0	5.8	22.3	71.9	14.3	1.40	68.1	76.2	68.1	75.7	13.3	0.65	10.3	62
B	1	92	83.1	44.8	5.8	29.3	64.9	14.5	1.37	68.0	76.0	68.0	75.1	13.4	0.65	9.0	70
C	1	92	82.8	45.7	6.0	24.6	69.4	15.2	1.38	67.2	75.3	67.8	75.6	13.1	0.61	10.7	85
D	1	91	82.8	43.2	7.6	33.2	59.2	14.3	1.47	66.3	74.7	67.8	75.0	14.2	0.62	10.3	77
E	1	95	83.0	46.6	4.9	26.3	68.9	14.9	1.39	66.9	75.6	67.0	74.7	13.9	0.62	9.4	66
F	1	93	81.5	47.7	3.5	22.4	74.2	15.5	1.45	66.1	73.4	67.0	74.2	14.4	0.63	10.8	57
G	1	96	83.0	44.7	7.4	32.3	60.3	14.7	1.44	65.2	73.4	65.7	73.4	13.7	0.69	11.5	56
Mean		93.3	82.8	45.8	5.9	27.2	67.0	14.8	1.42	66.8	74.9	67.3	74.8	13.8	0.64	10.3	69
Range		5	2.1	4.8	4.1	10.9	15.0	1.2	0.10	2.9	2.8	2.4	2.3	1.3	0.08	2.5	29

Note: small kernels: through No.6 slotted sieve (2.38 × 19.05 mm); medium kernels: passing No.7 but remained above No.6 slotted sieve; large kernels: remained above No.7 slotted sieve (2.78 × 19.05 mm). HVK = hard vitreous kernels; TWT = test weight; TKW = thousand kernel weight; KSD = kernel size distribution; SY = semolina yield; TMY = total milling yield; TYP = total yellow pigment.

Table 2. Significance of kernel size, genotype, and their interaction on major durum wheat quality parameters as measured by F values and percentage of variability assignable to each factor and their interaction.

	F Values			Percentage of Variability Assignable to Each Factor			
	Kernel Size	Genotype	Interactions	Kernel Size	Genotype	Interactions	Error
Wheat properties							
TWT	502 ****	215.2 ****	66.8 ****	82.6	10.6	6.6	0.2
TKW	9365.2 ****	8.3 ***	11.2 ****	98.9	0.3	0.7	0.1
Protein	555.5 ****	2159.1 ****	68.9 ****	7.4	86.6	5.5	0.4
Ash	800.9 ****	281.2 ****	33.9 ****	42.6	44.9	10.8	1.7
Milling quality							
Semolina yield	13177.7 ****	546.9 ****	36.0 ****	87.6	10.9	1.4	0.1
Total milling yield	7392.8 ****	531.9 ****	24.4 ****	80.8	17.4	1.6	0.1
Semolina ash	2143.3 ****	131.0 ****	30.6 ****	77.9	14.3	6.7	1.1
Semolina color							
Semolina TYP	2897.9 ****	1276.6 ****	24.5 ****	42.1	55.6	2.1	0.2
L*	608.8 ****	10.4 ****	9.6 ****	84.7	4.3	8.0	2.9
a*	238.2 ****	5.7 ***	6.6 ****	75.5	5.4	12.5	6.7
b*	96.0 ****	177.8 ****	13.5 ****	13.1	73.0	11.0	2.9

Table 2. *Cont.*

	F Values			Percentage of Variability Assignable to Each Factor			
	Kernel Size	Genotype	Interactions	Kernel Size	Genotype	Interactions	Error
Pasta color							
L*	18998 ****	184.9 ****	293.1 ****	89.1	2.6	8.1	0.1
a*	30824.8 ****	452.5 ****	150.0 ****	93.1	4.0	2.7	0.1
b*	9625.5 ****	406.4 ****	59.0 ****	85.9	10.9	3.2	0.1

, * indicate significance level of 0.001 and 0.0001. TWT = test weight; TKW = thousand kernel weight; TYP = total yellow pigment.

Table 3. Impact of genotype on wheat, milling, semolina, and pasta quality parameters of selected genotypes at three different level of kernel sizes.

	Wheat Properties			Milling Quality					Semolina Color			Pasta Color		
	TWT (kg/hL)	TKW (g)	Protein (%)	Ash (%)	SY (%)	TMY (%)	Semolina Ash (%)	Semolina TYP	L*	a*	b*	L*	a*	b*
							Large kernels							
Mean	83.7	51.0	14.8	1.41	68.0	75.3	0.57	9.4	83.7	−2.7	33.5	74.5	3.2	65.4
SD	0.7	1.8	0.5	0.04	0.9	0.9	0.02	0.8	0.2	0.1	1.1	1.3	0.4	1.5
Range	2.2	4.5	1.3	0.11	2.7	2.5	0.06	2.3	0.5	0.2	3.1	3.9	1.1	4.0
F values	130 ****	23.81 ***	1343.5 ****	130.6 ****	203.7 ****	220.4 ****	52. 7****	511 ****	9.27 ***	5.7 **	49.5 ***	658 ****	299 ****	223 ****
							Medium kernels							
Average	82.2	36.1	14.6	1.41	66.7	74.0	0.60	10.6	83.4	−2.7	34.5	72.6	4.6	66.2
SD	0.6	0.9	0.5	0.04	0.7	0.7	0.02	0.8	0.2	0.1	1.0	0.4	0.3	1.4
Range	1.5	2.2	1.3	0.08	1.8	1.9	0.05	2.5	0.5	0.3	2.8	1.1	0.9	3.3
F values	73.3 ****	4.5 *	1100.1 ****	96.0 ****	170.6 ****	160.8 ****	73.2 ****	1464 ****	11.5 ***	11.5 ***	68 ***	56 ****	171 ****	205 ****
							Small kernels							
Average	77.6	23.9	14.9	1.49	63.6	71.9	0.67	11.1	81.7	−2.1	33.9	69.4	6.5	59.4
SD	2.0	0.4	0.4	0.05	0.5	0.6	0.03	1.0	0.6	0.2	1.3	0.4	0.4	1.0
Range	6.1	1.2	1.2	0.15	1.3	1.5	0.07	3.0	1.7	0.6	4.1	1.1	1.3	2.3
F values	465.6 ****	1.9 ns	386.4 ****	121.0 ****	128.6 ****	106.1 ****	70.3 ****	2149 ****	17.3 ****	9 ***	92 ***	57 ****	282 ****	96 ****

*, **, ***, **** indicate significance level of 0.05, 0.01, 0.001 and 0.0001, respectively. ns = not significant; SD = standard deviation; TWT = test weight; TKW = thousand kernel weight; SY = semolina yield; TMY = total milling yield; TYP = total yellow pigment.

TWT can be affected by wheat moisture, kernel density, kernel shape, and packing factors, which were not directly associated with milling yield [14–18]. Simmons and Meredith attributed the difference in TWT to bran surface roughness, distribution of kernel size, shape, volume, and kernel density [19]. Troccoli and di Fonzo found that kernel shape such as rectangular aspect ratio (kernel width/kernel length) and circularity shape factor ($4\pi \times$ area/perimeter^2) were positively related to TWT [20]. More recently, Wang and Fu reported that durum wheat with a high proportion of SK could exhibit TWT comparable to the wheat samples of larger kernel size but exhibited much lower milling yields [5]. The relationship appears to be genotype dependent. The great variation in TWT of the SK fraction could likely be attributable to large differences in kernel shape and packing density. Due to the potential strong impact of genotype, TWT can vary widely for small-sized kernels. Therefore, TWT may not be reliable as a direct indicator of the milling potential of durum wheat when SK is predominantly present. It is critical to monitor the KSD when a larger proportion of SK is present. Wang and Fu (2020) demonstrated that by accounting for the difference in KSD, greater relationships were found for TKW ($R^2 > 0.91$, $p < 0.001$) or the proportion of kernels passing the no.6 slotted sieve with milling yields than TWT alone ($R^2 = 0.75$, $p < 0.001$) by studying 21 wheat composites of four major CWAD varieties [5].

3.2. *Influence of Kernel Size and Genotype on Milling Quality of Durum Wheat*

From Table 2, a significant impact of kernel size, genotype and their interactions was found on durum milling performance (semolina and total milling yields and semolina ash content). Based on the ANOVA test, more than 80% of variation in milling yields was attributed to kernel size alone, with a greater impact of kernel size being noted for semolina yield than total milling yield (*F* value: 13177.7 vs. 7392.8). Figure 2 demonstrates the semolina and total milling yields in relation to TKW and TWT as affected by kernel size. Regardless of genotype selected, decrease of kernel size significantly reduced semolina and total milling yields. A drastic reduction of milling yields was evident for kernels passing no.6 slotted sieve (Table 3). On average, LK (68.0 ± 0.9%) had 1.3% higher SY than MK (66.7 ± 0.7%), and the latter was about 3.1% higher in SY than that of SK (63.6 ± 0.7%). Kernel size is clearly a better indicator of average milling yields for SK than the TWT (Figure 2). For LK and MK; however, both TWT and TKW provided strong indication of average milling quality. A similar adverse effect of SK on durum milling quality was reported by Wang and Fu (2020) and Dexter et al. (2007) by examining durum composites with a wide variation in kernel sizes [4,5].

From Figure 2, considering the response of genotype to the relationship between kernel size and milling quality, genotypes A and B appeared to be more susceptible to kernel size variations showing a greater decrease (~4.9%) in semolina yield from 68.9 to 64.0% than those of the inferior ones (e.g., G) from 66.2 to 62.9% (vs. 3.3%). A similar trend was found for total milling yield (3.5% vs. 2.7%). There were significant differences in semolina and total milling yields among the genotypes at all three kernel size fractions (Figure 2a,b). The difference in milling yields was greater for LK (2.7%) than MK (1.8%) and SK (1.3%) among the selected genotypes (Table 3).

When comparing milling quality of all kernel size fractions (Figure 2), semolina and total milling yields of MK segregated from genotypes with superior milling quality (A and C) were comparable or superior to the LK from genotypes of inferior or moderate milling quality (E, F, and G) despite the TKW of those MK (34.8 to 37.0 g) being significantly lower than LK counterparts (48.3 to 52.8 g). In addition, LK from genotypes with inferior milling quality showed lower milling yields. SK exhibited inferior milling quality to both MK and LK regardless of the genotypes selected (Table 3). SK is very detrimental to the overall milling quality but usually represents only a small proportion in commercial durum shipments. Analysis of variance by excluding SK revealed that genotype accounted for 52.0% of variation in semolina yield, followed by kernel size of 44.3% and their interaction of 3.4%. These results strongly suggest that the intrinsic kernel properties could play an important role in determining the milling quality of durum wheat. Selection of genotypes with superior milling quality could compensate the negative impact of SK which is usually present in higher percentage in dry and hot growing seasons. When a large proportion of small kernels was present; however, milling quality could be poor regardless of the genotypes selected.

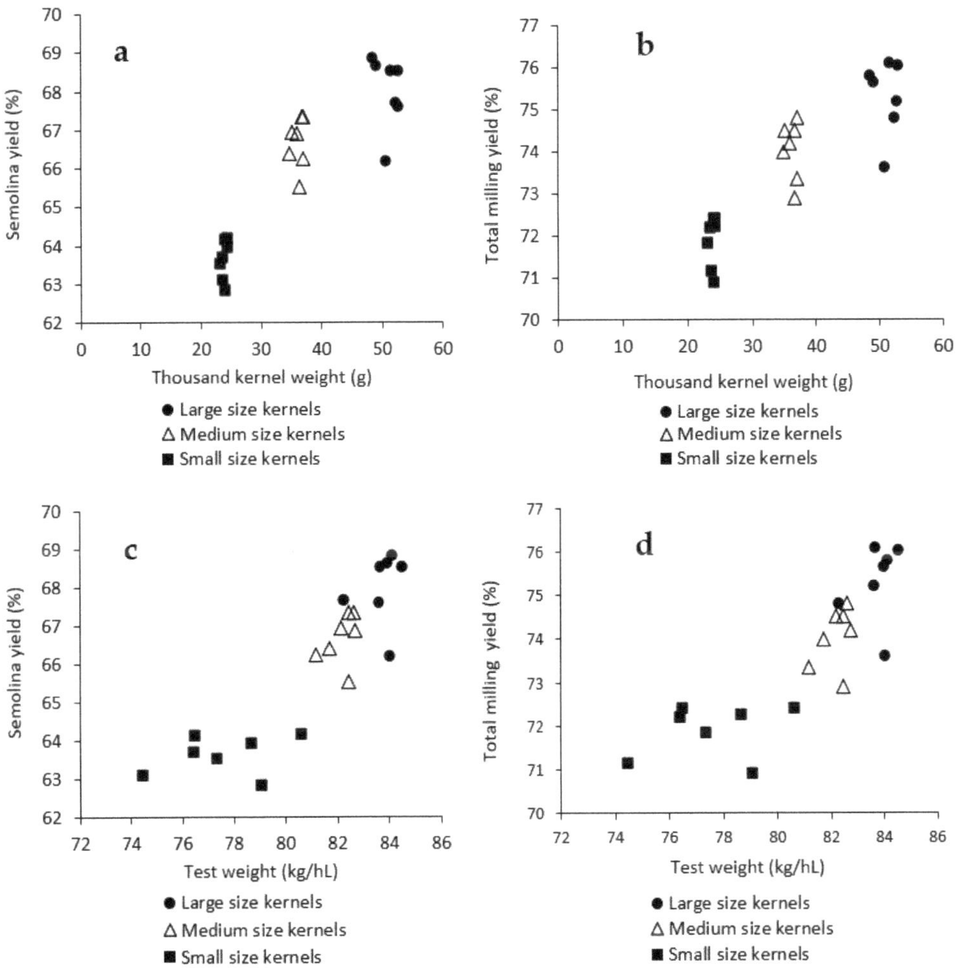

Figure 2. Impact of genotype and kernel size on semolina and total milling yield of selected durum samples in relation to thousand kernel weight (**a**,**b**) and test weight (**c**,**d**).

In addition to the milling yields, ash content is an important part of overall milling quality. The ash contents of wheat and semolina increased with the decrease of kernel size (Table 3). Coupled with the lower semolina yield of SK, its high semolina ash could further decrease the wheat milling potential when a constant degree of semolina refinement is required.

Milling quality of durum wheat is a complicated trait [10]. From Figure 2, a cooperative effect between kernel size and genotype on durum milling quality was evident when considering both MK and LK. The average milling yields of SK were lower and the impact of genotype was much less (Table 3). While the impact of some common kernel physical parameters (e.g., vitreousness, TWT, and KSD) on milling quality has been extensively investigated, the work on the intrinsic properties that contribute to varietal differences in milling quality of durum wheat are scarce [19,21–24]. Both kernel morphological parameters (e.g., length, width, thickness, size, shape, etc.) and kernel physical properties (e.g., hardness, vitreousness, TWT) could affect milling quality. Simmons and Meredith (1979) summarized three major factors that contribute to the difference in milling quality: the amount of endosperm contained in the grain (endosperm-to-bran ratio); the separability of the endosperm from the aleurone and bran layers (structure dissociates on fracture and milling); and endosperm hardness, which determines how the kernel fragments during the milling process [19]. Novaro et al. (2001) reported ellipsoidal volume was the best predictor of semolina yield among other grain morphological parameters evaluated [25]. Haraszi et al. found that the rheological phenotype phases of an average crush response profile obtained from a single kernel characterization system provided good predictions of the laboratory milling potential of durum wheats [26].

Due to the relatively large kernel size of the original unsorted samples (Table 1) and the similar TKW of the segregated kernel fractions (Table 3), the varietal differences in milling quality among selected genotypes could be attributed to their intrinsic kernel properties. Information on hardness, endosperm-to-bran ratio, and kernel fracture behavior could shed some light on the genotypic variation in milling quality. A study is currently being conducted in our laboratory to investigate the underlying factors, which could affect the milling quality of durum genotypes with a similar size of wheat kernels.

3.3. Influence of Kernel Size and Genotype on Semolina and Pasta Color Parameters

Both genotype and kernel size significantly affected semolina TYP (Table 2). Figure 3 presents the semolina TYP of three kernel size fractions segregated from the selected genotypes. The decrease of kernel size led to significant increase in semolina TYP for all genotypes. Alvarez et al. (1999) reported a similar negative relationship between kernel weight and yellow pigment concentration [27]. A greater difference in TYP was shown between MK and LK (1.0–1.6 ppm) than between small and medium ones (0.2–0.9 ppm). The degree of increase in semolina TYP as shown in Figure 3 was comparable to the level previously reported by Wang and Fu, who found that semolina TYP of SK was about 1.5 ppm higher than that of LK segregated from a bulk CWAD cargo composite [5]. Large genetic variations in semolina TYP from 2.3 to 3.0 ppm were noted for the genotypes used in this study across three different kernel sizes.

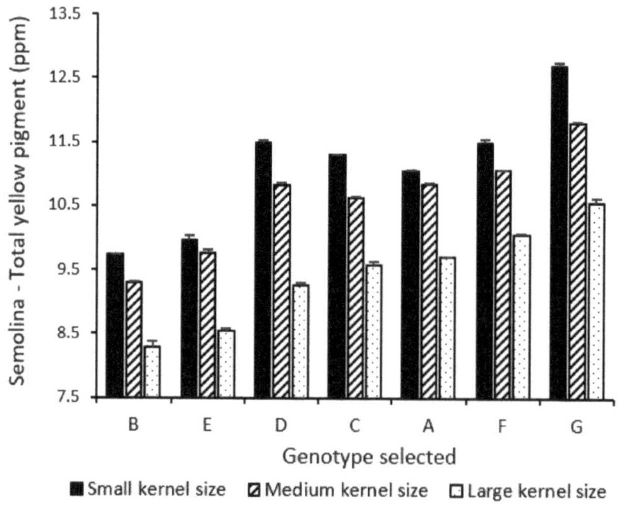

Figure 3. Impact of genotype and kernel size on semolina total yellow pigment content of selected genotypes.

The colour of semolina and pasta made from the size fractions are summarized in Figure 4. Brightness and redness of semolina were greatly influenced by kernel size, while the genotype had a large impact on semolina yellowness (Table 2). In general, semolina prepared from MK and LK was much brighter (Figure 4a) and less dull (Figure 4c) compared to that prepared from SK. Much greater variation in brightness and redness was also shown for SK fractions than MK and LK ones (Figure 2 and Table 3).

With the decrease of kernel size from LK to MK, significant increases in semolina TYP and yellowness were shown (Figure 4e). However, except for genotypes D and G, reduction of kernel size from MK to SK did not lead to further increase in semolina yellowness despite the TYP being significantly higher in SK. The drastic decrease in semolina brightness and increase in redness for small kernels might mask semolina yellowness.

Table 2 showed a large impact of kernel size on pasta color. The decrease in kernel size led to a significant reduction in pasta brightness (Figure 4b) and an increase in pasta redness (Figure 4d). Superior yellowness was seen for pasta prepared from medium and large kernel fractions. However, a drastic decrease in pasta yellowness of about 7 units was noticed for SK despite its semolina TYP being significantly higher (Figure 4f). By plotting semolina yellowness against TYP for three different kernel size fractions of the selected genotypes, it was shown that semolina b* linearly increased about 1.2 units with each ppm increase in TYP (Figure 5a). The degree of increase in semolina yellowness in relation to TYP was similar for all three size fractions. For a given TYP, however, semolina prepared from LK and MK consistently showed superior yellowness than that of SK, inferring the negative impact of SK on semolina yellowness. This negative impact was much more profound for pasta yellowness (Figure 5b). As far as SK fraction is concerned, the increase in semolina TYP resulted in little increase in pasta yellowness. This is in contrast to the MK and LK fractions evaluated in this study.

Pasta brightness and yellowness decrease with the increase of semolina ash content [28,29]. Although SK have lower semolina and total milling yields, the higher ash content suggests inclusion of a greater proportion of external tissues, which could lead to pasta browning due to high enzymatic activities [28]. Maillard reaction between amino acid and reducing sugars could lead to the undesirable reddish color of pasta dried at high temperature [30,31]. Although the protein content was not significantly higher for SK as compared with MK and LK, pasta prepared from SK was much redder (6.2–7.3 in a*) than that made from LK (2.7–3.7 in a*), suggesting other underlying factors such as amino acid composition or reducing sugar content may favor the development of the reddish coloration of pasta prepared from small kernels. Joubert et al. revisited the role of particle size, ash, and protein on pasta color and viscoelasticity [32]. By combining the milling fractions of five durum wheat patches, a series of formulated mixes of semolina/flour were prepared so that the effect of protein, ash, and particle size distribution (PSD) could be evaluated in an unbiased manner. The authors found that pasta brightness and yellowness decreased while redness increased with the increase of semolina ash content regardless of protein content and PSD. The authors attributed the increase in pasta redness to the elevation of reducing sugars accompanied by the high ash content in the semolina. A significant correlation was found between the ash content and total arabinoxylans in semolina, which were known to concentrate in the outer layers of the grain [33]. The extrusion process can significantly increase the reducing sugars due to shearing stress [34]. It is likely that the SK contains a high level of arabinoxylan, which could result in a high level of reducing sugar during extrusion and increase the potential of Maillard reactions [32]. The elevated redness/brownness and decrease in pasta brightness could subsequently mask pasta yellowness. Wang and Fu proposed that the drastic elevation in pasta redness due to the Maillard reaction under high-temperature (85 °C) drying conditions could adversely impact pasta yellowness regardless of the level of TYP [5].

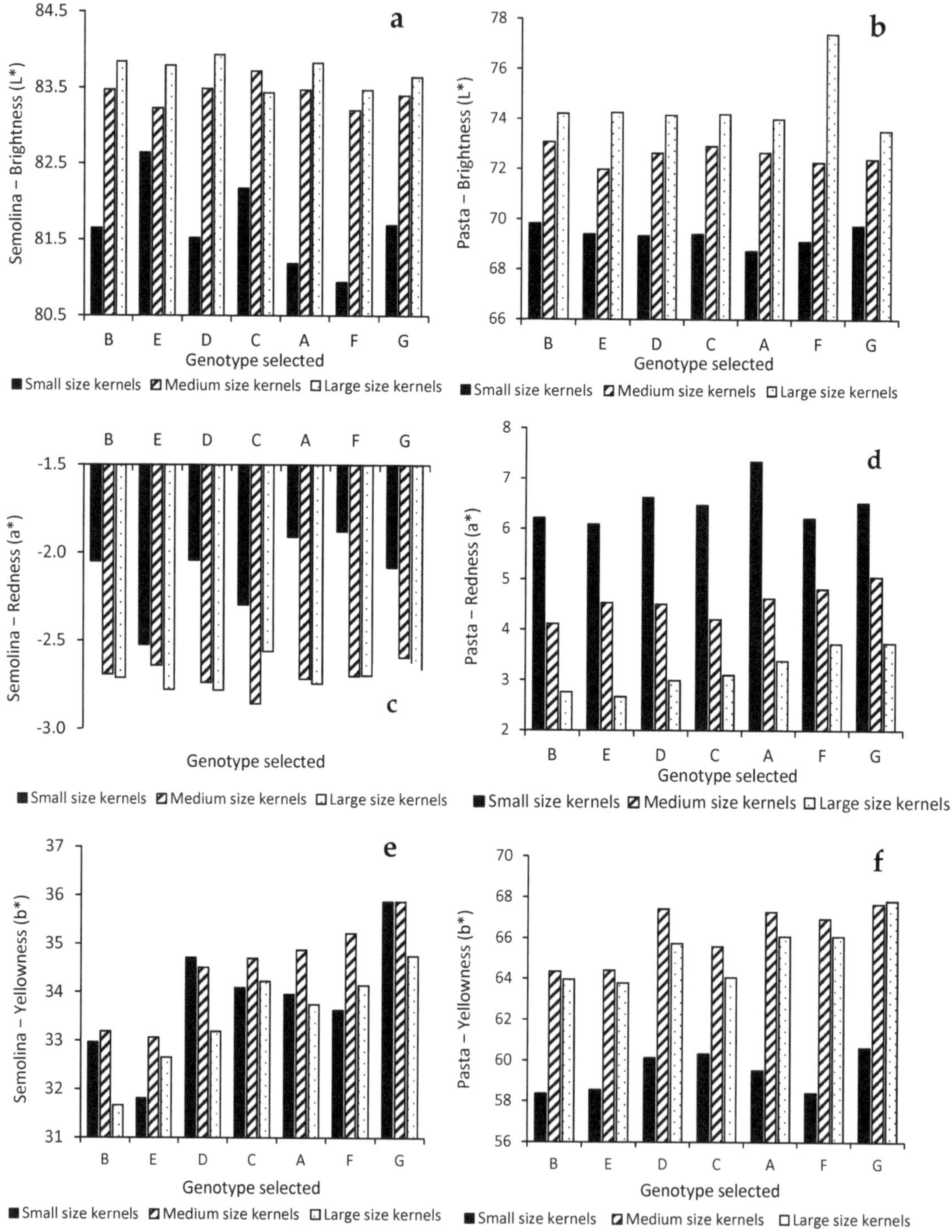

Figure 4. Impact of genotype and kernel size on semolina (**a**,**c**,**e**) and pasta color (**b**,**d**,**f**).

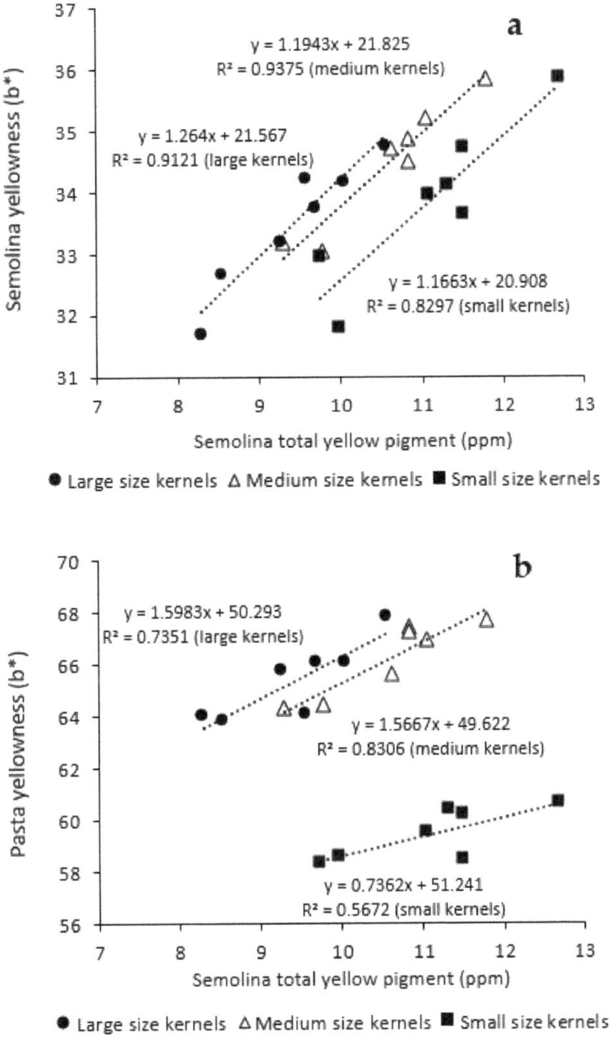

Figure 5. Combined effect of kernel size and genotype on yellowness of semolina (**a**) and pasta (**b**).

4. Conclusions

By segregating durum samples of selected genotypes into three kernel size fractions, the impact and relative importance of kernel size, genotype, and their interaction on major quality parameters were characterized in this study. For LK and MK fractions, TWT and kernel size are closely related. However, a greater influence of genotype on TWT of SK was evident. Regardless of the genotype, the SK fraction is detrimental to durum milling performance as shown by low semolina yield, high semolina ash content, and poor semolina color. The degree of impact of genotype on the durum milling performance appears to be related to kernel size. A greater impact was shown for LK than MK and SK, based on seven genotypes evaluated in this study. When the SK fraction is excluded, the genotype or intrinsic property of the durum kernel played an important role in contributing to overall milling quality. Genotype is a dominant factor in determining semolina TYP and yellowness despite TYP increases with the decrease of kernel size. Semolina and pasta prepared from MK and LK fractions were much brighter and less dull than those made from SK. Regardless of the genotype, the SK fraction exerted a strong detrimental effect on pasta yellowness, despite the higher level of TYP in SK. To meet the milling and end-product quality expectation of domestic and international durum buyers, it is critical to monitor the presence of SK (through a no.6 slotted sieve) in commercial durum samples, particularly in hot and dry growing seasons. More research is needed to confirm the potential interactions between genotype and kernel size and their effects on durum quality by using wheat samples from various genotypes and different growing conditions.

Author Contributions: Conceptualization, K.W. and B.X.F.; methodology, D.T., J.S., Y.C. and K.W.; software, K.W.; validation, K.W.; writing—original draft preparation, K.W.; writing—review and editing, K.W., B.X.F. and J.S.; supervision, B.X.F.; project administration, K.W. All authors have read and agreed to the published version of the manuscript.

Funding: This research received no external funding.

Institutional Review Board Statement: Not applicable.

Informed Consent Statement: Not applicable.

Acknowledgments: The authors are grateful to Dave Turnock for milling the durum samples in the standard milling protocol and providing assistance in the micro-scale purification process. The technical assistance of staff at the CGC Analytical Services and Industry Services is gratefully acknowledged.

Conflicts of Interest: The authors declare no conflict of interest.

References

1. Feillet, P.; Dexter, J.E. Quality requirements of durum wheat for semolina milling and pasta production. In *Pasta and Noodle Technology*, 1st ed.; Kruger, J.E., Matsuo, R.B., Dick, J.W., Eds.; AACC: St. Paul, MN, USA, 1996; pp. 95–131.
2. Sissions, M.; Abecassis, J.; Marchylo, B. Methods used to access and predict quality of durum wheat, semolina, and pasta. In *Durum Wheat Chemistry and Technology*; Sissions, M., Abecassis, J., Marchylo, B., Carcea, M., Eds.; AACC International: St. Paul, MN, USA, 2012; pp. 213–234.
3. Dexter, J.E.; Matsuo, R.R.; Martin, D.G. The relationship of durum wheat test weight to milling performance and spaghetti quality. *CFW* **1987**, *32*, 772–777.
4. Dexter, J.E.; Symons, S.J. Impact of durum wheat test weight, kernel size, kernel weight and protein content on semolina milling quality. *Int. Miller.* **2007**, *1*, 27–33.
5. Wang, K.; Fu, B.X. Inter-relationships between test weight, thousand kernel weight, kernel size distribution and their effects on durum wheat milling, semolina composition and pasta processing quality. *Foods* **2020**, *9*, 1308. [CrossRef]
6. Fu, B.; Wang, K.; Dupuis, B.; Taylor, D.; Nam, S. Kernel vitreousness and protein content: Relationship, interaction and synergistic effects on durum wheat quality. *J. Cereal Sci.* **2017**, *78*, 2–9. [CrossRef]
7. Baasandorj, T.; Ohm, J.; Manthey, F.; Simek, S. Effect of kernel size and mill type on protein, milling yield, and baking quality of hard red spring wheat. *Cereal Chem.* **2015**, *92*, 81–87. [CrossRef]
8. Canadian Grain Commission: Procedure and Equipment for Determining Test Weight. Available online: https://www.grainscanada.gc.ca/en/grain-quality/grain-grading/grading-factors/grading-factors-wheat/test-weight.html (accessed on 12 July 2021).
9. Dexter, J.E.; Matsuo, R.R.; Kruger, J.R. The spaghetti making quality of commercial durum wheat samples of variable amylase activity. *Cereal Chem.* **1990**, *67*, 405–412.
10. Wang, K.; Taylor, D.; Pozniak, C.; Fu, B.X. Developing a high-throughput micromilling protocol for evaluating durum wheat milling performance and semolina quality. *Cereal Chem.* **2019**, *96*, 802–814. [CrossRef]
11. Williams, P.; Sobering, D.; Antoniszyn, J. Protein testing methods. In *Wheat Protein Production and Marketing*; Fowler, D.B., Geddes, W.E., Johnston, A.M., Preston, K.R., Eds.; University Extension Press: Saskatoon, SK, Canada, 1998; pp. 37–47.
12. AACC International. *Approved Method of Analysis*, 11th ed.; Methods 76.31.01 and Methods 38-12.02; AACC International: St. Paul, MN, USA, 2010.
13. Fu, B.; Schlichting, L.; Pozniak, C.J.; Singh, A.K. Pigment loss from semolina to dough: Rapid measurement and relationship with pasta color. *J. Cereal Sci.* **2013**, *57*, 560–566. [CrossRef]
14. Hook, S.C. Specific weight and wheat quality. *J. Sci. Food Agric.* **1984**, *35*, 1136–1141. [CrossRef]
15. Tkachuk, R.; Kuzina, F.D. Wheat: Relationship between some physical and chemical properties. *Can. J. Plant Sci.* **1979**, *59*, 15–20. [CrossRef]

16. Cabral, A.L.; Jordan, M.C.; Larson, G.; Somers, D.J.; Humphreys, D.G.; McCartney, C.A. Relationships between QTL for grain shape, grain weight, test weight, milling yield, and plant height in the spring wheat cross RL4452/'AC Domain'. *PLoS ONE* **2018**, *13*, e0190681. [CrossRef]
17. Yabwalo, D.N.; Berzonsky, W.A.; Brabec, D.; Pearson, T.; Glover, K.D.; Kleinjan, J.L. Impact of grain morphology and genotype by environment interactions on test weight of spring and winter wheat (*Triticum aestivum* L.). *Euphytica* **2018**, *2014*, 125–141. [CrossRef]
18. Samaan, J.; El-Khayat, G.H.; Manthey, F.A.; Fuller, M.P.; Brennan, C.S. Durum wheat quality: II. The relationship of kernel physicochemical composition to semolina quality and end product utilization. *Int. J. Food Sci. Technol.* **2006**, *41*, 47–55. [CrossRef]
19. Simmons, L.; Meredith, P. Width, weight, endosperm, and bran of the wheat grain as determinants of flour milling yield in normal and shrivelled wheats. *N. Z. J. Sci.* **1979**, *22*, 1–10.
20. Troccoli, A.; di Fonzo, N. Relationship between kernel size features and test weight in Triticum durum. *Cereal Chem.* **1999**, *76*, 45–49. [CrossRef]
21. Dziki, D.; Laskowski, J. Wheat kernel physical properties and milling process. *Acta Agrophysica* **2005**, *6*, 59–71.
22. Lyford, C.P.; Kidd, W.; Rayas-Duarte, P.; Deyoe, C. Prediction of flour extraction rate in hard red winter wheat using the single kernel characterization. *J. Food Qual.* **2005**, *28*, 279–288. [CrossRef]
23. Chaurand, M.; Lempereur, I.; Roulland, T.M.; Autran, J.C.; Abecassis, J. Genetic and agronomic effects on semolina milling value of durum wheat (*Triticum durum* Desf.). *Crop. Sci.* **1999**, *2*, 277.
24. Peyron, S.; Surget, A.; Mabille, F.; Autran, J.C.; Rouau, X.; Abecassis, J. Evaluation of tissue dissociation of durum wheat grain (*Triticum durum* Desf.) generated by milling process. *J. Cereal Sci.* **2002**, *36*, 199–208. [CrossRef]
25. Novaro, P.; Colucci, F.; Venora, G.; D'Egidio, M.G.D. Image analysis of whole grains: A noninvasive method to predict semolina yield in durum wheat. *Cereal Chem.* **2001**, *78*, 217–221. [CrossRef]
26. Haraszi, R.; Sissons, M.; Juhasz, A.; Kadkol, G.; Tamas, L.; Anderssen, R. Using rheological phenotype phases to predict rheological features of wheat hardness and milling potential of durum wheat. *Cereal Chem.* **2016**, *93*, 369–376. [CrossRef]
27. Alvarez, J.B.; Martin, L.M.; Martin, A. Genetic variation for carotenoid pigment content in the amphiploid *Hordeum chilense x Triticum turgidum* conv. *durum*. *Plant Breed.* **1999**, *118*, 187–189. [CrossRef]
28. Borrelli, G.M.; Troccoli, A.; Di Fonzo, N.; Fares, C. Durum wheat lipoxygenase activity and other quality parameters that affect pasta color. *Cereal Chem.* **1999**, *76*, 335–340. [CrossRef]
29. Matsuo, R.R.; Dexter, J.E. Relationship between some durum wheat physical characteristics and semolina milling properties. *Can. J. Plant Sci.* **1980**, *60*, 49–53. [CrossRef]
30. Acquistucci, R. Influence of Maillard reaction on protein modification and color development in pasta. Comparison of different drying conditions. *LWT* **2000**, *33*, 48–52. [CrossRef]
31. Anese, M.; Nicoli, M.C.; Massini, R.; Lerici, C.R. Effects of drying processing on the Maillard reaction in pasta. *Food Res. Int.* **1999**, *32*, 193–199. [CrossRef]
32. Joubert, M.; Morel, M.; Lullien-Pellerin, V. Pasta color and viscoelastivity: Revisiting the role of particle size, ash, and protein content. *Cereal Chem.* **2018**, *95*, 386–398. [CrossRef]
33. Lempereur, I.; Rouau, X.; Abecassis, J. Genetic and agronomic variation in arabinoxylan and ferulic acid contents of durum wheat (Triticum durum.) grain and its milling fractions. *J. Cereal Sci.* **1997**, *25*, 103–110. [CrossRef]
34. Lintas, C.; D'Appolonia, B.L. Effect of spaghetti processing on semolina carbohydrates. *Cereal Chem.* **1973**, *50*, 563–570.

foods

MDPI

Article

The Use of Durum Wheat Oil in the Preparation of *Focaccia*: Effects on the Oxidative Stability and Physical and Sensorial Properties

Francesca Vurro, Carmine Summo, Giacomo Squeo, Francesco Caponio and Antonella Pasqualone *

Department of Soil, Plant and Food Science (DISSPA), University of Bari Aldo Moro, via Amendola 165/a, I-70126 Bari, Italy
* Correspondence: antonella.pasqualone@uniba.it

Abstract: Durum wheat oil is an innovative oil that could be considered the "second life" of durum wheat milling by-products. In this study, we proposed the use of this oil in the reformulation of a traditional Italian greased flat bread, namely *focaccia*, whose typical sensorial features are due to the presence of relevant amounts of oil in its formulation. The chemical, physical, and sensorial features of *focaccia* with durum wheat oil (DWO) were compared with those of *focaccia* prepared with olive oil (OO) and sunflower oil (SO). The results showed the prevalence of polyunsaturated fatty acids in DWO, followed by SO. DWO was more resistant to oxidation than SO (induction time 86.2 and 66.3 min, respectively), due to its higher content of tocotrienols (1020 and 70.2 mg/kg in DWO and SO, respectively), but was less resistant than OO, richer in monounsaturated fatty acids, and contained phenolic compounds. The volatile oxidation markers, namely hexanal and nonanal, were less prevalent in OO and DWO than in SO. Texture and color were positively influenced by the use of durum wheat oil, allowing the nutritional improvement of this flat bread in a sustainable and circular manner.

Keywords: flat bread; durum wheat oil; acidic composition; tocotrienols; tocopherols; volatile compounds; texture profile analysis; sensorial properties

Citation: Vurro, F.; Summo, C.; Squeo, G.; Caponio, F.; Pasqualone, A. The Use of Durum Wheat Oil in the Preparation of *Focaccia*: Effects on the Oxidative Stability and Physical and Sensorial Properties. *Foods* **2022**, *11*, 2679. https://doi.org/10.3390/foods11172679

Academic Editor: Mike Sissons

Received: 31 July 2022
Accepted: 29 August 2022
Published: 2 September 2022

1. Introduction

Bread is a traditional staple food consumed by people worldwide, with hundreds of different examples [1]. Among them, flat breads are the oldest, and they are still very popular due to their high versatility, organoleptic properties, and convenience [2,3]. These reasons justify the relevant growth of their market, especially in relation to the modern style of life and new eating preferences [4].

Italy has a long tradition of garnished flat breads, some of which are also renowned abroad, such as pizza. *Focaccia* is another typical garnished flat bread widely consumed in several Italian regions, where it originated [5]. *Focaccia* is oven-baked in a low pan and, prior to cooking, is topped with fresh tomato and oregano; onions and potatoes; cheese; or salt and rosemary, etc., providing a myriad of diverse and nuanced varieties [6]. This old and traditional food product, which has been included in the list of Italian Traditional Agri-Food Products (TAP) [7], is similar to pizza but has some distinct characteristics [2,5]. The consumption patterns of *focaccia* and pizza are different: the first is a quick snack consumed at any time [5], while the second is usually preferred for dinner or lunch (except for the "*pizza a portafoglio*", the typical street food of Naples) [8]. The difference between pizza and *focaccia*, however, is not limited to their consumption patterns, as their formulation, appearance, and sensory characteristics are also different. The preparation of pizza and *focaccia* starts in the same way, by kneading flour, water, yeast, and salt. Then, only in the preparation of *focaccia*, abundant oil is incorporated into the dough and poured onto its surface to confer the typical greasiness [2,5]. Conversely, in the preparation of the

"Traditional Specialty Guaranteed" (TSG) Neapolitan pizza, only a very small amount of oil is used [9]. Specifically, only extra virgin olive oil can be used in TAP-labeled *focaccia* and TSG-labeled pizza, which are high-quality niche products. However, most of the commonly marketed *focaccia* and pizza, which are not labeled as TAP or TSG, contain olive oil or sunflower oil.

Pizza has been the object of several investigations aimed at improving its nutritional features [10–13], without neglecting gluten-free versions [14,15], but very few studies are available on *focaccia*. The reformulation of *focaccia* using *Apulian black chickpea* flour, which provides anthocyanins and increases antioxidant activity, was investigated by Pasqualone et al. [12], while Delcuratolo et al. [16,17] evaluated the role of *focaccia* toppings on the oxidation stability and content of polar compounds arising from triacylglycerol oxidation and hydrolysis, which are responsible for negative health implications [18]. However, only a single study has investigated the use of fat replacers to reduce the oil content of *focaccia* [19], and no studies have compared the effects of different vegetable oils on its nutritional and sensory characteristics.

Italy is not only famous for pizza and *focaccia*, but also for pasta and special baked goods made of durum wheat semolina [20–22]. However, the milling process for obtaining semolina involves the production of by-products (bran, germ, and various middlings) [23], which should be upcycled and reintroduced into the food system to comply with the principles of a circular economy [24,25]. These by-products have a proven potential for oil extraction [25]. A previous study has evaluated the effect of using durum wheat oil in the preparation of biscuits [26], whose long shelf life can be affected by rancidity onset. The substitution of sunflower oil with durum wheat oil significantly increased the resistance of biscuits to oxidation due to the abundance of tocols in durum wheat oil, especially tocotrienols [26].

At this historical moment, the war in Ukraine is causing problems for the supply of sunflower oil [27], the fourth most consumed oil in the world [28]. Moreover, since 2013, the "silent war" of *Xylella fastidiosa* has been changing the Italian landscape, causing a decrease in the production of olive oil [29]. Therefore, alternatives to these largely consumed oils should be considered. The use of durum wheat oil in a traditional product such as *focaccia* could valorize the entire durum wheat supply chain and, at the same time, could offer producers and consumers an alternative to the currently used oils.

Therefore, the aim of this study was to evaluate the effect of durum wheat oil on the oxidation stability and physical–sensory characteristics of *focaccia*, in comparison with olive oil and sunflower oil.

2. Materials and Methods

2.1. Materials

Durum wheat oil, prepared as reported in Squeo et al. [25], was provided by Casillo Next Food Srl (Corato, Italy). Wheat flour type 0 (Casillo Spa, Corato, Italy) (carbohydrate 72 g/100 g; proteins 11 g/100 g; fat 2 g/100 g; fiber 2 g/100 g); sunflower oil (Olearia De Santis, Bitonto, Italy); olive oil (Olearia De Santis, Bitonto, Italy); yeast (*Saccharomyces cerevisiae*, Mulino Caputo, Naples, Italy); and sea salt (Atisale Spa, Margherita di Savoia, Italy) were purchased from local retailers.

2.2. Sample Preparation

Three different types of *focaccia* were prepared: (i) *focaccia* with sunflower oil (SO); (ii) *focaccia* with olive oil (OO); (iii) *focaccia* with durum wheat oil (DWO), according to the formulation reported in Table 1. The *focaccia* samples were prepared as described in Pasqualone et al. [12]. Flour, water, and yeast were kneaded for 6 min using a spiral kneader (Bosh MFQ40304, München, Germany). Then, salt and oil (70% of the total oil amount) were added, and kneading was continued for 6 min. The first fermentation was carried out for 1 h and 30 min in controlled conditions at 35 °C, RH = 20% (Memmert proofer, EN.CO. Srl, Spinea, Italy). The leavened dough was divided into portions, which were manually shaped into discs with a thickness of 1.5 cm and a diameter of about 30 cm. The discs were placed into metal pans, previously greased with oil (10% of the total oil amount), and left to rise again in the same conditions. The *focaccia* surface was then greased by pouring oil over it (20% of the total oil amount), followed by baking in an electric oven (Oem Ali Group Srl, Bozzolo, Italy) at 200 °C for 25 min.

Table 1. Formulation of the experimental *focaccia* samples. SO = *focaccia* with sunflower oil; OO = *focaccia* with olive oil; DWO = *focaccia* with durum wheat oil.

Ingredients	SO (g)	OO (g)	DWO (g)
Wheat flour type 0	600	600	600
Water	420	420	420
Sunflower oil	85	-	-
Olive oil	-	85	-
Durum wheat oil	-	-	85
Salt	15	15	15
Yeast	5	5	5

2.3. Determination of the Resistance to Oxidation

A RapidOxy oxidation stability tester (Anton Paar, Blankenfelde-Mahlow, Germany) was used, as described in AOCS Method Cd 12c-16 [30]. Two grams of the samples (*focaccia*, finely crushed, or oil) was oxidized at a temperature of 140 °C with an oxygen pressure of 700 kPa until the pressure decreased by 10%. The samples were tested in triplicate.

2.4. Determination of Fatty Acid Composition

The fatty acid composition of the oils used in the *focaccia* preparation was analyzed as described by Squeo et al. [25]. A gas chromatograph (mod. 7890A, Agilent Technologies, Santa Clara, CA, USA) equipped with an FID detector (set at 220 °C) and an SP2340 capillary column of 60 m × 0.25 mm × 0.2 mm film thickness (Supelco Park, Bellefonte, PA, USA) was used to separate the fatty acid methyl esters. Comparison with the retention time of the standard mixture (C_4–C_{24}) (Sigma-Aldrich, St. Louis, MO, USA) was used for the identification of each fatty acid present in the sample. The analyses were carried out in triplicate.

2.5. Lipid Extraction

The lipid fraction of the *focaccia* was extracted by the Soxhlet apparatus (SER 148 extraction system, Velp Scientifica Srl, Usmate, Italy). The solvent used for the extraction was diethyl ether (Carlo Erba, Milan, Italy).

2.6. Determination of Tocopherols and Tocotrienols

The tocopherols and tocotrienols of oils and of the lipid fraction extracted were determined by HPLC (Agilent 1100 Series, Agilent Technologies, Santa Clara, CA, USA). Primarily, 0.02–0.03 g of sample was dissolved in 1 mL of 2-propanol. The samples were filtered by a 0.45 µm polytetrafluoroethylene (PTFE) filter and injected into an HPLC system consisting of a Waters 600E quaternary pump (Milford, MA, USA), a 7725i Rheodyne injector (20-µL sample loop), and a fluorescent detector (excitation wavelength 292 nm, emission wavelength 330 nm). The stationary phase was an Acclaim™ 120 Å C18 column, with a particle size of 3 µm and 3 mm × 150 mm in length (Thermo Fisher Scientific, Waltham, MA, USA); the mobile phase was 96:4 (*v*/*v*) methanol and water at a flow rate of 1 mL/min. The software used was Chromeleon (Thermo Fisher Scientific, Waltham, MA, USA). The single tocol was determined by the external standard method based on a previously set calibration curve. The content of tocopherols and tocotrienols was expressed as mg/kg of the total weight of oil. The analyses were carried out in triplicate.

2.7. Determination of Polar Compounds of the Lipid Fraction of Focaccia

The polar compounds of the oil extracted from *focaccia* samples were separated by silica gel column chromatography and quantified using high-performance size-exclusion chromatography (HPSEC) according to Difonzo et al. [31]. The content of polar compounds was expressed as g/100 g of oil extracted. The analyses were carried out in triplicate.

2.8. Determination of Antioxidant Activity

The extraction of antioxidant compounds was conducted as described by Troilo et al. [32], with the only changes being that the ratio of sample and extraction solvent and the number of washes were altered from 1:5 and two, respectively.

The sample extracts were submitted to a radical scavenging assays using 2,2′-azino-bis-3-ethylbenzthiazoline-6-sulphonic acid (ABTS) and 1,1-diphenyl-2-picrylhydrazyl (DPPH) radical, according to Difonzo et al. [33]. The absorbance was read using a Cary 60 spectrophotometer (Cernusco, Milan, Italy). The results were expressed as μmol Trolox equivalents (TE)/g. The determinations were carried out in triplicate.

2.9. Determination of Volatile Compounds

Volatile compounds of the *focaccia* were analyzed by headspace solid-phase micro-extraction (HS-SPME) coupled with gas chromatography/mass spectrometry (GC–MS), as described by Difonzo et al. [31]. The identification of the volatile compounds was carried out using an LRI and by computer matching with the reference mass spectra of the National Institute of Standards and Technology (NIST) and Wiley libraries. The quantification of the volatile compounds was performed considering the standardization of the respective peak areas with the peak area of the 1-propanol used as internal standard. The results were expressed as μg/g of sample. The analyses were carried out in triplicate.

2.10. Texture Profile Analysis

Texture profile analysis (TPA) was executed as described in Pasqualone et al. (2019) [12], with the only modification being the use of a cylindrical probe (36 mm diameter). The following parameters were calculated from the TPA graphic: hardness (N), chewiness (N), cohesiveness, and springiness. Six replications were carried out.

2.11. Color Determination

The color of *focaccia* (crumb and crust) was measured using the CM-600d colorimeter (Konica Minolta, Tokyo, Japan) supported by SpectraMagic NX software (Konica Minolta, Tokyo, Japan). The color properties were determined in the CIE (International Commission on Illumination) color space. Lightness (L^*, from black to white), red index (a^*, from green to red), and yellow index (b^*, from blue to yellow) were determined. The total color difference (ΔE) was calculated as follows:

$$\Delta E = \sqrt{[(L^* - L_0^*)^2 + (a^* - a_0^*)^2 + (b^* - b_0^*)^2}$$

where L^*_0, a^*_0, and b^*_0 are the color coordinates of *focaccia* with olive oil (OO), while L^*, b^*, and a^* are the color coordinates of the other *focaccia* samples. The calculation considered the mean values. The following ΔE scale was considered for the evaluation of the results: 0–0.5 = no relevant difference; 0.5–1.5 = a slight difference; 1.5–3.0 = difference recognizable by an experienced observer; 3.0–6.0 = an appreciable difference; 6.0–12.0 = a large difference; and >12.0 = a very evident difference [14].

Nine replications were carried out.

2.12. Determination of Dimensional Parameters

The diameter (D) and thickness (T) of *focaccia* before and after baking were determined as described in Pasqualone et al. [12]. A caliper was used. The percentage variation due to baking was calculated as follows:

$$\%\text{ of variation of D (or T)} = \frac{[\text{D (or T) after baking} - \text{D (or T) before baking}]}{\text{D (or T) before baking}} \times 100$$

The analyses were carried out in triplicate.

2.13. Determination of Sensory Properties

The quantitative descriptive analysis (QDA) of *focaccia* was conducted according to the International Standardization Organization (ISO) standard 13299 [34] by a trained panel of eight members, previously selected for their reliability, consistency, and discriminating ability. The panel was composed of four men and four women, ranging in age from 23 to 55 years, who expressed written consent according to the ethical guidelines of the laboratory of Food Science and Technology of the Department of Soil, Plant, and Food Science of the University of Bari (Italy). The panelists were regular consumers of baked products and had no food allergies or intolerances. A pre-test session was conducted, as outlined in the ISO Standard 11132 [35]. A total of 12 descriptors were selected for the consideration of the *focaccia* samples made with different oils: 3 descriptors for visual appearance (surface color intensity, inner color intensity, crumb porosity); 3 descriptors for the odor (focaccia odor, roasted odor, oxidized odor); 3 for texture perceived during the tasting (crumb elasticity, softness, crumb moisture); and 3 for taste (saltiness, sweetness, greasiness). The intensity of every attribute was expressed on a 10 cm unstructured linear scale (contractual units—c.u.). The scale anchors for focaccia odor, roasted odor, oxidized odor, saltiness, sweetness, and greasiness were: 0 c.u. = minimum intensity; 10 c.u. = maximum intensity. The scale anchors for surface and inner color intensity were: 0 c.u. = ivory; 10 c.u. = brown. The scale anchors for the crumb porosity were: 0 c.u. = dense structure with very few pores; 10 c.u. = open structure, very porous. The scale anchors for crumb moisture were: 0 c.u. = dry; 10 c.u. = wet. The scale anchors for crumb softness were: 0 c.u. = very hard; 10 c.u. = very soft. The scale anchors for crumb elasticity were: 0 c.u. rigid =; 10 c.u. = very elastic.

The samples were randomized and presented to the panelists in white dishes marked with alphanumeric codes. The testing was performed at ambient room temperature (20 ± 2 °C). In accordance with the ISO 8589 [36] standard, the sensory analysis was carried out by physically separating panelists during analysis. Three replicates were carried out.

2.14. Statistical Analysis

Statistical analysis was carried out by Minitab Statistical Software (Minitab Inc., State College, PA, USA). The results were all expressed as mean ± standard deviation (SD). The Anderson–Darling test was applied to evaluate the normal distribution of the data, and the Levene test was used to evaluate the homoscedasticity of variances. The significant differences ($\alpha = 0.05$) were verified through the application of parametric one-way analysis of variance (ANOVA), followed by the Tukey HSD test, considering the type of oil as the independent variable.

3. Results and Discussion

3.1. Oxidation Stability, Fatty Acid Composition, and Tocols Content

Lipid oxidation is a negative event affecting many food products, particularly when their fat content is relevant and the processing or storage conditions are favourable to degradative reactions. Oxidative events cause a change in taste, texture, and appearance, as well as the production of toxic compounds and the loss of nutritional value [26].

A RapidOxy oxidation stability tester was used to evaluate the effect of varying the oil type on the oxidative stability of the *focaccia* samples. This instrument, which does not need solvents, enforces pro-oxidising conditions and measures the induction time (IT) of the lipid fraction, which is known to be positively related to the resistance to oxidation [37]. The OO *focaccia* showed the highest IT, followed by DWO and SO (Table 2).

Table 2. Resistance to forced oxidation of the oils and *focaccia*. SO = *focaccia* with sunflower oil; OO = *focaccia* with olive oil; DWO = *focaccia* with durum wheat oil.

Sample	Induction Time (min)
Focaccia	
OO	134 ± 2.06 ^{a}
SO	66.3 ± 4.81 ^{c}
DWO	86.2 ± 2.53 ^{b}
Oils	
Olive oil	59.5 ± 0.07 ^{a}
Sunflower oil	30.7 ± 0.64 ^{c}
Durum wheat oil	39.8 ± 0.26 ^{b}

Data are presented as means ± SD of three replicates. Different letters for the same sample type indicate significant differences at $p < 0.05$.

This trend mirrored the differences observed for the oils, which, in turn, could be explained in terms of different fatty acid composition and content of antioxidant compounds.

Polyunsaturated fatty acids (PUFAs) were the most abundant class in durum wheat oil and sunflower oil (Table 3). Particularly notable was the content of linolenic acid (n-3 PUFA) observed in durum wheat oil, accounting for 5.06 ± 0.03%, while in sunflower oil the content of linolenic acid was significantly lower (0.90 ± 0.01%). This difference is relevant because studies suggest that n-3 PUFAs reduce the risk of inflammatory and cardiovascular diseases, steatohepatitis, obesity, and diabetes [38]. Although the concentration of linolenic acid in durum wheat oil was lower than in the typical sources, such as flaxseeds, it was higher than the values reported for the majority of commonly used oils, such as sunflower oil, olive oil, corn oil, and palm oil [39–41].

Table 3. Percentage of the main fatty acids in the olive oil, sunflower oil, and durum wheat oil used in the preparation of *focaccia* samples.

Fatty Acids (%)	Olive Oil	Durum Wheat Oil	Sunflower Oil
$C_{14:0}$	0.02 ± 0.00 ^{c}	0.07 ± 0.00 ^{b}	0.08 ± 0.00 ^{a}
$C_{16:0}$	12.9 ± 0.06 ^{b}	14.84 ± 0.04 ^{a}	6.48 ± 0.01 ^{c}
$C_{18:0}$	3.24 ± 0.04 ^{b}	1.38 ± 0.01 ^{c}	3.82 ± 0.01 ^{a}
$C_{18:1}$	71.0 ± 0.11 ^{a}	22.2 ± 0.05 ^{c}	30.3 ± 0.00 ^{b}
$C_{18:2}$	9.42 ± 0.08 ^{c}	55.1 ± 0.05 ^{b}	57.5 ± 0.01 ^{a}
$C_{18:3}$	0.90 ± 0.01 ^{b}	5.06 ± 0.03 ^{a}	0.23 ± 0.01 ^{c}
∑SFA	16.9 ± 0.03 ^{b}	17.0 ± 0.00 ^{a}	10.8 ± 0.01 ^{c}
∑MUFA	72.5 ± 0.11 ^{a}	22.6 ± 0.06 ^{c}	30.5 ± 0.00 ^{b}
∑PUFA	10.6 ± 0.08 ^{c}	60.5 ± 0.06 ^{a}	58.7 ± 0.01 ^{b}

∑SFA, sum of saturated fatty acids; ∑MUFA, sum of monounsaturated fatty acids; ∑PUFA, sum of polyunsaturated fatty acids. Data are presented as means ± SD of three replicates. Different letters in the same row indicate significant differences at $p < 0.05$.

Monounsaturated fatty acids (MUFAs) were the most represented fatty acids in olive oil, which contained an amount of oleic acid accounting for 71.0 ± 0.11%. The saturated fatty acids (SFAs) were significantly more abundant in durum wheat oil and olive oil than in sunflower oil.

As PUFAs are the most susceptible to oxidation, the observed fatty acid composition easily explains the finding that olive oil showed the highest resistance to oxidation, since it had the lowest content of PUFAs. Moreover, the higher resistance to oxidation observed in durum wheat oil compared to sunflower oil, though the former had a slightly higher PUFA content, could be attributable to its higher content of SFAs, as well as to the greater presence of antioxidant compounds, primarily tocols [25].

Tocotrienols and tocopherols are recognized as natural antioxidants typical of vegetable oils and are used as additives by the food industries to cope with the low oxidative stability of PUFAs [38]. The concentrations of tocols ascertained in the oils are shown in Figure 1A, while those of the lipid fraction extracted from the *focaccia* samples are reported in Figure 1B. Durum wheat oil and DWO *focaccia* were the richest in tocotrienols, while tocopherols were more present in sunflower oil and in the corresponding *focaccia* (SO). This difference was interesting because studies have suggested that tocotrienols exert greater antioxidant activity than tocopherols and have more relevant health benefits [42,43]. Durum wheat oil contained 1094 mg/kg of tocotrienols, significantly higher than the amount determined for sunflower oil and olive oil, wherein these antioxidants were very scarce or absent. Other authors have reported similar findings in sunflower and olive oil [44] and have observed that wheat flour provides only a minimal contribution to the content of tocotrienols in baked goods [45].

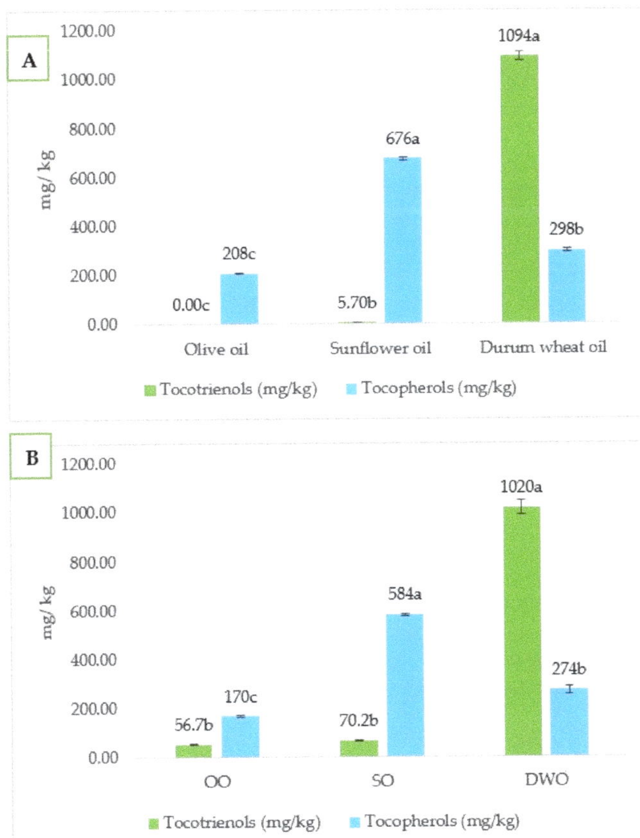

Figure 1. Tocopherols and tocotrienols content (mg/kg) of oil (**A**) and *focaccia* (**B**). SO = *focaccia* with sunflower oil; OO = *focaccia* with olive oil; DWO = *focaccia* with durum wheat oil. Data are presented as means ± SD of three replicates. Different letters in the same row indicate significant differences at $p < 0.05$.

3.2. Polar Compounds Content

The oxidative reactions that affect the lipid fraction of food determine the formation of compounds characterized by a higher polarity than unaltered triacylglycerols. In particular, oxidized triacylglycerols (ox-TAGs) are composed of triacylglycerols with an oxidized fatty acyl group, while triacylglycerol oligopolymers (TAGPs) are obtained from the latter with bonds that generate complex molecules, such as dimers and polymers. Finally, diacylglycerols (DAGs), monoacylglycerols (MAGs), and free fatty acids (FFAs) arise from the hydrolysis of triacylglycerols, as a result of lipolytic enzyme activity and moisture [18,46]. Recently, Chen et al. [18] compared the polar compounds of peanut, rapeseed, soybean, and linseed oils in different cooking conditions, observing that unsaturated fatty acids can lead to a high level of polar compounds. The SO *focaccia*, indeed, was richer in TAGPs and ox-TAGs (Figure 2) than the OO and DWO *focaccia*. These results were strongly associated with the oxidation stability; therefore, MUFA-rich oils and/or antioxidant-rich oils are able to limit the production of potentially adverse compounds [46].

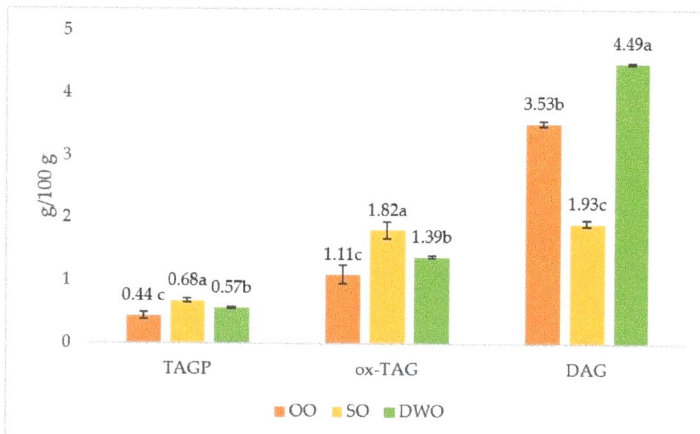

Figure 2. Polar compounds content (g/100 g of extracted fat) of *focaccia*. SO = *focaccia* with sunflower oil; OO = *focaccia* with olive oil; DWO = *focaccia* with durum wheat oil; TAGP = triacylglycerol oligopolymers; ox-TAG = oxidized triacylglycerols; DAG = diacylglycerols. Data are presented as means ± SD of three replicates. Different letters in the same row indicate significant differences at $p < 0.05$.

The DWO *focaccia* had a significantly higher content of DAGs than SO and OO, probably due to lipolytic events affecting the raw materials used for the extraction of oil. The milling industry should consider specific containment measures for these events [25], in spite of the fact that researchers have shown the health benefits of DAGs, especially in terms of body weight [47]. DAGs also have a function in the food industry due to their emulsifying properties [26,48,49].

The presence of polar compounds in *focaccia* has previously been reported. Delcuratolo et al. [16] studied the role of different toppings on the content of polar compounds, considering only the use of extra virgin olive oil. The type of *focaccia* topping influenced the exposition to thermal stress: potato-topped *focaccia*, which was moister than *focaccia* topped with onion and rosemary, was characterized by a less intense lipid degradation. Our study, on the other hand, highlighted that the type of oil influences the concentration of polar compounds in the final products. Our trials were conducted in the absence of toppings to avoid any interferences. However, the optimal combination of toppings and oil type could allow the dramatic limitation of the content of polar compounds in *focaccia*, thus avoiding consumer exposure.

3.3. Antioxidant Activity

ABTS and DPPH assays are based on the color change of a sample extract in connection with the capacity of an antioxidant to reduce a colored oxidant. The antioxidants derive from the sample, while the oxidants are in the solution that is prepared for the assay [50]. For both these assays, the DWO and OO *focaccia* showed a higher antioxidant activity than SO (Table 4). These findings reflected the high content of bioactive compounds in the durum wheat and olive oils, namely tocols in durum wheat oil and phenolic compounds in olive oil (accounting for 81.5 ± 3.15 mg GAE/kg oil—data not shown). Another study [51] compared wheat germ oil (from *T. aestivum*), sunflower oil, and olive oil. The latter, however, was chosen to be high-phenolic olive oil, accounting for 320 mg GAE/kg oil, which is remarkably high considering that the refining process reduces the levels of these compounds [52]. The authors detected the highest antioxidant activity in the high-phenolic olive oil, followed by wheat germ oil and sunflower oil [51].

Table 4. Total antioxidant capacity of experimental *focaccia*. OO = *focaccia* prepared with olive oil; DWO = *focaccia* prepared with durum wheat oil; SO = *focaccia* prepared with sunflower oil.

Focaccia Type	ABTS (μmol TE/g)	DPPH (μmol TE/g)
OO	0.64 ± 0.03 ab	0.46 ± 0.01 ab
DWO	0.70 ± 0.04 a	0.55 ± 0.02 a
SO	0.56 ± 0.04 b	0.38 ± 0.03 b

Expressed as μmol/g trolox equivalent. Data are presented as means ± SD of three replicates. Different letters in the same column indicate significant differences at $p < 0.05$.

3.4. Volatile Compounds

Bread is characterized by over 540 volatile compounds [53]: alcohols, aldehydes, esters, ketones, acids, pyrazines, furans, and sulfur compounds [54], although only a small number of them really influence the flavor profile [53]. Different volatile compounds may have different origins. Microorganisms ferment the sugars and produce ethanol, which is partly lost during baking, while some of them take part in secondary fermentation events, which lead to short-chain alcohols and fatty acids, esters, and carbonyl compounds [53,54]. The oxidation of lipids causes the production of aldehydes, such as hexanal, nonanal, octanal, heptanal, and 2-heptenal. The typical baking flavor is due to the Maillard reaction involving amino acids and sugars. The caramelization of sugars and the thermal degradation of sugars and amino acids form furans, acetic acid, acetaldehyde, and other compounds [54].

In the current study, the type of oil used in the preparation of *focaccia* significantly influenced the volatile profile (Table 5). Hexanal and nonanal, markers of lipid oxidation, were significantly higher in SO, followed by DW and OO, mirroring the other chemical determinations. Additionally, the 2-methylbutanal content varied among the different oils: SO and DWO were richer in 2-methylbutanal than OO, while the content of 3-methylbutanal was higher in DWO than in SO and OO. These compounds, due to the Maillard reaction [55], positively influence the aroma of the crust, conferring a malty and roasted odor [54]. Several authors have described the effect of the fatty acid composition on the intensity of the Maillard reaction and have found that its development is favored by a higher unsaturation level [56].

The Maillard reaction also generated benzaldehyde and furans; the content of the former was significantly lower in OO, while the content of the latter was significantly higher in DWO. This result could be attributed to the simultaneous presence, in DWO, of high concentrations of PUFA and diglycerides, which positively influence the presence of furans, as observed by Emektar et al. [57]. Pyrazines, with their olfactory properties, confer a pleasant roasted odor on baked goods [54,58] and are therefore used as additives to improve the organoleptic properties of bread and other bakery products [58]. DWO was significantly richer in pyrazines than SO and OO. These findings could be connected to the differing acidic compositions of the oils used, in particular to the PUFA content, which was the highest in durum wheat oil, followed by sunflower oil, then olive oil. In support of this,

Negroni et al. [59], studying the formation of pyrazines in glucose–lysine or xylose–lysine model systems added to olive oil, canola oil, and sunflower oil, suggested that higher unsaturation levels could lead to a higher presence of pyrazines.

Table 5. Volatile compounds of experimental *focaccia*. OO = *focaccia* prepared with olive oil; DWO = *focaccia* prepared with durum wheat oil; SO = *focaccia* prepared with sunflower oil.

Volatile Compounds (µg/g)	*Focaccia* Type		
	OO	DWO	SO
Aldehydes			
Hexanal	15.7 ± 0.01 c	22.2 ± 0.0 b	25.8 ± 1.07 a
Heptanal	1.00 ± 0.08 b	1.85 ± 0.02 a	2.01 ± 0.24 a
Nonanal	4.91 ± 0.33 b	4.83 ± 0.15 b	7.20 ± 0.01 a
2-Methylbutanal	12.0 ± 0.45 c	17.7 ± 0.64 b	20.5 ± 0.31 a
3-Methylbutanal	16.3 ± 0.63 c	25.5 ± 0.66 a	22.7 ± 0.08 b
Octanal	1.35 ± 0.02 ab	0.87 ± 0.16 b	1.71 ± 0.12 a
2-Heptenal	5.00 ± 0.01 b	4.63 ± 0.11 c	9.75 ± 0.15 a
2,4-Heptadienal	0.78 ± 0.06 c	1.56 ± 0.08 b	3.28 ± 0.11 a
Benzacetaldheyde	2.57 ± 0.04 b	4.58 ± 0.09 a	1.83 ± 0.08 c
Benzaldehyde	6.18 ± 0.55 b	7.55 ± 0.36 a	7.28 ± 0.18 a
Alcohols			
Ethyl alcohol	2.29 ± 0.32 a	2.10 ± 0.46 a	2.67 ± 0.22 a
2-Phenylethanol	8.54 ± 0.16 a	4.73 ± 0.12 c	7.57 ± 0.23 b
1-Hexanol	6.14 ± 0.00 b	2.82 ± 0.13 c	10.8 ± 0.08 a
Carboxylic acid			
Acetic acid	2.95 ± 0.00 a	1.41 ± 0.09 c	1.73 ± 0.05 b
Ketones			
Methyl ethyl ketone	1.72 ± 0.28 a	1.73 ± 0.17 a	1.48 ± 0.04 a
Furan compounds			
2-Furanmethanol	1.28 ± 0.25 c	9.67 ± 0.90 a	6.59 ± 0.27 b
Furan-2-pentyl	2.25 ± 0.26 c	4.80 ± 0.35 a	3.83 ± 0.09 b
2-Furancarboxaldehyde, 5-methyl-	0.50 ± 0.10 c	1.44 ± 0.07 a	0.63 ± 0.04 b
2-Furancarboxaldehyde (furfural)	1.46 ± 0.01 c	5.32 ± 0.06 a	5.15 ± 0.06 b
Pyrazines			
Methyl-pyrazine	2.72 ± 0.11 c	10.4 ± 0.82 a	8.20 ± 0.85 b
Ethyl-pyrazine	1.42 ± 0.12 c	2.66 ± 0.07 a	2.42 ± 0.02 b

Data are presented as means ± SD of three replicates. Different letters in the same row indicate significant differences at $p < 0.05$.

The fermentation of *focaccia*, caused by compressed yeast (*Saccharomyces cerevisiae*), produces ethyl alcohol. Despite its partial evaporation during baking, ethyl alcohol contributes to the aroma of baked goods [60], but its concentration was not influenced by the type of oil.

3.5. Physical Determinations

Ingredients and processing, especially baking, are principally responsible for the color of baked products: the golden-brown color of the crust is considered an important quality parameter [61]. Table 6 reports the colorimetric indices of the *focaccia* prepared with different oils, shown in Figure 3. DWO crumb and crust were significantly less luminous (lower L^*) and showed higher a^* (more intense red tone) than SO and OO, while no significant differences were observed for b^* (yellow index). These results agreed with the data for volatile compounds. In fact, higher levels of pyrazines and furans, both arising from thermal reactions which cause browning, were observed in DWO than in the other two *focaccia* types. These observations were reinforced by the calculation of the total color differences

(ΔE) of the crust and crumb, with OO taken as reference. The color differences of the crumb were lower than those observed in the crust, with the latter being more exposed to heat and more impacted by non-enzymatic browning. In particular, the difference in color between OO and DWO crumbs was considered recognizable only by an experienced observer ($1.5 < \Delta E < 3.0$). Instead, the difference in crust color was considered clearly recognizable ($3.0 < \Delta E < 6.0$). Other authors, studying bread, have reported the effect of the type of oil on the color of the crumb and crust [62,63].

Table 6. Physical determinations (color, texture, and dimensional variations during baking) of the experimental *focaccia* samples. OO = *focaccia* prepared with olive oil; DWO = *focaccia* prepared with durum wheat oil; SO = *focaccia* prepared with sunflower oil.

	***Focaccia* Type**		
	OO	**DWO**	**SO**
Color			
Crumb			
a^*	0.40 ± 0.08 ^{b}	0.87 ± 0.11 ^{a}	0.67 ± 0.08 ^{a}
b^*	18.2 ± 0.31 ^{a}	18.99 ± 0.46 ^{a}	21.0 ± 0.38 ^{a}
L^*	72.8 ± 1.37 ^{ab}	71.15 ± 1.58 ^{b}	74.5 ± 0.13 ^{a}
ΔE	-	1.88	3.28
Crust			
a^*	7.15 ± 0.34 ^{b}	10.1 ± 1.77 ^{a}	9.31 ± 0.66 ^{ab}
b^*	32.0 ± 1.14 ^{a}	33.4 ± 2.02 ^{a}	33.2 ± 2.39 ^{a}
L^*	67.5 ± 0.82 ^{a}	62.7 ± 1.89 ^{b}	64.7 ± 1.99 ^{ab}
ΔE	-	5.82	3.71
Texture			
Hardness (N)	7.69 ± 1.06 ^{b}	8.67 ± 1.13 ^{b}	12.1 ± 1.13 ^{a}
Springiness	0.94 ± 0.01 ^{a}	0.94 ± 0.02 ^{a}	0.95 ± 0.01 ^{a}
Chewiness (N)	5.73 ± 0.84 ^{b}	6.39 ± 0.44 ^{b}	9.80 ± 0.48 ^{a}
Cohesiveness	0.79 ± 0.01 ^{a}	0.82 ± 0.07 ^{a}	0.82 ± 0.01 ^{a}
Dimensional variations during baking			
Diameter variation (%)	-0.73 ± 0.01 ^{a}	-0.73 ± 0.01 ^{a}	-0.75 ± 0.01 ^{a}
Thickness variation (%)	117 ± 9.91 ^{a}	110 ± 11.7 ^{a}	118 ± 8.08 ^{a}

Data are presented as means $\pm$ SD of three replicates. Different letters in the same row indicate significant differences at $p < 0.05$.

Figure 3. (**a**) *Focaccia* with olive oil; (**b**) *focaccia* with durum wheat oil; (**c**) *focaccia* with sunflower oil.

Texture profile analysis (TPA) consisted of compressing a food sample twice in a reciprocating motion that mimicked the action of the jaw [64]. Four parameters were measured:

hardness, springiness, chewiness, and cohesiveness. The springiness and cohesiveness were very similar in all *focaccia* types, while the hardness and chewiness showed significant differences among samples (Table 6). In particular, the use of durum wheat oil and olive oil were related to a lower hardness and chewiness than sunflower oil. This result could be related to the content of DAGs, which was higher in these two oils than in the sunflower oil. DAGs, indeed, together with monoglycerides, are extensively used in breadmaking as emulsifiers to improve crumb softness. In addition, their presence can delay the firming process, due to the ability to form complexes with amylose and amylopectin [49,65].

During baking, an increase in volume occurs due to the thermal expansion of gases [12]. As a consequence, the thickness of the *focaccia* increased with baking, to a similar extent in all samples (Table 6). Meanwhile, the diameter decreased, but without an influence exerted by the type of oil. The effect of oil on the variation in the dimensional parameters during baking, therefore, was secondary, while other studies have reported a significant effect caused by the type of flour, due to differing fiber and gluten contents [12].

3.6. The Sensory Profile of Focaccia

The type of oil also significantly influenced the sensory properties of *focaccia* (Table 7).

Table 7. Sensory profile of experimental *focaccia*. OO = *focaccia* prepared with olive oil; DWO = *focaccia* prepared with durum wheat oil; SO = *focaccia* prepared with sunflower oil.

	***Focaccia* Type**		
Sensory Descriptor	**OO**	**DWO**	**SO**
Surface color	3.80 ± 0.35^{b}	5.55 ± 0.28^{a}	4.22 ± 0.30^{b}
Inner color	0.58 ± 0.12^{b}	0.85 ± 0.05^{a}	0.75 ± 0.00^{a}
Crumb porosity	4.27 ± 0.35^{b}	5.53 ± 0.12^{a}	3.57 ± 0.25^{b}
Focaccia odor	6.50 ± 0.05^{b}	7.67 ± 0.25^{a}	6.75 ± 0.05^{b}
Oxidized odor	0.00 ± 0.00^{b}	0.00 ± 0.00^{b}	0.63 ± 0.10^{a}
Roasted odor	1.15 ± 0.15^{b}	1.67 ± 0.22^{a}	1.13 ± 0.15^{b}
Crumb elasticity	5.38 ± 0.06^{a}	5.17 ± 0.38^{a}	5.60 ± 0.22^{a}
Softness	6.18 ± 0.29^{b}	7.08 ± 0.08^{a}	5.77 ± 0.08^{b}
Crumb moisture	5.55 ± 0.05^{a}	5.55 ± 0.79^{a}	5.48 ± 0.08^{a}
Greasiness	6.07 ± 0.19^{a}	5.12 ± 0.43^{b}	5.93 ± 0.19^{a}
Sweetness	1.15 ± 0.13^{a}	1.22 ± 0.28^{a}	1.42 ± 0.08^{a}
Saltiness	5.03 ± 0.20^{a}	4.37 ± 0.25^{b}	4.90 ± 0.09^{a}

Data are presented as means ± SD of three replicates. Different letters in the same row indicate significant differences at $p < 0.05$.

The perception of crumb and crust color varied with the type of oil, with DWO being darker than the others. The sensory evaluation of color agreed with the instrumental determination. Moreover, DWO was perceived as softer and more porous than the other *focaccia* types, while the elasticity of the crumbs was similar in all samples.

The type of oil did not affect the perception of sweetness and crumb moisture, while the panelists perceived DWO to be less salty and oily, which was interesting, considering the preference of consumers for *focaccia* that has not been excessively greased [12]. A hint of oxidized odor was detected only in SO, while none was observed in DWO and OO. A roasted odor, as well as the typical odor of *focaccia*, were both perceived significantly more intensely in DWO, due to its higher content of pyrazines.

4. Conclusions

Considering the significant nutritional, sensory, and health importance of the lipid fraction of *focaccia*, this study suggested that the choice of oil to be used in its preparation is not trivial. Although olive oil, rich in MUFAs, was proven to be the most resistant to oxidation, durum wheat oil, rich in PUFAs and tocols, was more stable than sunflower oil thanks to the greater presence of antioxidants. Moreover, the use of durum wheat oil demonstrated a positive effect on the physical and sensory characteristics of the end product. Therefore, the reformulation of bakery products with this oil will increase the value of the by-products generated by the durum wheat milling industries, while respecting the principles of the circular economy. This oil could offer a healthier alternative to consumers while combining the tradition of *focaccia* making with a viable strategy for product innovation and, at the same time, increasing the sustainability of the durum wheat chain.

Durum wheat oil could also respond to the need to find new alternatives to sunflower oil, the supply of which is facing considerable difficulties due to the war in Ukraine. It should be noted, however, that durum wheat oil is a high-quality niche product with a relatively high price (5.00 €/kg, compared to 2.50–3.00 €/kg for olive oil and 1.50–2.00 €/kg for sunflower oil). Its price is justified by the high nutritional value related to the remarkably high concentration of tocols, especially tocotrienols, and favorable levels of n-3 PUFAs. Currently, there is a single producer of durum wheat oil, with a productive capacity of 4000 tons/year. Therefore, there is presently not enough durum wheat oil to make up for potential losses in olive and sunflower oil, but there is good development potential because other companies will probably start producing it in the future.

Future investigations, however, are needed to deepen our knowledge of the effect of this oil on products' shelf lives and to widen its application in the food sector and beyond. In particular, the performance of durum wheat oil during *focaccia* storage should be investigated by conducting shelf-life studies in comparison with other refined oils. Furthermore, durum wheat oil could also be considered for interesting applications in pharmaceuticals, nutraceuticals, and the cosmetic sector, which could represent the main routes, alongside the food industry, for the valorization of cereal by-products.

Author Contributions: Conceptualization, A.P. and F.C.; formal analysis, A.P. and F.V.; investigation, F.V.; writing—original draft preparation, A.P. and F.V.; writing—review and editing, A.P., C.S., F.C., G.S., and F.V.; supervision, A.P. and F.C.; funding acquisition, A.P. and F.C. All authors have read and agreed to the published version of the manuscript.

Funding: This paper was supported by the PRIMA program under grant agreement No. 2031, project Flat Bread of Mediterranean area: Innovation and Emerging Process and Technology (Flat Bread Mine). The PRIMA program is an Art.185 initiative supported and funded under Horizon 2020, the European Union's Framework Program for Research and Innovation. The results and content found in this paper reflect only the authors' view. The PRIMA Foundation is not responsible for any use that may be made of the information this paper contains.

Institutional Review Board Statement: Not applicable.

Informed Consent Statement: For the sensory analysis, written informed consent was obtained from all the panelists.

Data Availability Statement: Data is contained within the article.

Acknowledgments: This research was made possible thanks to the contribution of experimental materials from Casillo Next Gen Food srl (Corato, Italy). The authors are grateful to Marcello Miani for the valuable discussion.

Conflicts of Interest: The authors declare no conflict of interest.

References

1. Rosell, C.M.; Bajerska, J.; El Sheikha, A.F. *Bread and Its Fortification: Nutrition and Health Benefits*; CRC Press: Boca Raton, FL, USA, 2015; pp. 1–10.
2. Pasqualone, A. Traditional flat breads spread from the Fertile Crescent: Production process and history of baking systems. *J. Ethn. Foods* **2018**, 5, 10–19. [CrossRef]
3. Boukid, F. Flatbread-a canvas for innovation: A review. *Appl. Food Biotechnol.* **2022**, 2, 100071. [CrossRef]
4. Research and Markets. Available online: https://www.researchandmarkets.com/reports/4987703/flatbread-market-to-2027-global-analysis-and#tag-pos-17 (accessed on 8 June 2022).
5. Pasqualone, A.; Delcuratolo, D.; Gomes, T. Focaccia Italian flat fatty bread. In *Flour and Breads and Their Fortification in Health and Disease Prevention*; Academic Press: Oxford, UK, 2011; pp. 47–58.

6. Pasqualone, A.; Vurro, F.; Summo, C.; Abd-El-Khalek, M.H.; Al-Dmoor, H.H.; Grgic, T.; Ruiz, M.; Magro, C.; Deligeorgakis, C.; Helou, C.; et al. The large and diverse family of Mediterranean flat breads: A database. *Foods* **2022**, *11*, 2326. [CrossRef] [PubMed]
7. Italian Ministry of Agricultural and Forest Policies. Decreto Ministeriale n. 350. Regolamento recante norme per l'individuazione dei prodotti tradizionali di cui all'articolo 8, comma 1, del D. Lgs. 30 aprile 1998, n. 173. *Ital. Off. J.* **1999**, *240*, 30–31.
8. Viganoni, L. Commercio e consumo nelle città che cambiano. In *Napoli, Città Medie, Spazi Esterni*; Franco Angeli: Milan, Italy, 2019; pp. 317–322.
9. European Commission. Commission Regulation (EU) No 97/2010 of 4 February 2010 entering a name in the register of traditional specialties guaranteed [Pizza Napoletana (TSG)]. *Off. J. Eur. Union* **2010**, *L34*, 7–16.
10. Bernklau, I.; Neußer, C.; Moroni, A.V.; Gysler, C.; Spagnolello, A.; Chung, W.; Becker, T. Structural, textural and sensory impact of sodium reduction on long fermented pizza. *Food Chem.* **2017**, *234*, 398–407. [CrossRef]
11. Camacho Flinois, J.; Dando, R.; Padilla-Zakour, O.I. Yogurt acid whey utilization for production of baked goods: Pancakes and pizza crust. *Foods* **2019**, *8*, 615. [CrossRef]
12. Pasqualone, A.; De Angelis, D.; Squeo, G.; Difonzo, G.; Caponio, F.; Summo, C. The effect of the addition of Apulian black chickpea flour on the nutritional and qualitative properties of durum wheat-based bakery products. *Foods* **2019**, *8*, 504. [CrossRef]
13. Sagar, N.A.; Pareek, S. Dough rheology, antioxidants, textural, physicochemical characteristics, and sensory quality of pizza base enriched with onion (*Allium cepa* L.) skin powder. *Sci. Rep.* **2020**, *10*, 18669. [CrossRef]
14. Pasqualone, A.; Costantini, M.; Faccia, M.; Difonzo, G.; Caponio, F.; Summo, C. The effectiveness of extruded-cooked lentil flour in preparing a gluten-free pizza with improved nutritional features and a good sensory quality. *Foods* **2022**, *11*, 482. [CrossRef]
15. Kumar, R.; Chandrashekar, S.; Kathalsar, A.K.; Prabhasankar, P. Multi-Fabaceae flour for gluten-free pizza base: Rheology, processing, microstructural, and physico-sensory characteristics. *J. Food Process. Preserv.* **2022**, *46*, 16239.
16. Delcuratolo, D.; Gomes, T.; Paradiso, V.M.; Nasti, R. Changes in the oxidative state of extra virgin olive oil used in baked Italian focaccia topped with different ingredients. *Food Chem.* **2008**, *106*, 222–226.
17. Delcuratolo, D.; Paradiso, V.M.; Nasti, R.; Gomes, T. Comparative study and quality evaluation of Italian Focaccias seasoned with extra virgin olive oil. *Int. J. Food Prop.* **2011**, *14*, 988–995.
18. Chen, J.; Zhang, L.; Li, Y.; Zhang, N.; Gao, Y.; Yu, X. The formation, determination and health implications of polar compounds in edible oils: Current status, challenges and perspectives. *Food Chem.* **2021**, *364*, 130451. [PubMed]
19. Bavaro, A.R.; Di Biase, M.; Conte, A.; Lonigro, S.L.; Caputo, L.; Cedola, A.; Del Nobile, M.A.; Logrieco, A.F.; Lavermicocca, P.; Valerio, F. *Weissella cibaria* short-fermented liquid sourdoughs based on quinoa or amaranth flours as fat replacer in focaccia bread formulation. *Int. J. Food Sci. Technol.* **2021**, *56*, 3197–3208.
20. Pasqualone, A. *Italian Durum Wheat Breads*; Nova Science Publisher Inc.: Hauppauge, NY, USA, 2012; pp. 57–79.
21. European Commission and Council. Commission Regulation (EU) No. 1291/2003 of 18 July 2003 on the protection of geographical indications and designations of origin for agricultural products and foodstuffs [Pane di Altamura (PDO)]. *Off. J. Eur. Union* **2003**, *L181*, 12–19.
22. Brescia, M.A.; Sacco, D.; Sgaramella, A.; Pasqualone, A.; Simeone, R.; Peri, G.; Sacco, A. Characterisation of different typical Italian breads by means of traditional, spectroscopic and image analyses. *Food Chem.* **2007**, *104*, 429–438.
23. Pasqualone, A.; Laddomada, B.; Centomani, I.; Paradiso, V.M.; Minervini, D.; Caponio, F.; Summo, C. Bread making aptitude of mixtures of re-milled semolina and selected durum wheat milling by-products. *LWT* **2017**, *78*, 151–159.
24. Skendi, A.; Zinoviadou, K.G.; Papageorgiou, M.; Rocha, J.M. Advances on the valorisation and functionalization of by-products and wastes from cereal-based processing industry. *Foods* **2020**, *9*, 1243.
25. Squeo, G.; Silletti, R.; Napoletano, G.; Greco Miani, M.; Difonzo, G.; Pasqualone, A.; Caponio, F. Characterization and effect of refining on the oil extracted from durum wheat by-products. *Foods* **2022**, *11*, 683.
26. Vurro, F.; Greco Miani, M.; Summo, C.; Caponio, F.; Pasqualone, A. Effect of durum wheat oil on the physico-chemical and sensory features of biscuits. *Foods* **2022**, *11*, 1282. [PubMed]
27. Glauben, T.; Svanidze, M.; Götz, L.; Prehn, S.; Jamali Jaghdani, T.; Đurić, I.; Kuhn, L. The war in Ukraine, agricultural trade and risks to global food security. *Intereconomics* **2022**, *57*, 157–163.
28. Statista 2022. Available online: https://www.statista.com/statistics/263937/vegetable-oils-global-consumption/ (accessed on 8 June 2022).
29. Ali, B.M.; van der Werf, W.; Lansink, A.O. Assessment of the environmental impacts of *Xylella fastidiosa* subsp. Pauca in Puglia. *J. Crop Prot.* **2021**, *142*, 105519.
30. AOCS. *Official Methods and Recommended Practices of the American Oil Chemist's Society Cd 12c-16 "Accelerated Oxidation Test for the Determination of the Oxidation Stability of Foods, Oils and Fats Using the Oxitest Oxidation Test Reactor"*; AOCS: Urbana, IL, USA, 2017.
31. Difonzo, G.; Pasqualone, A.; Silletti, R.; Cosmai, L.; Summo, C.; Paradiso, V.M.; Caponio, F. Use of olive leaf extract to reduce lipid oxidation of baked snacks. *Food Res. Int.* **2018**, *108*, 48–56.
32. Troilo, M.; Difonzo, G.; Paradiso, V.M.; Pasqualone, A.; Caponio, F. Grape Pomace as Innovative Flour for the Formulation of Functional Muffins: How Particle Size Affects the Nutritional, Textural and Sensory Properties. *Foods* **2022**, *11*, 1799. [PubMed]
33. Difonzo, G.; Russo, A.; Trani, A.; Paradiso, V.M.; Ranieri, M.; Pasqualone, A.; Summo, C.; Tamma, G.; Silletti, R.; Caponio, F. Green extracts from Coratina olive cultivar leaves: Antioxidant characterization and biological activity. *J. Funct. Foods* **2017**, *31*, 63–70.
34. *ISO 13299*; Sensory Analysis. Methodology. General Guidance for Establishing a Sensory Profile. ISO: Geneva, Switzerland, 2016.

35. *ISO 11132*; Sensory Analysis. Methodology. Guidelines for the Measurement of the Performance of a Quantitative Descriptive Sensory Panel. ISO: Geneva, Switzerland, 2021.
36. *ISO 8589*; Sensory Analysis. General Guidance for the Design of Test Rooms. ISO: Geneva, Switzerland, 2007.
37. Caruso, M.C.; Galgano, F.; Colangelo, M.A.; Condelli, N.; Scarpa, T.; Tolve, R.; Favati, F. Evaluation of the oxidative stability of bakery products by OXITEST method and sensory analysis. *Eur. Food Res. Technol.* **2017**, *243*, 1183–1191.
38. Saini, R.K.; Prasad, P.; Sreedhar, R.V.; Akhilender Naidu, K.; Shang, X.; Keum, Y.S. Omega-3 Polyunsaturated Fatty Acids (PUFAs): Emerging plant and microbial sources, oxidative stability, bioavailability, and health benefits—A Review. *Antioxidants* **2021**, *10*, 1627.
39. Codex Standards for Fats and Oils from Vegetable Sources. Available online: https://www.fao.org/3/y2774e/y2774e04.htm#bm4.2 (accessed on 22 July 2022).
40. Giakoumis, E.G. Analysis of 22 vegetable oils' physico-chemical properties and fatty acid composition on a statistical basis, and correlation with the degree of unsaturation. *Renew. Energy* **2018**, *126*, 403–419.
41. Dorni, C.; Sharma, P.; Saikia, G.; Longvah, T. Fatty acid profile of edible oils and fats consumed in India. *Food Chem.* **2018**, *238*, 9–15.
42. Peh, H.Y.; Tan, W.D.; Liao, W.; Wong, W.F. Vitamin E therapy beyond cancer: Tocopherol versus tocotrienol. *Pharmacol. Ther.* **2016**, *162*, 152–169. [PubMed]
43. Tan, S.M.Q.; Chiew, Y.; Ahmad, B.; Kadir, K.A. Tocotrienol-rich vitamin E from palm oil (tocovid) and its effects in diabetes 608 and diabetic nephropathy: A pilot phase II clinical trial. *Nutrients* **2018**, *10*, 1315.
44. Rossi, M.; Alamprese, C.; Ratti, S. Tocopherols and tocotrienols as free radical-scavengers in refined vegetable oils and their 610 stability during deep-fat frying. *Food Chem.* **2007**, *102*, 812–817.
45. Mignogna, R.; Fratianni, A.; Niro, S.; Panfili, G. Tocopherol and tocotrienol analysis as a tool to discriminate different fat ingredients in bakery products. *Food Control* **2015**, *54*, 31–38.
46. Chen, H.; Chen, Y.; Wan, C.; Deng, Q. Free-radical-mediated formation mechanism of polar polymeric triglycerides in vegetable oil studied by electron spin resonance and high-performance liquid chromatography. *J. Agric. Food Chem.* **2021**, *69*, 9034–9042.
47. Atehli, D.; Wang, J.; Yu, J.; Ali, F.; Wang, Y. Effects of mono-and diglycerides of fatty acids on the milk fat globule membrane after heat treatment. *Int. J. Dairy Technol.* **2020**, *73*, 667–673.
48. Azizi, M.H.; Rajabzadeh, N.; Riahi, E. Effect of mono-diglyceride and lecithin on dough rheological characteristics and quality of flat bread. *LWT-Food Sci. Technol.* **2003**, *36*, 189–193.
49. Tebben, L.; Shen, Y.; Li, Y. Improvers and functional ingredients in whole wheat bread: A review of their effects on dough properties and bread quality. *Trends Food Sci. Technol.* **2018**, *81*, 10–24.
50. Xiao, F.; Xu, T.; Lu, B.; Liu, R. Guidelines for antioxidant assays for food components. *Food Front.* **2020**, *1*, 60–69.
51. Liu, R.; Lu, M.; Zhang, T.; Zhang, Z.; Jin, Q.; Chang, M.; Wang, X. Evaluation of the antioxidant properties of micronutrients in different vegetable oils. *Eur. J. Lipid Sci. Technol.* **2020**, *122*, 1900079.
52. Durmaz, G.; Gökmen, V. Effect of refining on bioactive composition and oxidative stability of hazelnut oil. *Int. Food Res. J.* **2019**, *116*, 586–591. [CrossRef] [PubMed]
53. Dong, Y.; Karboune, S. A review of bread qualities and current strategies for bread bioprotection: Flavor, sensory, rheological, and textural attributes. *Compr. Rev. Food Sci. Food Saf.* **2021**, *20*, 1937–1981. [CrossRef] [PubMed]
54. Pico, J.; Bernal, J.; Gómez, M. Wheat bread aroma compounds in crumb and crust: A review. *Food Res. Int.* **2015**, *75*, 200–215. [CrossRef]
55. Pasqualone, A.; Bianco, A.M.; Paradiso, V.M.; Summo, C.; Gambacorta, G.; Caponio, F.; Blanco, A. Production and characterization of functional biscuits obtained from purple wheat. *Food Chem.* **2015**, *180*, 64–70. [CrossRef]
56. Zamora, R.; Hidalgo, F.J. Coordinate contribution of lipid oxidation and Maillard reaction to the nonenzymatic food browning. *Crit. Rev. Food Sci. Nutr.* **2005**, *45*, 49–59. [CrossRef]
57. Emektar, K.; Kantekin-Erdogan, M.N.; Tekin, A. Furan formation in some vegetable oils during heat treatments. *Food Chem.* **2022**, *386*, 132744. [CrossRef]
58. Mortzfeld, F.B.; Hashem, C.; Vranková, K.; Winkler, M.; Rudroff, F. Pyrazines: Synthesis and industrial application of these valuable flavor and fragrance compounds. *Biotechnol. J.* **2020**, *15*, 2000064. [CrossRef]
59. Negroni, M.; D'Agostin, A.; Arnoldi, A. Effects of olive, canola, and sunflower oils on the formation of volatiles from the Maillard reaction of lysine with xylose and glucose. *J. Agric. Food Chem.* **2001**, *49*, 439–445. [CrossRef]
60. De Luca, L.; Aiello, A.; Pizzolongo, F.; Blaiotta, G.; Aponte, M.; Romano, R. Volatile organic compounds in breads prepared with different sourdoughs. *Appl. Sci.* **2021**, *11*, 1330. [CrossRef]
61. Melini, V.; Melini, F.; Acquistucci, R. Nutritional characterization of an Italian traditional bread from ancient grains: The case study of the durum wheat bread "Pane di Monreale". *Eur. Food Res. Technol.* **2021**, *247*, 193–200. [CrossRef]
62. Debonne, E.; De Leyn, I.; Verwaeren, J.; Moens, S.; Devlieghere, F.; Eeckhout, M.; Van Bockstaele, F. The influence of natural oils of blackcurrant, black cumin seed, thyme and wheat germ on dough and bread technological and microbiological quality. *LWT-Food Sci.* **2018**, *93*, 212–219. [CrossRef]
63. Pycia, K.; Kapusta, I.; Jaworska, G. Walnut oil and oilcake affect selected the physicochemical and antioxidant properties of wheat bread enriched with them. *J. Food Process. Preserv.* **2020**, *44*, 14573. [CrossRef]

64. Funami, T.; Nakauma, M. Instrumental food texture evaluation in relation to human perception. *Food Hydroc.* **2022**, *124*, 107253. [CrossRef]
65. Keskin, S.Ö.; Sumnu, G.; Sahin, S. Usage of enzymes in a novel baking process. *Food/Nahrung* **2004**, *48*, 156–160. [CrossRef] [PubMed]

foods

MDPI

Article

Effect of Durum Wheat Oil on the Physico-Chemical and Sensory Features of Biscuits

Francesca Vurro [1], Marcello Greco Miani [2], Carmine Summo [1], Francesco Caponio [1] and Antonella Pasqualone [1,*]

[1] Department of Soil, Plant and Food Science (DISSPA), University of Bari Aldo Moro, Via Amendola, 165/a, 70126 Bari, Italy; francesca.vurro@uniba.it (F.V.); carmine.summo@uniba.it (C.S.); francesco.caponio@uniba.it (F.C.)

[2] Casillo Next Gen Food srl, Laboratory R&D Casillo Group, Via A. Sant'Elia, 70033 Corato, Italy; marcello.miani@casillogroup.it

* Correspondence: antonella.pasqualone@uniba.it

Abstract: Lipids play an important role in defining the overall quality of biscuits, particularly in terms of resistance to oxidation, as well as for their influence on textural and sensorial properties. The aim of this work was to investigate the effects of durum wheat oil on the physico-chemical and sensory features of biscuits. Control biscuits (C) prepared with the commonly used sunflower oil were compared with samples prepared with durum wheat oil at 50% (D50) and 100% replacement levels (D100). The reformulated biscuits were very rich in tocols, especially tocotrienols (982.9, 635.2, and 64.1 mg/kg on lipid fraction weight in D100, D50, and C, respectively). The higher content of antioxidants extended the resistance to the oxidation of biscuits (induction time = 53.61, 70.87, and 79.92 h in C, D50, and D100, respectively). D100 showed the lowest amounts of triacylglycerol oligopolymers and oxidized triacylglycerols, and the lowest amounts of the volatile markers of lipid oxidation (hexanal and nonanal). The use of durum wheat oil did not affect the sensorial and textural properties, compared to C. This study suggests that durum wheat oil could be effectively used in biscuit-making to decrease the oxidative phenomena and increase the bioactives of the end-products.

Keywords: biscuits; durum wheat oil; by-products; tocotrienols; tocopherols; induction time; oxidation; sensory properties; volatile compounds

Citation: Vurro, F.; Greco Miani, M.; Summo, C.; Caponio, F.; Pasqualone, A. Effect of Durum Wheat Oil on the Physico-Chemical and Sensory Features of Biscuits. *Foods* **2022**, *11*, 1282. https://doi.org/10.3390/foods11091282

Academic Editor: Marina Carcea

Received: 29 March 2022
Accepted: 26 April 2022
Published: 28 April 2022

1. Introduction

Biscuits are one of the most popular bakery items [1] and their global demand has grown by 31.6% during the outbreak of COVID-19 [2,3]. These products are widely consumed, due to their pleasant sensory characteristics, affordability, and long shelf-life [4,5].

The production process is relatively simple and consists of mixing the ingredients, shaping, baking, and packaging [6,7]. The usual ingredients are flour, fat or oil, sugar, water, and chemical leavening agents, and the optional minor ingredients are salt, eggs, emulsifiers, milk, and flavoring compounds [4,6,8]. Though not being perceived as fatty foods, biscuits show high contents of lipids, from 7.5% to 25% [9,10], which play an important technological role. During mixing, indeed, lipids lubricate flour and, due to their hydrophobic nature, inhibit hydration and gluten development [8]. As a consequence, lipids positively influence the physical and sensory characteristics of biscuits [11,12], conferring the typical 'melt in the mouth' and crumbly texture, as well as flavor [1,8]. Lipids, however, are susceptible to hydrolytic degradation, oxidation processes and thermal polymerization. The compounds originated by the oxidative degradation, namely triglyceride oligopolymers and oxidized triglycerides, are among the major biscuit contaminants related to storage [13] and are considered "nutritionally suspect", with potential implications on human health [14]. Lipid oxidation causes the development of off-odors and flavors that negatively influence the palatability and shelf-life of biscuits. This aspect, together with the great impact on the texture, makes the choice of fat crucial [5,11,15].

The formulation of low-fat biscuits or the use of healthier fats represents a good opportunity for bakery companies to release new products [1]. In particular, vegetable oils or hydrocolloids and oleogels with lubricant and flow properties similar to those of fats have been proposed to replace them totally or partially [1,16–18]. Other studies proposed to replace fat with flour from tomato seeds, poppy seeds, or apricot kernels [19–21]. However, the problem of keeping the biscuit consistency unaltered remains of relevant importance, because fat replacement has a great impact on the textural attributes [8].

At a global level, wheat is the second cultivated cereal crop, with a production of over 775 million tons in 2021 and a similar production forecast for 2022 [22,23]. The majority is represented by soft wheat (*Triticum aestivum* L.), whereas durum wheat (*Triticum turgidum* L. var *durum* Desf.), which is fundamental in the production of pasta, bulgur, couscous, freekeh, and some types of bread [24–26], accounts for 5% [27]. Canada is the main producer of durum wheat globally, while Italy is the main producer in the European Union, contributing about 4 million tons [28]. The first Italian region for durum wheat production is Apulia, where more than 83% of the cereal-producing area is represented by this crop [29]. Germ, bran, and de-branning fractions are the main by-products of the wheat milling industry, mostly destined for animal feeding [30]. These by-products contain about 80% of the wheat lipids, roughly 65% in the germ, and 15% in bran, with a total lipid content accounting for 2.4–3.8% of wheat kernel weight on a dry basis [31]. The industrial extraction of oil from the wheat milling by-products is well established in the bread wheat chain, while the availability of durum wheat oil is still limited. This situation, however, is going to change because of ongoing investments—in particular in the Apulia region—aimed at exploiting the durum wheat milling by-products for extracting oil. Such investments have been prompted by the need of implementing the principles of the circular economy in the durum wheat chain [32]. A recent paper has evidenced the interesting nutritional properties of durum wheat oil, extracted from a 40/60 *w*/*w* mixture of milling and de-branning by-products (bran, germ, and de-branning fractions). Even after the refining process following the solvent extraction, this oil showed an outstanding content of phytosterols (20.9 g/kg; mainly composed of β-sitosterol, followed by sitostanol, campestanol, and campesterol) and policosanols (754 mg/kg) [33]. Furthermore, it showed a very high content of tocotrienols (about 1100 mg/kg) [33]. No studies, however, have considered the use of durum wheat oil in the preparation of biscuits so far.

After a campaign calling for the substitution of palm oil in bakery products, including biscuits with other oils, sunflower oil has become the most used in Italy. However, the very recent crisis in Ukraine, one of the major producers of sunflower oil (over 5 million tons in 2019) [22], caused a shortage of this kind of oil. Proposing a larger use in bakery products, namely biscuits, of the oil extracted from the by-products of durum wheat milling and de-branning could therefore: (i) significantly contribute to the expected transition from the linear economy to a circular economy in the durum wheat chain; (ii) improve the quality and nutritional value of the end-products; (iii) allow for a valid alternative in the case of shortage of other oils more frequently used.

The aim of this work was, therefore, to investigate the effects of durum wheat oil on the physico-chemical and sensory features of biscuits in comparison with sunflower oil.

2. Materials and Methods

2.1. Materials

Wheat flour type 00 (Despar Italia, Casalecchio di Reno, Italy) (carbohydrates 72 g/100 g; protein 10 g/100 g; fat 1.7 g/100 g; fiber 1.4 g/100 g), sugar (sucrose) (Despar Italia, Casalecchio di Reno, Italy), partially skimmed milk (Granarolo, Bologna, Italy) (carbohydrates 5 g/100 g, protein 3.4 g/100 g; fat 1.6 g/100 g), baking powder (R. Barra s.a.s., Crispiano, Italy), refined sunflower oil (Olearia De Santis, Bitonto, Italy) were purchased from local retailers. Durum wheat oil, produced as reported in Squeo et al. [33], was provided by Casillo Next Gen Food srl (Corato, Italy).

2.2. Biscuit Preparation

Three biscuit types were prepared with: 100% sunflower oil (C); 50% sunflower oil and 50% durum wheat oil (D50); and 100% durum wheat oil (D100), according to the formulation reported in Table 1. Biscuit formulation was defined by means of preliminary trials. The process consisted of (i) mixing wheat flour, sugar, and oil using a spiral kneader (Bosh MFQ40304, München, Germany) for 5 min; (ii) adding partially skimmed milk and baking powder; and kneading for 12 min; (iii) rolling the dough and shaping as rectangular biscuits (6 cm length; 2.5 cm width; 1 cm thickness); and (iv) baking in an electric oven (Smeg SI 850 RA-5 oven, Smreg S.p.A., Guastalla, Italy) for 16 min at 160 °C. Biscuits were placed in the baking tray according to a randomized block distribution to take into account possible border effects.

Table 1. Formulation of the experimental biscuits. C = Biscuits prepared with 100% sunflower oil; D50 = biscuits prepared with sunflower oil (50%) and durum wheat oil (50%); D100 = biscuits prepared with 100% durum wheat oil.

Ingredients	C (g)	D50 (g)	D100 (g)
Wheat flour	400	400	400
Sunflower oil	112	56	–
Durum wheat oil	–	56	112
Sugar	112	112	112
Partially skimmed milk	128	128	128
Baking powder	4.8	4.8	4.8

2.3. Determination of the Resistance to Oxidation

The resistance to oxidation was determined by RapidOxy (Anton Paar, Blankenfelde-Mahlow, Germany). The samples (1 g) were analyzed at 140 °C and under 700 kPa O_2 pressure. The induction time, i.e., the time needed for a 10% drop of the O_2 pressure, was recorded. The analysis was carried out in triplicate.

2.4. Determination of Tocopherols and Tocotrienols

The tocopherols and tocotrienols of the oils and of the fatty fraction of biscuits—the latter extracted by Soxhlet method, using diethyl ether (SER 148 extraction system, Velp Scientifica srl, Usmate, Italy)—were determined by RP-UHPLC-FLD (Dionex Ultimate 3000 RSLC, Waltham, MA, USA). In particular, 0.02–0.03 g of sample was dissolved in 1 mL of 2-propanol. The samples were filtered by a 0.45 μm polytetrafluoroethylene (PTFE) filter and injected into a UHPLC system consisting of an HPG-3200 RS Pump, a WPS-3000 autosampler, a TCC-3000 column compartment, and an FLD-3400RS fluorescent detector (excitation wavelength 295 nm, emission wavelength 325 nm). The stationary phase was a Dionex Acclaim 120 C18 analytical column (Thermo Scientific, Waltham, MA, USA) with 3 μm particle size, 120 Å, 3 × 150 mm; the mobile phase was 1:1 (v/v) methanol and acetonitrile at a flow rate of 1 mL/min in isocratic elution. The software was Chromeleon (Dionex-ThermoFisher Scientific, Waltham, MA, USA). The single tocopherols were determined by external standard method on the basis of a previously set calibration curves obtained for α-tocotrienols and α-tocopherols. The content of tocopherols and tocotrienols were expressed as mg/kg on lipid fraction weight. The determinations were carried out in triplicate.

2.5. Determination of Polar Compounds of the Lipid Fraction

The polar compounds were recovered from the lipid fraction of biscuits by silica gel column chromatography and analyzed by high-performance size-exclusion chromatography (HPSEC) as described by Caponio et al. [34] with the only modification of using tetrahydrofuran (THF) as eluant, instead of dichloromethane. The analyses were carried out in duplicate.

2.6. Determination of Volatile Compounds

The volatile compounds of biscuits were determined by headspace solid-phase microextraction (HS-SPME) coupled with gas chromatography/mass spectrometry (GC–MS) as previously reported [35]. The quantification was carried out by standardizing the peak areas of the volatile compounds with the peak area of the internal standard (1-propanol). The analyses were carried out in duplicate.

2.7. Texture Profile Analysis

The textural properties of biscuits were determined by a 3-point bending test as described in Pasqualone et al. [35], with few modifications. The force (N) required to fracture the sample was recorded as biscuit hardness. A Texture Analyzer (Z1.0 TN, Zwick GmbH & Co., Ulm, Germany), equipped with a 1000 N load cell, was used. The distance between the support bars was 3 cm. The probe, the speed of which was set at 5 mm/min, moved downward until the biscuit was broken. Six replicated analyses were carried out.

2.8. Color Measurement

The color of the biscuits was analyzed in the CIE $L^*a^*b^*$ scale, under a D65 illuminant, by using a CM-600d colorimeter (Konica Minolta, Tokyo, Japan). Lightness (L^*), redness (a^*), and yellowness (b^*) were measured. Ten replicated analyses were carried out.

The total color variation (ΔE) was calculated to compare the differences between C biscuits and the two types of biscuits containing durum wheat oil (D50 and D100), according to the equation:

$$\Delta E = [(L^* - L_0^*)^2 + (a^* - a_0^*)^2 + (b^* - b_0^*)^2]^{1/2}$$

where L_0^*, a_0^*, and b_0^* were the color coordinates for the reference biscuits (C), whereas L^*, b^*, a^* were the color coordinates of the other samples. The mean values were considered in the calculation. The obtained results were then evaluated according to the following ΔE scale: 0–2.0 = unrecognizable difference; 2.0–3.5 = difference recognizable by an experienced observer; <3.5 = clear difference [36].

2.9. Determination of Dimensional Parameters

The dimensional parameters of biscuits (thickness, width, and length) were measured before and after baking by a calliper, and the increase induced by baking was calculated by difference. Six replicated analyses were carried out.

2.10. Determination of Sensory Properties

The quantitative descriptive analysis (QDA) of the sensory properties of biscuits was carried out according to the International Standardization Organization (ISO) standard 13299 [37], by a trained panel of 8 members. Panelists (4 men and 4 women) ranged in age from 23 to 55 years. Panelists, regular consumers of biscuits, had neither food allergies nor intolerances. They were informed about the study aims and provided written consent to perform the sensory analysis, according to the ethical guidelines of the laboratory of Food Science and Technology of the Department of Soil, Plant and Food Science of the University of Bari (Italy). Pre-test sessions were made to define the list of descriptors and to verify the discriminating ability, consistency, and reliability of panelists, as in the ISO Standard 11132 [38]. The sensory terms are defined in detail in Table 2. The intensity of every attribute was expressed on a 10 cm unstructured linear scale. The samples were randomized and presented to the panelists in white dishes marked with alphanumeric codes. The sensory properties were evaluated in a conference room, where temporary partitions were used to set up isolated tasting booths for separating the panelists during the analysis, in agreement with the ISO standard 8589 [39]. The testing was performed at ambient room temperature (20 ± 2 °C).

Table 2. Descriptive terms used for the sensory profiling of biscuits.

Descriptor	Definition	Scale Anchors	
		Min = 0 (c.u.) *	Max = 10 (c.u.)
	Visual–tactile characteristics		
Porosity	Presence of pores	Absent	Very intense
	The way the biscuit fractures when broken by fingers	Breaks with difficulty	Crumbly, breaks easily
Breakability	The way the biscuit fractures when broken by fingers	Breaks with difficulty	Crumbly, breaks easily
	Odor notes		
Caramel	Typical odor associated with caramel	Absent	Very intense
Oxidized oil	Typical odor associated with oxidized oil	Absent	Very intense
Shortbread	Typical odor associated with biscuits	Absent	Very intense

* c.u. = contractual units.

2.11. Statistical Analysis

The results were expressed as the mean ± standard deviation (SD). Significant differences were determined at $p < 0.05$, according to the one-way analysis of variance (ANOVA), followed by a Tukey test for multiple comparisons. Statistical analysis was carried out by the Minitab Statistical Software (Minitab Inc., State College, PA, USA).

3. Results and Discussion

3.1. Tocopherols and Tocotrienols Content, Resistance to the Oxidation, and Polar Compounds

Table 3 shows the content of tocopherols and tocotrienols in the oils used in the experimental trials, and in the obtained biscuits. Durum wheat oil was remarkably rich in tocotrienols, which instead were assessed in very low amounts in sunflower oil. On the contrary, significantly higher levels of tocopherols were observed in sunflower oil than in the durum wheat one. Overall, the total sum of tocopherols + tocotrienols in durum wheat oil accounted for 1425.2 mg/kg, roughly doubling the total amount of sunflower oil. The wheat germ, which was present in the by-product mixture used to extract the durum wheat oil used in these biscuit-making trials, is known to be an important source of tocopherols and tocotrienols. These compounds are a group of eight isomers, collectively known as tocols or vitamin E, synthesized only by plants and photosynthetic microorganisms [40]. The different forms of tocopherols and tocotrienols (α, β, γ, δ) depend on the number and location of methyl groups in the hydrophilic head of 6-chromanol [40,41]. Tocopherols and tocotrienols both act as natural antioxidants [42–44]. Furthermore, tocotrienols have been reported to be effective in the prevention of cancer-related processes, cardiovascular pathologies, and Alzheimer's disease [45–47].

The different concentrations of tocopherols and tocotrienols of the two oils influenced the content of these compounds in biscuits: those prepared with 100% durum wheat oil showed a higher tocotrienol concentration and lower tocopherol level than biscuits prepared with total or partial replacement of sunflower oil. The content of tocotrienols observed in all the biscuits was also positively influenced by the contribution of wheat flour, known to contain more tocotrienols than tocopherols [48,49], thus explaining the presence of tocotrienols observed in the 100% sunflower oil-containing biscuits.

Table 3. Total tocopherols and tocotrienols of the oils and biscuits. C = Biscuits prepared with 100% sunflower oil; D50 = biscuits prepared with sunflower oil (50%) and durum wheat oil (50%); D100 = biscuits prepared with 100% durum wheat oil.

Sample	Tocopherols (mg/kg)	Tocotrienols (mg/kg)
Oils		
Sunflower oil	677.9 ± 7.1 [a]	7.4 ± 1.4 [b]
Durum wheat oil	305.6 ± 7.6 [b] **	1119.6 ± 19.5 [a] **
Biscuits		
C	601.8 ± 10.1 [a]	64.1 ± 11.8 [c]
D50	418.9 ± 11.1 [b]	635.2 ± 38.7 [b]
D100	280.6 ± 8.3 [c]	982.9 ± 11.2 [a]

** From [33]. Different letters in the same column, for the same sample type, indicate significant differences at $p < 0.05$.

Table 4 reports the resistance to the oxidation of the oils and the experimental biscuits. The results are expressed as induction time (IT), i.e., the 'stability time' before fat oxidation, which corresponds to a 10% decrease of the O_2 pressure in the testing device due to the consumption of oxygen by the sample being oxidized [50].

Table 4. Resistance to oxidation of the oils and biscuits. C = Biscuits prepared with 100% sunflower oil; D50 = biscuits prepared with sunflower oil (50%) and durum wheat oil (50%); D100 = biscuits prepared with 100% durum wheat oil.

Sample	IT * (h)
Oils	
Sunflower oil	31.50 ± 0.42 [b]
Durum wheat oil	39.80 ± 0.09 [a]
Biscuits	
C	53.61 ± 1.87 [c]
D50	70.87 ± 2.94 [b]
D100	79.92 ± 2.21 [a]

* IT = Induction time. Different letters for the same sample type indicate significant differences at $p < 0.05$.

Durum wheat oil was more stable than sunflower against the onset of rancidity and oxidative deterioration, as indicated by its higher value of IT. The analysis of biscuits evidenced that replacing sunflower oil with durum wheat oil progressively and significantly increased the IT according to the percentage of replacement. This result was due to the high level of total tocols of the durum wheat oil. Similarly, Sharif et al. [51] observed that biscuits prepared with rice bran oil had an extended shelf-life due to high levels of tocopherols, tocotrienols, and oryzanols.

To have a better insight into the effect of oil substitution on biscuit oxidation, which starts during the production process and goes ahead during storage [13], the analysis of polar compounds was also carried out (Table 5). This analysis enabled the separation and quantification of the different classes of substances due to both oxidation (triacylglycerol oligopolymers and oxidized triacylglycerols) and hydrolysis (diacylglycerols) of any lipid [9,52]. The oxidation products, in particular, are most suspected of altering the nutritional properties of foods and causing adverse physiological effects [14,53,54]. The biscuits prepared with sunflower oil showed significantly higher contents of triacylglycerol oligopolymers and oxidized triacylglycerols than biscuits with durum wheat oil. This result was in line with the lower level of antioxidants observed in the former.

Table 5. Polar compounds of the lipid fraction of the experimental biscuits. C = Biscuits prepared with 100% sunflower oil; D50 = biscuits prepared with sunflower oil (50%) and durum wheat oil (50%); D100 = biscuits prepared with 100% durum wheat oil.

Compound (g/kg)	Sample Type		
	C	D50	D100
TAGP	0.43 ± 0.03 [a]	0.31 ± 0.02 [b]	0.17 ± 0.02 [c]
ox-TAG	3.52 ± 0.13 [a]	2.63 ± 0.43 [b]	1.81 ± 0.19 [b]
DG	1.30 ± 0.02 [c]	3.03 ± 0.05 [b]	4.93 ± 0.17 [a]

TAGP = triacylglycerol oligopolymers; ox-TAG = oxidized triacylglycerols; DG = diacylglycerols Different letters in the same row indicate significant differences at $p < 0.05$.

The biscuits prepared with durum wheat oil, however, showed a higher level of lipid hydrolytic degradation than biscuits made with sunflower oil, mirroring the high content of diacylglycerols of the durum wheat oil used [33]. The latter, in turn, was probably due to lipolytic phenomena occurred in the starting wheat milling/de-branning by-products, against which containment measures should be taken. On the other hand, the most detrimental for quality are the compounds derived from the oxidative degradation. In a comparative study involving the use of different oils (extra virgin olive oil, olive oil, olive-pomace oil, and refined palm oil) in the preparation of dry bakery products similar to biscuits [55], it appeared that the choice of lipid was very influential on quality because refined oils showed high levels of oxidized triacylglycerols and polymerization compounds, which further raised during processing. Similarly, the levels of lipid degradation compounds ascertained in an early survey on the quality of Italian biscuits, where refined oils and margarines were mostly used, were found to be high [9]. On the contrary, durum wheat oil, despite the detrimental effect of the refining process [33], is rich in antioxidants which help limiting the formation of the polar compounds, particularly the oxidation-related ones.

3.2. Volatile Compounds

The volatile compounds were significantly different among biscuits (Table 6). In particular hexanal, hexenal, 2-heptenal, nonanal were lower in biscuits with durum wheat oil than in the biscuits with sunflower oil. These compounds are related to the typical rancid off-flavor and are considered markers of lipid oxidation [56]. Hexanal derives from the oxidation of linoleic acid, while hexenal comes from the linolenic acid. In general, the aldehydes deriving from the action of lipoxygenase are responsible for undesirable odors, especially hexanal. The lower presence of these compounds in the biscuits prepared with the durum wheat oil indicated a lower level of oxidation and was in line with the ascertained levels of polar compounds. The high level of tocols present in the durum wheat oil reduced the oxidation therefore limiting the presence of volatile oxidation markers in biscuits. This was in agreement with Kishimoto et al. [57], who observed that α-tocopherol added to extra virgin olive oil reduced the formation of hexanal and other oxidative markers during storage.

2-Methylbutanal and 3-methylbutanal were both more abundant in the biscuits prepared with durum wheat oil. They are Strecker aldheydes characterized by a malty odor, derived by the reaction of Maillard from the aminoacids isoleucine and leucine, respectively. Benzaldehyde, derived from the phenylalanine metabolism [58], showed the same trend. The Maillard reaction also determined the formation of pyrazines and furan compounds. Though the carboxen/polydimethylsiloxane SPME fiber used is not very sensitive to pyrazine, three of them were identified: pyrazine, methyl-pyrazine, and ethyl-pyrazine. They were significantly more abundant in biscuits with durum wheat oil than in those prepared with sunflower oil. In addition, 2-furanmethanol and 2-furancarboxaldehyde (or furfural), the latter being commonly detected in biscuits [35], were assessed in higher amounts in the biscuits with durum wheat oil.

Table 6. Volatile compounds of biscuits. C = Biscuits prepared with 100% sunflower oil; D50 = biscuits prepared with sunflower oil (50%) and durum wheat oil (50%); D100 = biscuits prepared with 100% durum wheat oil.

Volatile Compounds (μg/g)	Sample Type		
	C	D50	D100
Aldehydes			
Hexanal	40.88 ± 0.49 [a]	33.53 ± 0.26 [b]	25.97 ± 1.06 [c]
Hexenal	5.38 ± 0.35 [a]	2.20 ± 0.11 [b]	2.52 ± 0.31 [b]
2-Heptenal	4.69 ± 0.17 [a]	2.83 ± 0.03 [b]	0.42 ± 0.16 [c]
Nonanal	11.83 ± 0.61 [a]	6.36 ± 0.36 [b]	6.38 ± 0.22 [b]
2-Methylbutanal	10.26 ± 0.23 [b]	13.92 ± 1.42 [a]	16.88 ± 0.41 [a]
3-Methylbutanal	17.71 ± 0.86 [b]	25.41 ± 0.77 [a]	27.03 ± 0.89 [a]
Benzaldehyde	4.85 ± 0.38 [b]	7.12 ± 0.92 [a]	7.24 ± 0.15 [a]
Furan compounds			
2-Furanmethanol	4.60 ± 0.51 [b]	8.78 ± 0.74 [a]	9.95 ± 0.68 [a]
2-Furancarboxaldehyde (furfural)	3.67 ± 0.25 [b]	5.41 ± 1.99 [ab]	8.60 ± 0.44 [a]
Pyrazines			
Pyrazine	8.27 ± 1.19 [b]	13.10 ± 0.13 [a]	13.15 ± 0.20 [a]
Methyl-pyrazine	30.79 ± 1.31 [b]	30.90 ± 1.82 [b]	44.19 ± 1.71 [a]
Ethyl-pyrazine	7.08 ± 1.21 [b]	6.15 ± 0.48 [b]	10.53 ± 0.17 [a]

Different letters in the same row indicate significant differences at $p < 0.05$.

3.3. Physical Characteristics

Table 7 shows the physical characteristics of the experimental biscuits (color, texture, and dimensional variations during baking). Replacing the sunflower oil with the durum wheat oil did not determine statistically significant changes in the color coordinates a^* (redness), b^* (yellowness), and L^* (luminosity). Therefore, the color difference (ΔE) between the reference biscuits (C) and those containing durum wheat oil (D50 and D100) was unrecognizable (values lower than 2). All biscuits showed a similar golden-brown color (Figure 1), imputable to Maillard reaction and caramelization. These reactions develop the typical color, as well as intense flavor and taste, very important in baked goods [8].

Table 7. Physical characteristics (color, texture, and dimensional variations during baking) of the experimental biscuits. C = Biscuits prepared with 100% sunflower oil; D50 = biscuits prepared with sunflower oil (50%) and durum wheat oil (50%); D100 = biscuits prepared with 100% durum wheat oil.

Parameter	Sample Type		
	C	D50	D100
	Color		
a^*	6.20 ± 0.32 [a]	7.04 ± 0.68 [a]	7.04 ± 0.52 [a]
b^*	33.33 ± 1.51 [a]	32.47 ± 0.30 [a]	34.56 ± 1.40 [a]
L^*	74.16 ± 1.63 [a]	73.96 ± 0.52 [a]	73.37 ± 1.43 [a]
ΔE	-	1.23	1.46
	Texture		
Hardness (N)	19.10 ± 0.45 [a]	18.70 ± 0.51 [a]	16.62 ± 0.53 [b]
	Dimensional variations during baking		
Thickness increase (mm)	7.5 ± 0.5 [a]	7.0 ± 0.5 [a]	7.2 ± 0.6 [a]
Length increase (mm)	1.5 ± 0.1 [a]	1.2 ± 0.3 [a]	1.5 ± 0.1 [a]
Width increase (mm)	2.0 ± 0.1 [a]	1.5 ± 0.5 [a]	1.7 ± 0.6 [a]

Different letters in the same row indicate significant differences at $p < 0.05$.

(a) (b) (c)

Figure 1. (**a**) Biscuits prepared with 100% sunflower oil; (**b**) biscuits prepared with sunflower oil (50%) and durum wheat oil (50%); (**c**) biscuits prepared with 100% durum wheat oil.

Biscuit hardness was determined by fracturing the samples through a three-point bending test. Biscuits prepared with sunflower oil were significantly ($p < 0.05$) harder than those obtained by using exclusively durum wheat oil. The increased presence of diacylglycerols in the latter could have determined an easier breakability. Mono- and diacylglycerols are, indeed, commonly used in the bakery sector as emulsifiers [59,60] to ensure proper gas retention during dough mixing and produce a more aerated structure, which in biscuits means a less hard consistency [61]. Biscuit hardness has an influence on consumer acceptability, with less hard biscuits being easier to chew and more appreciated. Other authors, in reformulating biscuits to improve their nutritional quality, added softening ingredients, such as honey or oleogels, to avoid excessive hardness [62,63].

The dimensional increase, caused by the thermal expansion of gases during baking [64], was not found to be significantly different when the oil type changed. The absence of significant variations as a consequence of the oil substitution can be explained being sunflower and durum wheat oil similar in lubricant ability and flow properties. The biscuits increased more in thickness than in length and width, as observed in other studies [56,64].

3.4. Sensory Features

The sensory properties did not show significant differences among the examined biscuits, with the only exception of breakability (Table 8). A significantly more pronounced breakability of biscuits with durum wheat oil was indeed observed, compared to the samples containing sunflower oil at 50% and 100%. These findings were in agreement with the instrumental measurement of hardness, which is inversely related to breakability. All biscuits were finely porous and showed an intense typical shortbread odor. Only a mild caramel odor was perceived, while no oxidized oil odor was perceived.

Table 8. Sensory features of the experimental biscuits. C = Biscuits prepared with 100% sunflower oil; D50 = biscuits prepared with sunflower oil (50%) and durum wheat oil (50%); D100 = biscuits prepared with 100% durum wheat oil.

Sensory Descriptor (c.u.) *	Sample Type		
	C	D50	D100
	Visual–tactile characteristics		
Porosity	4.2 ± 0.4 [a]	4.4 ± 0.2 [a]	5.2 ± 0.5 [a]
Breakability	3.5 ± 0.2 [b]	4.3 ± 0.4 [a]	5.1 ± 0.3 [a]
	Odor notes		
Caramel	0.6 ± 0.2 [a]	0.7 ± 0.2 [a]	0.7 ± 0.1 [a]
Oxidized oil	0.1 ± 0.1 [a]	0.0 ± 0.0 [a]	0.0 ± 0.0 [a]
Shortbread	7.6 ± 0.5 [a]	7.3 ± 0.5 [a]	6.6 ± 0.7 [a]

* c.u. = contractual units. Different letters in the same row indicate significant differences at $p < 0.05$.

4. Conclusions

Results from the study demonstrated that durum wheat oil incorporation into biscuits improves the oxidative stability of the end-products, due to the high content of tocotrienols which characterizes this kind of oil. Moreover, the use of durum wheat oil did not influence negatively the physical and sensory characteristics of biscuits compared to the commonly used sunflower oil.

Wheat oil is currently produced mostly from soft wheat germ by solvent extraction and subsequent refining, being the by-products from durum wheat de-branning and milling still largely underused. The food industry should start exploiting its full potential, still partially undiscovered [65], also to face the possible unavailability of soft wheat. Innovative uses of durum wheat by-products would lead to an improvement of the nutritional characteristics of food products and satisfy customers' demands for healthy and functional foods. Further investigations will be therefore carried out to study the use of durum wheat oil in other bakery products.

Author Contributions: Conceptualization, A.P., F.C. and M.G.M.; formal analysis, A.P. and F.V.; investigation, F.V.; writing—original draft preparation, A.P. and F.V.; writing—review and editing, A.P., C.S., F.C., M.G.M. and F.V.; supervision, A.P.; funding acquisition, F.C. All authors have read and agreed to the published version of the manuscript.

Funding: This research received no external funding.

Institutional Review Board Statement: Not applicable.

Informed Consent Statement: For the sensory analysis, written informed consent was obtained from all the panelists.

Data Availability Statement: The data presented in this study are available upon request from the corresponding author. The data are not publicly available due to the absence of in institutional platform.

Acknowledgments: This research was made possible thanks to the contribution of experimental materials from Casillo Next Gen Food srl (Corato, Italy).

Conflicts of Interest: The author Marcello Greco Miani is an employee of a company, and he participated in the design of the experiments, but he did not receive any additional financial advantage for taking part in this study. The company provided a material for the experiments, namely the wheat oil.

References

1. Tarancón, P.; Sanz, T.; Fiszman, S.; Tárrega, A. Consumers' hedonic expectations and perception of the healthiness of biscuits made with olive oil or sunflower oil. *Food Res. Int.* **2014**, *55*, 197–206. [CrossRef]
2. Janssen, M.; Chang, B.P.; Hristov, H.; Pravst, I.; Profeta, A.; Millard, J. Changes in food consumption during the COVID-19 pandemic: Analysis of consumer survey data from the first lockdown period in Denmark, Germany, and Slovenia. *Front. Nutr.* **2021**, *8*, 635859. [CrossRef] [PubMed]
3. Cookie Report—TOP Agency. Available online: https://topagency.com/report/cookie-report/ (accessed on 20 December 2021).
4. Canali, G.; Balestra, F.; Glicerina, V.; Pasini, F.; Caboni, M.F.; Romani, S. Influence of different baking powders on physico-chemical, sensory and volatile compounds in biscuits and their impact on textural modifications during soaking. *J. Food Sci. Technol.* **2020**, *57*, 3864–3873. [CrossRef] [PubMed]
5. Romani, S.; Rodriguez-Estrada, M.T. Bakery products and electronic nose. In *Electronic Noses and Tongues in Food Science*; Academic Press: Oxford, UK, 2016; pp. 39–47.
6. Drakos, A.; Andrioti-Petropoulou, L.; Evageliou, V.; Mandala, I. Physical and textural properties of biscuits containing jet milled rye and barley flour. *J. Food Sci. Technol.* **2019**, *56*, 367–375. [CrossRef] [PubMed]
7. Konstantas, A.; Stamford, L.; Azapagic, A. Evaluation of environmental sustainability of biscuits at the product and sectoral levels. *J. Clean. Prod.* **2019**, *230*, 1217–1228. [CrossRef]
8. Arepally, D.; Reddy, R.S.; Goswami, T.K.; Datta, A.K. Biscuit baking: A review. *LWT Food Sci. Technol.* **2020**, *131*, 109726. [CrossRef]
9. Caponio, F.; Summo, C.; Delcuratolo, D.; Pasqualone, A. Quality of the lipid fraction of Italian biscuits. *J. Sci. Food Agric.* **2006**, *86*, 356–361. [CrossRef]
10. Cervenka, L.; Brožková, I.; Vytrasová, J. Effects of the principal ingredients of biscuits upon water activity. *J. Food Nutr. Res.* **2006**, *45*, 39–43.
11. Devi, A.; Khatkar, B.S. Physicochemical, rheological and functional properties of fats and oils in relation to cookie quality: A review. *J. Food Sci. Technol.* **2016**, *53*, 3633–3641. [CrossRef]
12. Xu, J.; Zhang, Y.; Wang, W.; Li, Y. Advanced properties of gluten-free cookies, cakes, and crackers: A review. *Trends Food Sci. Technol.* **2020**, *103*, 200–203. [CrossRef]
13. Pasqualone, A.; Haider, N.N.; Summo, C.; Coldea, T.E.; George, S.S.; Altemimi, A.B. Biscuit Contaminants, Their Sources and Mitigation Strategies: A Review. *Foods* **2021**, *10*, 2751. [CrossRef] [PubMed]
14. Billek, G. Health aspects of thermoxidized oils and fats. *Eur. J. Lipid Sci. Technol.* **2000**, *102*, 587–593. [CrossRef]
15. Balestra, F.; Verardo, V.; Tappi, S.; Caboni, M.F.; Dalla Rosa, M.; Romani, S. Chemical and physical changes during storage of differently packed biscuits formulated with sunflower oil. *J. Food Sci. Technol.* **2019**, *56*, 4714–4721. [CrossRef] [PubMed]
16. Goswami, M.; Sharma, B.D.; Mendiratta, S.K.; Pathak, V. Quality evaluation of functional carabeef cookies incorporated with guar gum (*Cyamopsis tetragonoloba*) as fat replacer. *Nutr. Food. Sci.* **2019**, *49*, 432–440. [CrossRef]
17. Santiago-García, P.A.; Mellado-Mojica, E.; León-Martínez, F.M.; López, M.G. Evaluation of *Agave angustifolia* fructans as fat replacer in the cookies manufacture. *LWT Food Sci.* **2017**, *77*, 100–109. [CrossRef]
18. Giarnetti, M.; Paradiso, V.M.; Caponio, F.; Summo, C.; Pasqualone, A. Fat replacement in shortbread cookies using an emulsion filled gel based on inulin and extra virgin olive oil. *LWT Food Sci.* **2015**, *63*, 339–345. [CrossRef]
19. Yashini, M.; Sahana, S.; Hemanth, S.D.; Sunil, C.K. Partially defatted tomato seed flour as a fat replacer: Effect on physicochemical and sensory characteristics of millet-based cookies. *J. Food Sci. Technol.* **2021**, *58*, 4530–4541. [CrossRef]
20. Yalcin, S. Determination of quality characteristic of biscuits including ground yellow poppy seed as fat replacer. *Int. J. Second Metab.* **2017**, *4*, 406–411. [CrossRef]
21. Seker, I.T.; Ozboy-Ozbas, O.; Gokbulut, I.; Ozturk, S.; Koksel, H. Utilization of apricot kernel flour as fat replacer in cookies. *J. Food Process. Preserv.* **2010**, *34*, 15–26. [CrossRef]
22. Food and Agriculture Organization of the United Nations (FAO). Data of Crops Production. 2020. Available online: https://www.fao.org/faostat/en/#data/QCL (accessed on 21 March 2022).
23. Food and Agriculture Organization of the United Nations (FAO). Available online: https://www.fao.org/worldfoodsituation/csdb/en/ (accessed on 21 March 2022).
24. Elias, E.M. Durum wheat products. *Options Mediterr.* **1995**, *40*, 23–31.
25. Pasqualone, A. Italian durum wheat breads. In *Bread Consumption and Health*; Pedrosa Silva Clerici, M.T., Ed.; Nova Science Publisher Inc.: Hauppauge, NY, USA, 2012; pp. 57–79.
26. Pasqualone, A. Traditional flat breads spread from the Fertile Crescent: Production process and history of baking systems. *J. Ethn. Foods* **2018**, *5*, 10–19. [CrossRef]
27. Consiglio per la Ricerca in Agricoltura e L'analisi Dell'economia Agraria. Available online: https://www.fidaf.it/wp-content/uploads/2020/05/CREA-Considerazioni-grano-duro-CREA-01.08.2016-AGGIORNAMENTO.pdf (accessed on 21 March 2022).
28. ISTAT. Available online: http://dati.istat.it/Index.aspx?DataSetCode=DCSP_COLTIVAZIONI (accessed on 21 March 2022).
29. ISTAT. Available online: https://www.istat.it/it/files//2021/04/Previsioni-coltivazioni-agricole.pdf (accessed on 20 December 2021).
30. Prueckler, M.; Siebenhandl-Ehn, S.; Apprich, S.; Hoeltinger, S.; Haas, C.; Schmid, E.; Kneifel, W. Wheat bran-based biorefinery 1: Composition of wheat bran and strategies of functionalization. *LWT Food Sci. Technol.* **2014**, *56*, 211–221. [CrossRef]

31. Lafiandra, D.; Masci, S.; Sissons, M.; Dornez, E.; Delcour, J.A.; Courtin, C.M.; Caboni, M.F. Kernel components of technological value. In *Durum Wheat Chemistry and Technology*, 2nd ed.; Sissons, M., Marchylo, B., Abecassis, J., Eds.; AACC International Inc.: St. Paul, MN, USA, 2012; pp. 85–124.
32. Osorio, L.L.D.R.; Flórez-López, E.; Grande-Tovar, C.D. The Potential of Selected Agri-Food Loss and Waste to Contribute to a Circular Economy: Applications in the Food, Cosmetic and Pharmaceutical Industries. *Molecules* **2021**, *26*, 515. [CrossRef] [PubMed]
33. Squeo, G.; Silletti, R.; Napoletano, G.; Greco Miani, M.; Difonzo, G.; Pasqualone, A.; Caponio, F. Characterization and Effect of Refining on the Oil Extracted from Durum Wheat By-Products. *Foods* **2022**, *11*, 683. [CrossRef]
34. Caponio, F.; Gomes, T.; Pasqualone, A.; Summo, C. Use of the high performance size exclusion chromatography analysis for the measurement of the degree of hydrolytic and oxidative degradation of the lipid fraction of biscuits. *Food Chem.* **2007**, *102*, 232–236. [CrossRef]
35. Pasqualone, A.; Makhlouf, F.Z.; Barkat, M.; Difonzo, G.; Summo, C.; Squeo, G.; Caponio, F. Effect of acorn flour on the physico-chemical and sensory properties of biscuits. *Heliyon* **2019**, *5*, e02242. [CrossRef]
36. Brochard, M.; Correia, P.; Barroca, M.J.; Guiné, R.P. Development of a New Pasta Product by the Incorporation of Chestnut Flour and Bee Pollen. *Appl. Sci.* **2021**, *11*, 6617. [CrossRef]
37. *ISO 13299*; Sensory Analysis. Methodology. General Guidance for Establishing a Sensory Profile. ISO: Geneva, Switzerland, 2016.
38. *ISO 11132*; Sensory Analysis. Methodology. Guidelines for the Measurement of the Performance of a Quantitative Descriptive Sensory Panel. ISO: Geneva, Switzerland, 2021.
39. *ISO 8589*; Sensory Analysis. General Guidance for the Design of Test Rooms. ISO: Geneva, Switzerland, 2007.
40. Szewczyk, K.; Chojnacka, A.; Górnicka, M. Tocopherols and Tocotrienols—Bioactive Dietary Compounds; What Is Certain, What Is Doubt? *Int. J. Mol. Sci.* **2021**, *22*, 6222. [CrossRef]
41. Ghafoor, K.; Özcan, M.M.; AL-Juhaımı, F.; Babıker, E.E.; Sarker, Z.I.; Ahmed, I.A.M.; Ahmed, M.A. Nutritional composition, extraction, and utilization of wheat germ oil: A review. *Eur. J. Lipid Sci. Technol.* **2017**, *119*, 1600160. [CrossRef]
42. Mishra, S.K.; Belur, P.D.; Iyyaswami, R. Use of antioxidants for enhancing oxidative stability of bulk edible oils: A review. *Int. J. Food Sci. Technol.* **2021**, *56*, 1–12. [CrossRef]
43. Buddrick, O.; Jones, O.A.; Morrison, P.D.; Small, D.M. Effect of fermentation and oil incorporation on the retention of E vitamers during breadmaking. *Cereal Chem.* **2015**, *92*, 327–331. [CrossRef]
44. Kamal-Eldin, A. Effect of fatty acids and tocopherols on the oxidative stability of vegetable oils. *Eur. J. Lipid Sci. Technol.* **2006**, *108*, 1051–1061. [CrossRef]
45. Nesaretnam, K.; Yew, W.W.; Wahid, M.B. Tocotrienols and cancer: Beyond antioxidant activity. *Eur. J. Lipid Sci. Technol.* **2007**, *109*, 445–452. [CrossRef]
46. Brigelius-Flohé, R. Bioactivity of vitamin E. *Nutr. Res. Rev.* **2006**, *19*, 174–186. [CrossRef] [PubMed]
47. Saldeen, K.; Saldeen, T. Importance of tocopherols beyond a tocopherol: Evidence from animal and human studies. *Nutr. Res.* **2005**, *25*, 877–889. [CrossRef]
48. Engelsen, M.M.L.; Hansen, Å. Tocopherol and tocotrienol content in commercial wheat mill streams. *Cereal Chem.* **2009**, *86*, 499–502. [CrossRef]
49. Labuschagne, M.T.; Mkhatywa, N.; Wentzel, B.; Johansson, E.; Van Biljon, A. Tocochromanol concentration, protein composition and baking quality of white flour of South African wheat cultivars. *J. Food Comp. Anal.* **2014**, *33*, 127–131. [CrossRef]
50. Imeneo, V.; Romeo, R.; Gattuso, A.; De Bruno, A.; Piscopo, A. Functionalized Biscuits with Bioactive Ingredients Obtained by Citrus Lemon Pomace. *Foods* **2021**, *10*, 2460. [CrossRef]
51. Sharif, M.K.; Butt, M.S.; Anjum, F.M.; Nasir, M.; Minhas, R.; Qayyum, M.N. Extension of cookies shelf life by using rice bran oil. *Int. J. Agric. Biol.* **2003**, *5*, 455.
52. Caponio, F.; Pasqualone, A.; Gomes, T. Effects of conventional and microwave heating on the degradation of olive oil. *Eur. Food Res. Technol.* **2002**, *215*, 114–117. [CrossRef]
53. Zhao, Z.; Huang, J.; Jin, Q.; Wang, X. Influence of oryzanol and tocopherols on thermal oxidation of rice bran oil during the heating process at Chinese cooking temperatures. *LWT Food Sci.* **2021**, *142*, 111022. [CrossRef]
54. Chen, J.; Zhang, L.; Li, Y.; Zhang, N.; Gao, Y.; Yu, X. The formation, determination and health implications of polar compounds in edible oils: Current status, challenges and perspectives. *Food Chem.* **2021**, *364*, 130451. [CrossRef]
55. Caponio, F.; Giarnetti, M.; Summo, C.; Gomes, T. Influence of the different oils used in dough formulation on the lipid fraction of taralli. *J. Food Sci.* **2011**, *76*, 549–554. [CrossRef] [PubMed]
56. Pasqualone, A.; Bianco, A.M.; Paradiso, V.M.; Summo, C.; Gambacorta, G.; Caponio, F.; Blanco, A. Production and characterization of functional biscuits obtained from purple wheat. *Food Chem.* **2015**, *180*, 64–70. [CrossRef] [PubMed]
57. Kishimoto, N. Evaluation of Photooxidation of Olive Oil by Determining the Concentration of Hexanal as an Oxidative Marker Using an Electronic Nose. *Chem. Eng. Trans.* **2021**, *85*, 181–186.
58. Pripis-Nicolau, L.; De Revel, G.; Bertrand, A.; Maujean, A. Formation of flavor components by the reaction of amino acid and carbonyl compounds in mild conditions. *J. Agric. Food Chem.* **2000**, *48*, 3761–3766. [CrossRef]
59. Kyacier, A.; Singh, R.K. Effect of mono-and diglyceride addition on baked tortilla chips and their storage properties. *J. Food Proc. Pres.* **2003**, *2*, 1–8. [CrossRef]

60. Kara, H.H.; Yasemin, B.O.R. A review on: Production, usage, health effect and analysis of mono-and di-glyceride of fatty acids. *J. Halal Ethical Res.* **2019**, *1*, 40–47.
61. Li, S.; Wu, G.; Jin, Q.; Wang, X.; Zhang, H. Relationship between the microstructure and physical properties of emulsifier based oleogels and cookies quality. *Food Chem.* **2022**, *377*, 131966. [CrossRef]
62. Guiné, R.P.; Souta, A.; Gürbüz, B.; Almeida, E.; Lourenço, J.; Marques, L.; Pereira, R.; Gomes, R. Textural properties of newly developed cookies incorporating whey residue. *J. Culin. Sci. Technol.* **2020**, *18*, 317–332. [CrossRef]
63. Li, S.; Wu, G.; Li, X.; Jin, Q.; Wang, X.; Zhang, H. Roles of gelator type and gelation technology on texture and sensory properties of cookies prepared with oleogels. *Food Chem.* **2021**, *356*, 129667. [CrossRef] [PubMed]
64. Pasqualone, A.; Laddomada, B.; Boukid, F.; Angelis, D.D.; Summo, C. Use of Almond Skins to Improve Nutritional and Functional Properties of Biscuits: An Example of Upcycling. *Foods* **2020**, *9*, 1705. [CrossRef] [PubMed]
65. Boukid, F.; Folloni, S.; Ranieri, R.; Vittadini, E. A compendium of wheat germ: Separation, stabilization and food applications. *Trends Food Sci. Technol.* **2018**, *78*, 120–133. [CrossRef]

foods

Article

Wheat Germ and Lipid Oxidation: An Open Issue

Silvia Marzocchi [1,*], Maria Fiorenza Caboni [1], Marcello Greco Miani [2] and Federica Pasini [1]

1 Department of Agricultural and Food Sciences and Technologies, University of Bologna, Piazza Goidanich 60, 47521 Cesena, Italy; maria.caboni@unibo.it (M.F.C.); federica.pasini5@unibo.it (F.P.)
2 Casillo Next Gen Food S.r.l., Via Sant'Elia Z.I., 70033 Corato, Italy; marcello.miani@casillogroup.it
* Correspondence: silvia.marzocchi4@unibo.it; Fax: +39-0547-382348

Abstract: Wheat germ (WG)'s shelf life after the milling process is incredibly short because of the presence of enzymes that aggravate the oxidation process; thus, stabilization is required in order to exploit the nutrients and bioactive compounds within WG. The critical point for the oxidation process is the mechanical treatment used to separate WG from the kernel, which exposes the lipid fraction to the air. Showing the connection between the quality of durum wheat, considering its storage management, and wheat germ oil (WGO), extracted with a cold press, solvent and supercritical CO_2, is the aim of the study. The acidity and peroxide values were analyzed to evaluate lipid oxidation, while fatty acids, tocols, sterols and policosanols were evaluated for WGO characterization. The first fundamental step to control lipid oxidation is raw material management. Subsequently, the tempering phase of durum wheat, which is applied before the degermination process, is the most critical point for oxidation to develop because of the increase in moisture in the caryopsis and the activation of lipase and lipoxygenase. This represents a paradox: in order to stabilize the germ with degermination, first it seems inevitable to carry out a process that destabilizes it. To retains its highest quality, this will lead to a better use of the whole grain by reducing WG and by-product waste.

Keywords: wheat germ; stabilization; lipid oxidation; oil quality

Citation: Marzocchi, S.; Caboni, M.F.; Greco Miani, M.; Pasini, F. Wheat Germ and Lipid Oxidation: An Open Issue. *Foods* **2022**, *11*, 1032. https://doi.org/10.3390/foods11071032

Academic Editor: Mike Sissons

Received: 10 February 2022
Accepted: 31 March 2022
Published: 1 April 2022

1. Introduction

Triticum aestivum is an hexaploid species that accounts for about 95% of wheat grown annually, and it is commonly used for bread production; on the other hand, durum wheat (*Triticum turgidum* ssp. *durum* Desf) is a tetraploid species most cultivated in the Mediterranean sea area due to the climate and conditions, and it is used mainly for pasta and couscous [1,2]. Durum wheat is an important source of bioactive compounds, with their relative health benefits, that are mainly contained in the grain bran and germ tissue [3]. Dietary fiber comprises carbohydrates (11.5–15.5% of dry wheat grain) that are able to reach the large intestine or colon [1]. The aleurone layer and embryo are a good source of micronutrients, such as selenium, iron and zinc, which are fundamental for the correct organism development during ingestion and for promoting gut and body health [3]. The most abundant metabolites in wheat grain are phenolic compounds, and bound phenolic acids in particular, that have antioxidant, anti-inflammatory and anti-carcinogenic properties [4].

The whole wheat kernel is composed of three structures: the endosperm (80–85% of dry wheat grain), which is formed of starch and proteins and surrounded by layers called "aleurone"; the bran (13–17%), which is formed of pericarp and testa, a hydrophobic tissue composed of lignin and lipidic compounds [5]; and the wheat germ.

In order to obtain the separation between flour and wheat germ (WG), caryopsis undergoes debranning and milling processes. The initial debranning is used for "covered" cereals; in fact, it removes the outer layers of caryopsis, allowing the recovery of intact kernels without causing damage to the endosperm region [6], producing some by-products. Debranning is followed by the common milling process of wheat, with which flour and co-products, such as coarse bran, fine bran and WG, are obtained [7].

Wheat germ represents 2–3% of the whole wheat kernel, and it is a precious milling co-product. Because of its concentration of high quality compounds, such as proteins, minerals, flavonoids, sterols, and vitamin E and B, WG is considered to be the most beneficial part of wheat grain; in fact, it has antioxidant, antihyperlipidemic, hypocholesterolemic and anticancer effects [8,9]. Meanwhile, the widespread use of WG is limited due to its rapid oxidation development because of the high presence of unsaturated fatty acids and hydrolytic and oxidative enzymes, such as lipoxygenase and lipase [10,11].

Removing WG during wheat milling is necessary to increase flour shelf life, but, on the other hand, WG shelf life is drastically shortened, and its stabilization is required in order to exploit its nutrients and bioactive compounds [12,13].

The WG separation process can be conducted with a direct and indirect approach: the former is represented by degermination, the latter by gradual separation, with an evolution of the milling process over the years leading up to the debranning process [9].

On the other hand, it is well known that WG stabilization can be conducted with different strategies, namely, physical, chemical and biological approaches. Briefly, physical technologies are represented by heat treatments, microwave, infrared and gamma radiation [9] and thermal/mechanical treatments. Heat treatments include steaming, which leads to the complete inactivation of lipase, while the lipoxygenase is inactivated to an extent of 80–92%, after 15 min treatment at 125–130 °C [12]; fluidization, which involves mass transfer between the material and hot air in the fluid bed, and which leads to complete lipase inactivation and 13.5% residual lipoxygenase [14]; and roasting, which is also used for stabilizing WG because it reduces moisture and enzyme activity [15]. Li and collaborators [16] found the optimal infrared irradiation effect to be at 90 °C for 20 min, obtaining residual lipase and lipoxygenase activity of about 19% for both enzymes; Jha et al. [17] stabilized WG through γ-irradiation, with a 31% inactivation of lipase. The chemical stabilization of WG can be caried out with enzyme denaturation via acidification using hydrochloric acid or acetic acid [18] or oil removal through organic solvents or supercritical CO_2 extraction [9]. The supercritical CO_2 extraction simplifies the oil refining process and eliminates the solvent distillation stage. In addition, CO_2 is nontoxic, nonflammable, noncorrosive and recyclable [19]. Finally, biological stabilization consists of a fermentation of WG by lactic acid bacteria in order to obtain enzyme denaturation as a consequence of acidification [20].

The critical point for the oxidation process is the mechanical treatment used to separate the WG from the kernel, which involves a necessary tempering phase and, consequently, lipase and lipoxygenase activation.

To the best of our knowledge, there are no studies that consider what happens in this preliminary step; thus, the aim of this study is to demonstrate the connection between the quality of durum wheat and wheat germ oil (WGO). Additionally, we sought to find a way to use WG with the highest quality, allowing producers to use the whole grain and reduce waste and by-products. As reported above, during the debranning and milling processes, different by-products are obtained, so the characterization of their acidity parameters is fundamental for our aims, as is the lipid characterization of WGO, all of which can be used to evaluate the quality of the extracted oil. In addition, WGO was extracted with different technologies, including mechanical extraction, extraction with solvents and extraction with supercritical CO_2, to evaluate if technology can affect its quality.

2. Materials and Methods

2.1. Tempering Phase of Wheat

Before the milling process, the raw durum wheat underwent a laboratory tempering process in water. Different treated water was used: water as control (C), water with 3% NaCl (W3%) and water with 5% NaCl (W5%) in order to see which has the least impact on the acidity of the wheat.

2.2. Samples

Winter durum wheat (*Triticum turgidum* ssp. *durum*) was supplied by an Italian mill, Molino Casillo S.p.A., and was used for wheat germ oil (WGO) extraction with different technologies. The durum wheat was sowed between the beginning of November and the first half of December, and the following thermal requirements were met: 2–3 °C for germination and tillering, 10 °C for rising, 15 °C for flowering and 20 °C for ripening. Durum wheat was harvested when it reached full ripening and its humidity was less than 13% (in the third week of May).

Two debranning by-products (DB1 and DB2) and one milling by-product (MB) were supplied by Molino Casillo S.p.A. A total of four samples for each by-product (DB1, DB2 and MB) were collected from four different plant systems.

Wheat germ for WGO was separated from durum wheat endosperm with the initial debranning and milling process, followed by sieving in order to obtain pure wheat germ.

2.3. Mechanical Extraction of WGO

The cold press extraction of WGO was performed using an SK60/1 press (Karl Strähle GmbH & Co., Dettingen unter Teck, Germany). Wheat germ was separated from durum wheat caryopsis and other impurities, such as metal pieces and stones, and wheat germ oil was extracted with a cold press (15 kg/h capacity).

2.4. Solvent Extraction of WGO

The standard protocol according to the AOAC Official Method [21] was used for the Soxhlet extraction of WGO; about 60 g of ground wheat germ was placed in a cellulose extraction thimble and the process was carried out for 2 h for complete extraction using a refluxing hexane. The residual solvent was evaporated using a rotary evaporator, Laborota 4001-efficient (Heidolph). Each extraction was performed twice.

2.5. Supercritical CO_2 Extraction of WGO

The supercritical fluid extraction of about 1 kg ground wheat germ was carried out for 3 h at 380 bar and 55 °C under a constant CO_2 flow rate of 20 kg/h.

2.6. Analytical Methods

2.6.1. Moisture Content

Moisture content (%) was determined according to the European Standard Method UNI EN ISO 712:2010 [22]. Each determination was calculated twice.

2.6.2. Lipid Determination of Durum Wheat and By-Products

Soxhlet extraction, according to the method of AOAC 920.39B [23], was conducted for lipid extraction in all samples. Each extraction was performed twice.

2.6.3. Free Acidity Determination

Acidity was determined by means of volumetric titration according to the UNI EN ISO 660:2009 [22] standard method. Each determination was calculated twice for each lipid extraction ($n = 4$).

2.6.4. Peroxide Value

The peroxide value was determined by means of volumetric titration according to UNI EN ISO 3960:2010 [22]. Each determination was calculated twice for each lipid extraction ($n = 4$).

2.7. Wheat Germ Oil Characterization

2.7.1. Fatty Acid Composition

The fatty acid composition was determined according to ISO 12966-2:2017 +ISO 12966-4:2015 [22]. The analyses were carried out using a gas chromatograph (Shimadzu, Tokyo,

Japan) equipped with a flame ionization detector (GC-FID), using a capillary column (CP-Sil 88-l = 100 m, 0.32 mm ID, film thickness 0.25 μm; Supelco, Bellefonte, PA, USA). Each determination was calculated twice.

2.7.2. Tocopherol Composition

Tocopherols were evaluated according to ISO 9936:2016 [22]. A sample amount was diluted in hexane and injected in a HPLC system (Agilent 1200 series, Palo Alto, CA, USA) operating in direct phase with a silica column 4.6 mm ID × 250 mm length (Luna Hilic Phenomenex). Reference tocopherols (Sigma Aldrich, Milano, Italy) were used for calibration curve construction for quantification. Each analysis were performed twice.

2.7.3. Unsaponifiable Matter

For the unsaponifiable matter, WGO samples extracted with different technologies were treated with alcoholic KOH solution, according to ISO 3956:2000 [22]. Each extraction was carried out twice.

2.7.4. Sterols Composition

The procedure for sterols content and composition was performed according to NGD 71-1989+NGD 72-1989 [22] using a gas chromatograph (Shimadzu, Tokyo, Japan) equipped with a flame ionization detector with a CPSil 8CB (Supelco, Bellefonte, PA, USA) capillary column (l = 30 m, 0.32 mm ID, film thickness 0.25 μm). Each determination was calculated twice for each insaponifiable extraction (n = 4).

2.8. Statistical Analysis

A one-way analysis of variance (ANOVA, with Tukey's honest significant difference multiple comparison) was evaluated using Statistica 8 software (2006, StatSoft, Tulsa, OK, USA). p-values lower than 0.05 were considered to be statistically significant.

3. Results and Discussion

3.1. Moisture, Lipid Content, Free Acidity and Peroxide Value

It is interesting to see that after the tempering phase, a fundamental step for the milling process, a significant acidity increase has been registered. In durum wheat before the tempering phase, the acidity was 9.5%, which then significantly ($p < 0.05$) increased after treatment. In fact, it increased by about 18–30 %; using different treated water, values of acidity reached 13.3, 13.8 and 11.6% using C, W3% and W5%, respectively. Therefore, this preliminary step before degermination is a critical point with regard to lipase and lipoxygenase activity, because, by adding water, the lipase cleaves triglycerides in fatty acids and the lipoxygenase catalyzes the oxidation of polyunsaturated fatty acids [24,25].

Table 1 shows the moisture content (%), lipid content (%), free acidity (%) and peroxide value (meqO_2/kg of fat) determined during the milling process after the tempering phase. It is possible to see that moisture content does not change along the whole milling process; in fact, it is in a range between 13.3 and 13.9%.

Table 1. Moisture content, lipid content, free acidity and peroxide value (PV) of the different milling by-products.

	Moisture Content (%)	Lipid Content (%)	Free Acidity (%)	PV (meqO_2/kg of Fat)
DW	13.7 ± 0.4 a	2.5 ± 0.2 c	4.8 ± 0.0 b	1.8 ± 0.0 d
DB1	13.8 ± 0.3 a	4.4 ± 0.2 b	6.5 ± 1.2 a	5.2 ± 0.2 b
DB2	13.9 ± 0.1 a	6.4 ± 0.3 a	5.3 ± 0.4 a	3.9 ± 0.4 c
MB	13.3 ± 0.6 a	4.1 ± 0.5 b	7.0 ± 0.7 a	6.5 ± 0.6 a

Abbreviation: DW, durum wheat after tempering phase; DB1, by-products after first debranning; DB2, by-product after second debranning; MB, by-product after milling. Results of the analysis of variance with Tukey's test are shown: $p < 0.05$; letters in the same column show significantly different values within each parameter.

Lipid content, instead, is affected by the milling process. In the intact durum wheat, lipid content was 2.5%, and it increased in the by-products after debranning, reaching a content of 4.4 and 6.4% in DB1 and DB2, respectively. This is due to the concentration of germ and aleurone in the by-products; in fact, is well known that oil in wheat is mainly concentrated in this part of the caryopsis [26,27]. Finally, the lipid content in the milling by-product was 4.1% after the process.

As regards the free acidity, this was 4.8% in the initial mixture of durum wheat, and during the debranning and milling process, it significantly increased, reaching values of 6.5, 5.3 and 7.0% in DB1, DB2 and MB, respectively. This can be due to the treatment conditions to which the wheat is exposed (such as temperature, heat, oil-water interface and water) [28].

The PV value trend is closely linked to the acidity one; in fact, in the initial durum wheat, it was 1.8 meqO_2/kg of fat, and it increased significantly during the milling process, but without exceeding the legal limit (20 meqO_2/kg of fat). After the first and second debranning, the PV reached a value of 5.2 and 3.9 meqO_2/kg of fat, respectively, and in the final by-product, it reached the most significant and highest value (6.5 meqO_2/kg of fat).

Considering Table 1, it is possible to see that acidity following the milling process decreases with respect to the one registered after the tempering phase.

Table 2 shows the yield, acidity and peroxide value of the WGO extracted from the separated germ with the three different technologies. The yield, calculated on a dry basis, and acidity were affected by the extraction technology, while the peroxide values were not. In fact, WGO extraction with a solvent registered a yield of 16%, significantly higher ($p < 0.05$) than the ones registered for mechanical and supercritical CO_2 extraction, which were 6.6 and 6.4%, respectively. The WGO extracted with supercritical CO_2 had a value of acidity of about 34%, significantly higher ($p < 0.05$) than the one obtained with mechanical (25.8%) and solvent (16%) extraction. Compared to the acidity registered during the milling process (Table 1), it is evident that it increased after the WGO extraction due to the technological process. The peroxide value was not affected by the extraction technology, and the results were in line with the ones registered following the milling process (Table 1).

Table 2. Yield (dry basis), free acidity and peroxide value of WGO extracted with different technologies.

	Yield (%)	Free Acidity (%)	PV (meqO_2/kg of Fat)
Mechanical extraction	6.6 ± 0.9 b	25.8 ± 1.1 b	3.5 ± 0.8 a
Solvent extraction	16.0 ± 2.1 a	16.0 ± 0.9 c	4.1 ± 0.7 a
Supercritical CO_2 extraction	6.4 ± 0.4 b	34.0 ± 2.1 a	3.6 ± 0.6 a

Results of the analysis of variance with Tukey's test are shown: $p < 0.05$; letters in the same column show significantly different values within each parameter.

3.2. Characterization of WGO Extracted with Different Technologies

Table 3 reports the fatty acid composition (%) of the WGO extracted with different technologies. Its composition was not affected by the extraction technology, and the results are in line with the literature [11,29], which reported linoleic acid (C18:2, ~53–58%) to be the major fatty acid in WGO, followed by oleic acid (C18:1, ~18–23%), palmitic acid (C16:0, ~13–17%) and linolenic acid (C18:3, ~3–6%).

Unsaponifiable matter was recorded to be at a significantly higher content in the WGO extracted with solvents and supercritical CO_2, at 4.1 and 5%, respectively, than the one registered in the WGO extracted with mechanical extraction (3.6%). The extracting conditions could affect the extraction efficiency because of the mixture of polar and non-polar compounds, which characterize the unsaponifiable matter [29]. Unsaponifiable matter contains tocopherols, tocotrienols and phytosterols, whose compositions are shown below; they are an important constituent of vegetable oils due to their health benefits [30].

Table 3. Fatty acid composition (%) of WGO extracted with different technologies.

Fatty Acid	*Mechanical Extraction*	*Solvent Extraction*	*Supercritical CO_2 Extraction*
C16:0	14.8 ± 1.4 ab	14.8 ± 0.9 b	17.1 ± 1.2 a
C18:0	1.3 ± 0.3 a	1.3 ± 0.4 a	1.4 ± 0.3 a
C18:1	20.3 ± 1.5 a	22.0 ± 2.3 a	21.0 ± 1.6 a
C18:2	56.7 ± 1.8 a	56.0 ± 2.0 a	53.5 ± 0.3 b
C18:3	4.6 ± 0.9 a	3.9 ± 0.5 a	4.0 ± 0.3 a

Results of the analysis of variance with Tukey's test are shown: $p < 0.05$; letters in the same row show significantly different values within each fatty acid.

Tocol amounts in durum wheat reach approximatively 60 mg/100 g db according to the literature [26], and, of their composition (Table 4) in WGO, β-tocotrienol was the most preponderant (60–88%), and the extraction technology that affected it above all was supercritical CO_2. In fact, the WGO extracted with supercritical CO_2 showed a significantly ($p < 0.05$) higher percentage of β-tocotrienol (88%) than the WGO extracted with a cold press and with a solvent, which presented a percentage of about 60%. On the other hand, α-tocopherol, β-tocopherol and α-tocotrienol were present at a significantly ($p < 0.05$) higher percentage in the WGO extracted with a cold press (12, 5 and 21%, respectively) and solvent (15, 4 and 18%, respectively) than in the WGO extracted with supercritical CO_2 (6.5, 2 and 2%, respectively). The others tocols were extracted in traces and did not show any significantly differences. In general, our results are in line with the literature [31–34]; the few differences identified can be due to the cultivar investigated and the cultivation technology.

Table 4. Tocol composition (%) of WGO extracted with different technologies.

Tocol	*Mechanical Extraction*	*Solvent Extraction*	*Supercritical CO_2 Extraction*
α-Tocopherol	12.9 ± 2.3 a	15.3 ± 2.2 a	6.5 ± 0.9 b
β-Tocopherol	5.3 ± 0.4 a	4.4 ± 0.7 a	1.9 ± 0.3 b
γ-Tocopherol	0.4 ± 0.0 a	0.4 ± 0.0 a	0.5 ± 0.0 a
δ-Tocopherol	0.1 ± 0.0 a	0.2 ± 0.0 a	0.1 ± 0.0 a
α-Tocotrienol	21.1 ± 6.0 a	18.1 ± 2.3 a	2.3 ± 0.2 b
β-Tocotrienol	60.1 ± 8.2 b	61.1 ± 3.7 b	88.3 ± 3.6 a
γ-Tocotrienol	0.1 ± 0.0 a	0.2 ± 0.0 a	0.1 ± 0.0 a
δ-Tocotrienol	0.1 ± 0.0 a	0.2 ± 0.0 a	0.2 ± 0.0 a

Results of the analysis of variance with Tukey's test are shown: $p < 0.05$; letters in the same row show significantly different values within each sterol.

It is well known that phytosterol consumption can reduce cardiovascular disease risk and blood LDL cholesterol levels [35]. The total sterol amount in durum wheat is in a range between 70 and 95 mg/100 g db [36]. Eight sterols (Table 5) were quantified in the WGO, with a preponderance of β-sitosterol (31–35%), followed by campestanol (15–18%), sitostanol (16–17%) and campesterol (12–13%). In general, WGO extracted with supercritical CO_2 showed a significantly ($p < 0.05$) higher concentration of sterols, in particular of campesterol, stigmasterol and β-sitosterol. Campestanol, sitostanol and Δ_7-avenasterol, instead, had a significantly ($p < 0.05$) higher concentration in the WGO extracted with a cold press and solvents. These results are in accordance with, or slightly lower than, the literature [29–34], but cultivar, the origin of the wheat germ and cultivation and environmental factors must be considered.

Table 5. Single sterol composition (%) of WGO extracted with different technologies.

Sterol	*Mechanical Extraction*	*Solvent Extraction*	*Supercritical CO_2 Extraction*
Campesterol	12.6 ± 0.9 ab	12.5 ± 0.4 b	13.8 ± 0.7 a
Campestanol	17.6 ± 1.0 a	18.5 ± 0.6 a	15.2 ± 0.8 b
Stigmasterol	3.2 ± 0.5 b	3.3 ± 0.3 b	4.1 ± 0.2 a
β-Sitosterol	32.7 ± 2.0 ab	31.5 ± 1.5 b	35.3 ± 1.0 a
Sitostanol	17.1 ± 1.4 ab	18.8 ± 0.8 a	16.3 ± 0.9 b
Δ_5-Avenasterol	7.4 ± 1.0 a	8.1 ± 0.7 a	8.0 ± 0.7 a
Δ_7-Stigmasterol	1.2 ± 0.2 b	1.6 ± 0.3 ab	1.8 ± 0.1 a
Δ_7-Avenasterol	2.1 ± 0.3 a	2.2 ± 0.2 a	1.3 ± 0.2 b

Results of the analysis of variance with Tukey's test are shown: $p < 0.05$; lowercase letters in the same row show significantly different values within each sterol.

4. Conclusions

For the different wheat germ oil extraction technologies, differences were registered with regard to wheat germ oil yield, while its composition was not affected by the extraction. Raw material management is the first critical point for oil quality, considering free fatty acids and the oxidation level, which may depend on cultivation, storage and degermination conditions; the tempering phase of durum wheat that is applied before the degermination process is the most critical point for the development of lipid hydrolysis and oxidation. This is because the increase in the amount of moisture in the caryopsis causes the activation of lipase and lipoxygenase. This represents a paradox: in order to stabilize the germ with degermination, first, it seems inevitable to carry out a process that destabilizes it. Therefore, the whole stabilization process must take account of the effective quality of the germ at that moment and, at the same time, of the final quality of the flour and germ by-products obtained. Further studies are necessary and interesting in order to obtain WG with high quality after its separation from the caryopsis; in this way, is it possible to use to use the whole grain, reducing waste and by-products.

Author Contributions: Conceptualization, S.M., M.F.C., M.G.M. and F.P.; Data curation, S.M. and M.F.C.; Formal analysis, M.G.M.; Investigation, S.M. and F.P.; Writing—original draft, S.M. and M.F.C. All authors have read and agreed to the published version of the manuscript.

Funding: This study was funded by the Agrifood PON i&Cc 2014-2020 project number F/200044/01-03/X45 "Produzione di uno sfarinato con proprietà nutraceutiche per il trattamento della Sindrome Metabolica—SIMBA".

Data Availability Statement: Data is contained within the article.

Acknowledgments: The authors also thank Molino Casillo S.p.a., who supplied the samples of this investigation, Giulia Napoletano (R&D assistant, Molino Casillo S.p.a) and Giovanni di Gioia (wheat junior trader, Molino Casillo Spa).

Conflicts of Interest: The authors declare no conflict of interest. The sponsors had no role in the design, execution, interpretation or writing of the study.

References

1. Shewry, P.; Hey, S. Do "ancient" wheat species differ from modern bread wheat in their contents of bioactive components? *J. Cereal Sci.* **2015**, *65*, 236–243. [CrossRef]
2. Peña-Bautista, R.J.; Hernandez-Espinosa, N.; Jones, J.M.; Guzmán, C.; Braun, H.J. Series on carbohydrates, wheat, grains, and health: Wheat-based foods: Their global and regional importance in the food supply, nutrition, and health. *Cereal Foods World* **2017**, *62*, 231–249. [CrossRef]
3. Hernandez-Espinosa, N.; Laddomada, B.; Payne, T.; Huerta-Espino, J.; Govindan, V.; Ammar, K.; Ibba, M.I.; Pasqualone, A.; Guzman, C. Nutritional quality characterization of a set of durum wheat landraces from Iran and Mexico. *LWT Food Sci. Technol.* **2020**, *124*, 109198. [CrossRef]

4. Laddomada, B.; Durante, M.; Mangini, G.; D'Amico, L.; Lenucci, M.S.; Simeone, R.; Piarulli, L.; Mita, G.; Blanco, A. Genetic variation for phenolic acids concentration and composition in a tetraploid wheat (*Triticum turgidum* L.) collection. *Genet. Resour. Crop Evol.* **2017**, *64*, 587–597. [CrossRef]
5. Hemery, Y.M.; Anson, N.M.; Havenaar, R.; Haenen, G.R.M.M.; Noort, M.W.J.; Rouau, X. Dry-fractionation of wheat bran increases the bioaccessibility of phenolic acids in breads made from processed bran fractions. *Int. Food Res. J.* **2010**, *43*, 1429–1438. [CrossRef]
6. Bottega, G.; Caramanico, R.; Lucisano, M.; Mariotti, M.; Franzetti, L.; Pagani, M.A. The debranning of common wheat (*Triticum aestivum* L.) with innovative abrasive rolls. *J. Food Eng.* **2009**, *94*, 75–82. [CrossRef]
7. Engelsen, M.M.; Hansen, Å. Tocopherol and tocotrienol content in commercial wheat mill streams. *Cereal Chem.* **2009**, *86*, 499–502. [CrossRef]
8. Zhu, K.X.; Wang, X.P.; Guo, X.N. Isolation and characterization of zinc-chelating peptides from wheat germ protein hydrolysates. *J. Funct. Foods* **2015**, *12*, 23–32. [CrossRef]
9. Boukid, F.; Folloni, S.; Ranieri, R.; Vittadini, E. A compendium of wheat germ: Separation, stabilization and food applications. *Trends Food Sci. Technol.* **2018**, *78*, 120–133. [CrossRef]
10. Tolouie, H.; Mohammadifar, M.A.; Ghomi, H.; Hashemi, M. Argon and nitrogen cold plasma effects on wheat germ lipolytic enzymes: Comparison to thermal treatment. *Food Chem.* **2021**, *346*, 128974. [CrossRef]
11. Liao, M.; Damayanti, W.; Zhao, Y.; Xu, X.; Zheng, Y.; Wu, J.; Jiao, S. Hot air-assisted radio frequency stabilizing treatment effects on physicochemical properties, enzyme activities and nutritional quality of wheat germ. *Food Bioprocess Technol.* **2020**, *13*, 901–910. [CrossRef]
12. Srivastava, A.K.; Sudha, M.L.; Baskaran, V.; Leelavathi, K. Studies on heat stabilized wheat germ and its influence on rheological charactaeristics of dough. *Eur. Food Res. Technol.* **2006**, *224*, 365–372. [CrossRef]
13. Geng, P.; Harnly, J.M.; Chen, P. Differentiation of whole grain from refined wheat (*T. aestivum*) flour using lipid profile of wheat bran, germ and endosperm with UHPLC-HRAM mass spectrometry. *J. Agric. Food Chem.* **2015**, *63*, 6189–6211. [CrossRef]
14. Sudha, M.L.; Srivastava, A.K.; Leelavathi, K. Studies on pasting an structural characteristics of thermally treated wheat germ. *Eur. Food Res. Technol.* **2007**, *225*, 351–357. [CrossRef]
15. Fawzy Ramadan, M.; El-Said Showky, H.; Mohamed Sulieman, A.E.R. Comparison between the effect of γ-irradiation and roasting on the profile and antioxidant activity of wheat germ lipids. *Grasas Aceites* **2008**, *59*, 166–173. [CrossRef]
16. Li, B.; Zhao, L.; Chen, H.; Sun, D.; Deng, B.; Li, J.; Liu, Y.; Wang, F. Inactivation of lipase and lipoxygenase of wheat germ with temperature-controlled short wave infrared radiation and its effect on storage stability and quality of wheat germ oil. *PLoS ONE* **2016**, *11*, e0167330. [CrossRef]
17. Jha, P.K.; Kudachikar, V.B.; Kumar, S. Lipase inactivation in wheat germ by gamma irradiation. *Rad. Phys. Chem.* **2013**, *86*, 136–139. [CrossRef]
18. Sharma, S.; Kaur, S.; Dar, B.N.; Singh, B. Storage stability and quality assessment of processed cereal brans. *J. Food Sci. Technol.* **2013**, *51*, 583–588. [CrossRef]
19. Gomez, A.; de la Ossa, E. Quality of wheat germ oil extracted by liquid and supercritical carbon dioxide. *JAOCS* **2000**, *77*, 969–974.
20. Marti, A.; Torri, L.; Casiraghi, M.C.; Franzetti, L.; Limbo, S.; Morandin, F.; Quaglia, L.; Pagani, M.A. Wheat germ stabilization by heat-teatment or sourdrough fermentation: Effects on dough rheology and bread properties. *LWT Food Sci. Technol.* **2014**, *59*, 1100–1106. [CrossRef]
21. AOAC International Website. Available online: https://www.aoac.org (accessed on 5 February 2022).
22. European Standard Methods. Available online: https://www.en-standard.eu (accessed on 5 February 2022).
23. AOAC. *Official Methods of Analysis*, 16th ed.; Association of Official Analysis Chemists: Washington, DC, USA, 2000.
24. Malekian, F. *Lipase and Lipoxygenase Activity, Functionality, and Nutrient Losses in Rice Bran during Storage*; LSU AgCenter: Baton Rouge, LA, USA, 2000; pp. 3–56.
25. Kapranchikov, V.S.; Zherebtsov, N.A.; Popova, T.N. Purification and characterization of lipase from wheat (*Triticum aestivum* L.) germ. *Appl. Biochem. Microbiol.* **2004**, *40*, 84–88. [CrossRef]
26. Liu, K. Comparison of Lipid Content and Fatty Acid Composition and Their Distribution within Seeds of 5 Small Grain Species. *J. Food Sci.* **2011**, *76*, 334–342. [CrossRef]
27. Suresh Kumar, G.; Gopala Krishna, A.G. Studies on the nutraceuticals composition of wheat derived oils wheat bran oil and wheat germ oil. *J. Food Sci. Technol.* **2015**, *52*, 1145–1151. [CrossRef]
28. Cardenia, V.; Sgarzi, F.; Mandrioli, M.; Tribuzio, G.; Rodriguez-Estrada, M.T.; Gallina Toschi, T. Durum wheat bran by-products: Oil and phenolic acids to be valorized by industrial symbiosis. *Eur. J. Lipid Sci. Technol.* **2018**, *120*, 1700209. [CrossRef]
29. Harrabi, S.; Ferchichi, A.; Fellah, H.; Feki, M.; Hosseinian, F. Chemical composition and in vitro anti-inflammatory activity of wheat germ oil depending on the extraction procedure. *J. Oleo Sci.* **2021**, *70*, 1051–1058. [CrossRef]
30. Salehi, B.; Rescigno, A.; Dettori, T.; Calina, D.; Docea, A.O.; Singh, L.; Cebeci, F.; ÖzÇelik, B.; Bhia, M.; Beirami, A.D.; et al. Avocado-soybean unsaponifiables: A panoply of potentialities to be exploited. *Biomolecules* **2020**, *10*, 130. [CrossRef]
31. Panfili, G.; Fratianni, A.; Irano, M. Normal Phase High-Performance Liquid Chromatography Method for the Determination of Tocopherols and Tocotrienols in Cereals. *J. Agric. Food Chem.* **2003**, *51*, 3940–3944. [CrossRef] [PubMed]
32. Eisenmenger, M.; Dunford, N.T. Bioactive components of commercial and supercritical carbon dioxide processed wheat germ oil. *J. Am. Oil Chem. Soc.* **2008**, *85*, 55–61. [CrossRef]

33. Özcan, M.M.; Rosa, A.; Dessi, M.A.; Marongu, B.; Piras, A.; AL-Juhai, F.Y.I. Quality of wheat germ oil obtained by cold pressind and supercritical carbon dioxide extraction. *Czech J. Food Sci.* **2013**, *31*, 236–240. [CrossRef]
34. Ghafoor, K.; Özcan, M.M.; AL-Juhaimi, F.; Babiker, E.E.; Sarker, Z.I.; Ahmed, I.A.M.; Ahmed, M.A. Nutritional composition, extraction, and utilization of wheat germ oil: A review. *Eur. J. Lipid Sci. Technol.* **2017**, *119*, 1600160. [CrossRef]
35. Woyengo, T.A.; Ramprasath, V.R.; Jones, P.J.H. Anticancer effects of phytosterols. *Eur. J. Clin. Nutr.* **2009**, *63*, 813–882. [CrossRef]
36. Iafelice, G.; Verardo, V.; Marconi, E.; Caboni, M.F. Characterization of Total, Free and Esterified Phytosterols in Tetraploid and Hexaploid Wheats. *J. Agric. Food Chem.* **2009**, *57*, 2267–2273. [CrossRef]

foods

Review

Impact of Quality Improvement and Milling Innovations on Durum Wheat and End Products

Ashok Sarkar [1,*] and Bin Xiao Fu [2]

1 Cereals Canada, 1000-303 Main Street, Winnipeg, MB R3C 3G7, Canada
2 Grain Research Laboratory, Canadian Grain Commission, 303 Main St., Winnipeg, MB R3C 3G8, Canada; binxiao.fu@grainscanada.gc.ca
* Correspondence: asarkar@cerealscanada.ca

Abstract: There are long-standing established intrinsic quality requirements of end products made from durum wheat semolina, with color, textural, and cooking properties of pasta and couscous representing persistent key attributes for consumers. Over time, traditional efforts to advance development in these areas with respect to raw material, equipment, and process improvements have been influenced by growing awareness of issues around food safety, health and nutrition, and climate change, necessitating that development strategies incorporate specific considerations relating to safety, traceability, and sustainability. We examined improvements in durum wheat quality and innovations in milling and the resulting impact on product quality in light of these considerations, which are now fundamental to the planning and development of any food process, as required by consumers and regulators alike.

Keywords: durum wheat; semolina milling; semolina quality; granulations; digitalization; sustainability

Citation: Sarkar, A.; Fu, B.X. Impact of Quality Improvement and Milling Innovations on Durum Wheat and End Products. *Foods* **2022**, *11*, 1796. https://doi.org/10.3390/foods11121796

Academic Editor: Mike Sissons

Received: 5 May 2022
Accepted: 11 June 2022
Published: 18 June 2022

1. Introduction

Durum wheat is an important crop that serves as a staple and a good source of nutrition for consumers around the world. Durum wheat is widely utilized in diverse traditional food products consumed in the Mediterranean basin [1]. Globally, durum wheat semolina is considered to be the most suitable raw material of choice for pasta, couscous, and a variety of breads due to its natural pigment color, hard kernel texture, and good protein content and quality. These properties are directly attributable to the intrinsic quality of the durum wheat that is milled into semolina. Furthermore, these end-use properties are influenced by various milling processes, and quality optimization requires an understanding of the specific effects of these processes. Other factors, including the successes of the breeding program, have also contributed to quality improvements in recent years.

In an environment of increasing awareness around issues of food safety and sustainability, desirable characteristics of semolina products now extend beyond quality and functional attributes, with increasing emphasis on food safety issues (e.g., cadmium and DON), as well methods for sustainable production. Processing equipment advances such as the use optical sorters and pearling systems have improved the elimination of undesirable and toxic extraneous materials found in incoming wheat, while the widespread adoption of fine semolina granulation and design advances relating to roller mills, sifters, and purifiers have allowed for improved energy conservation without sacrificing quality or safety. Additionally, online quality control of wheat and products and application of advanced automation through digitalization hold considerable promise toward further improving productivity, energy efficiency, process optimization, and reduced wastage.

In this paper, we examine and discuss important factors relating to the current state of durum wheat production, milling, and processing, with particular attention to advances in the areas of product quality, safety, process optimization, and sustainability.

2. Advances in the Quality of Durum Wheat

Durum wheat is widely used in various products, including long and short dried pasta, fresh and sheeted pasta, couscous, and baked bread. High-quality durum wheat has superior milling quality, producing a high yield of semolina with low ash content and speck count, and has high yellow pigment content necessary to produce products with a bright yellow color. A high protein content and strong gluten characteristics ensure superior pasta-cooking quality and good performance in certain products, such as durum bread. New durum varieties expressing low cadmium uptake are required to meet the food safety regulations in many markets.

For high-quality production of semolina, the selection of appropriate quality of the raw material—durum wheat—is vitally important. Recent advances in the availability of improved durum wheat varieties have been a great success story in raising the standard of quality, which is higher than ever before, particularly with respect to the following key quality attributes.

2.1. Protein Content

Durum wheat with a high protein and good physical condition will generally yield semolina of uniform particle size with a minimum number of starchy particles [2]. Protein in semolina facilitates hydration during mixing and provides the structure for pasta. A high protein concentration is the prerequisite for superior pasta-cooking quality. The protein content is a major determinant of the value for durum milling and pasta processing [3].

Both genetic factors and environmental conditions influence the protein concentration in durum wheat. Growing varieties with a high protein potential is an effective way to maintain grain protein content at the highest possible level in a low-input production system characterized by low rates of nitrogen fertilizer application [4].

2.2. Gluten Strength

Gluten strength has been widely considered as an important secondary prerequisite for superior pasta-cooking quality [5]. The continuity and strength of the protein matrix formed during extrusion are important in determining the textural characteristics of the pasta. The relationship between gluten strength and pasta-cooking quality is complex and inconclusive. Furthermore, there is strong evidence that, under high-temperature drying conditions, gluten strength has less influence on the pasta's cooking quality than under low-temperature drying conditions [6,7]. High-temperature drying is predominant in today's pasta industry.

Regardless of whether gluten strength might be overestimated as a cooking quality prerequisite, particularly for pasta dried at a high temperature, there has been increasing emphasis placed on strength as a durum wheat quality specification. Efforts in improving durum wheat gluten quality have resulted in a significant increase in the gluten strength in the last 20 years (Table 1). It is important to have strong and extensible gluten to be suitable for most products, including long and short goods, fresh pasta, sheeted pasta, couscous, baked bread, etc. [8]. Durum wheat with inextensible gluten has limited applications. They can cause processing problems (e.g., sheeted pasta) or result in poor product quality (e.g., bread).

Table 1. Selected quality parameters of Canadian durum export aggregates.

Parameters	2006–2010	2011–2015	2016–2020
Total Yellow Pigment Content (semolina), ppm	7.5–9.0	8.5–10	9.0–11
Gluten Index (Semolina), %	40–55	60–85	65–85
Cadmium (Grain), ppb	140–80	85–65	85–65

Source: www.grainscanada.gc.ca (accessed on 1 February 2022).

2.3. Pigment Content and Pigment Loss

One of the most important quality factors in durum wheat is the potential of producing semolina and pasta products with a bright yellow color. Semolina and pasta yellowness is affected by various factors: the yellow pigment content of the grain; the oxidative degradation of pigments by lipoxygenase (LOX) during pasta processing; and the processing conditions such as drying temperature, extrusion die design, and type [9,10].

Over the past few decades, efforts to improve the yellow color of durum semolina in durum breeding programs have resulted in the release of cultivars with high pigment levels, as reflected by the Canadian durum export cargo quality monitoring (Table 1). Market feedback has been very positive for this improvement. There is usually a slight elevation of pasta redness with the increase of total yellow pigments in durum wheat. However, the increase in pasta yellowness more than compensated for the elevation in redness in the overall appearance of pasta. Millers also use high pigment durum for blending with wheat of low pigment to improve the color of semolina and pasta products.

Some of the pigments in semolina will be degraded and lose yellow color during pasta processing through oxidation induced by LOX. Significant progress has been made in the genetics of LOX to facilitate durum wheat improvement by developing new cultivars with low LOX activity [11]. The allelic variation for a deletion of the *Lpx-B1.1* was associated with a significant reduction in LOX activity and improved pasta color due to reduced pigment degradation during pasta processing [12]. While high pigment content is the primary factor for superior color of semolina and pasta, a reduction in LOX activity will further improve the color of pasta products.

2.4. Milling Performance

Milling performance is the most important factor that determines the industrial value of durum wheat. The key indicators of milling quality are yields (total and semolina), ash content, and speck counts in the finished granular product. Yield is a key indicator of profit for durum mill. There is a legal limit for semolina ash content in some EU countries. The speck count is a deciding factor of consumer acceptance for many durum products. Durum wheat with superior milling quality is characterized by high test weight, large kernels, and high percentage of hard vitreous kernel (HVK). Table 2 listed tolerances of some key grading factors for Canadian durum wheat.

Table 2. Tolerances of selected grading factors for CWAD wheat.

Grading Factors	No. 1	No. 2	No. 3	No. 4
Test Weight, kg/hL	80	79	78	75
HVK, %	80	60	40	No Minimum
Ergot, %	0.02	0.02	0.04	0.04
Fusarium Damage, %	0.5	0.5	2.0	2.0

Source: www.grainscanada.gc.ca (accessed on 1 February 2022).

The physical defects associated with surface discoloration of kernels are important because bright speck-free semolina is required to give the aesthetic appearance of premium semolina and pasta products. Kernels with surface discoloration are tolerated in very low amounts in high-quality durum wheat. Ergot sclerotia, smudge, black point, mildew, and midge are the main physical defects associated with surface discoloration [13]. Resistance to disease and insect damage not only prevents loss of yield but also protects the grade and quality of the grain, especially for durum wheat, because of the major impact on speck count.

To ensure that Canada Western Amber Durum (CWAD) wheat meets the quality and safety expectation of customers, stringent tolerances are set for various milling grades.

2.5. Cadmium Level

High levels of Cd in cereal grains are a health concern. Durum wheat normally accumulates more Cd than other commonly grown cereals, with concentrations ranging from less than 30 to more than 300 μg/kg [14]. A low Cd concentration in durum wheat is mainly controlled by a single dominant gene (Cdu-B1) that is highly heritable [15]. Incorporation of the low Cd allele into cultivars reduces the grain Cd by about 50%. Low Cd is mandatory for registration of durum wheat cultivars in Canada [3]. The environment and soil conditions, however, play an important role in Cd content in the grain, even for low-Cd varieties.

The Codex General Standard for Contaminants and Toxins in Food (Rev. 5, 2009) lists the maximum limit for Cd in wheat grain as 200 ppb. EU has instituted a new standard for Cd at 180 ppb for durum wheat since August 2021 [16]. The current guide of <100 ppb for registering new durum varieties in Canada can effectively manage the Cd level.

2.6. Future Trends in Durum Quality Improvement

Fusarium head blight (FHB) is a major fungal disease. Durum is notorious for its extreme susceptibility to FHB and breeding for FHB resistance is difficult. FHB has a great economic impact on durum crops due to the reduced seed quality and agnomical yield. Semolina yield, gluten strength, and pasta quality can adversely be affected by the presence of fusarium damaged kernels (FDKs), which are usually shriveled [13]. The current FDK tolerances for grading Canadian wheat (Table 2) can effectively ensure durum wheat milling performance and semolina and pasta quality. FDK is also highly problematic because of fusarium mycotoxins, which render the grain unfit for food and feed. Genetic resistance is the most cost-effective and environment friendly approach for controlling FHB and development of new cultivars with improved resistance is the major goal for improving durum wheat quality and safety [17].

Drought and heat are major abiotic stresses affecting durum wheat production and quality worldwide. In addition to the yield loss, the effects of drought and heat stresses have a very significant impact on durum quality, as indicated by a lower test weight, smaller kernels, lower milling yield, and higher semolina ash [18]. Enhanced heat and drought tolerance of wheat is not only to ensure a stable yield across both good and bad seasons while maintaining a high yield under optimal conditions, but also to protect quality by minimizing their impacts on kernel morphological properties, which, in turn, adversely affect milling quality.

Protein content is critical for kernel virtuosity and pasta-cooking quality [2]. Nitrogen fertilizer is the most used nutrient source in modern agriculture and represents significant environmental and production costs. The selection of new varieties with higher nitrogen-use efficiency has become of ever-increasing importance.

3. Raw Material and Semolina Milling

Semolina milling operations is a vital link between durum wheat and quality end products such as pasta, couscous, and popular Mediterranean baked goods. While semolina quality is strongly correlated to the intrinsic quality of the raw material—wheat—the processing environment plays a major role in determining the outcome of the final quality of semolina consistently over the long production periods.

Starting with high-quality raw material characterized by a good kernel size, high test weight, and high percentage of hard vitreous kernels (HVKs), along with all the appropriate physical and functional quality attributes, is the best approach to produce good quality semolina. The value of good and uniform kernel size has been emphasized in the literature for good milling performance [19]. The relationship of milling properties with physical properties of wheat kernel has been studied extensively. Djiki and Laskowski (2005) [20] stated that the 1000 kernel weight in durum wheat is associated with semolina yield and test weight. They also suggested that kernel size uniformity is very important in the milling process with respect to cleaning, conditioning, debranning, and grinding. Recent research on the relationship of kernel size and genotype on milling performance further substantiates the importance of these aspects for the selection of appropriate durum wheat [18]. Often the semolina quality specified requires the blending of two or more wheat types to achieve the desired quality of the blend. The motivation for blending may also be to reduce the cost of wheat mix. This may also introduce the differences in kernel sizes if the wheats in the blend are of variable sizes. Major durum-wheat-exporting countries maintain their quality standard by implementing a numerical grading system, with tolerance levels set for reflecting quality, as previously shown in Table 2.

The next critical step is in the milling process, which allows millers to generate desirable levels of semolina yields of specified quality. Developments in the processing of semolina has been influenced by the continued increased demand for finer semolina over the years. The stringent semolina-processing requirements of traditional coarse semolina (<630 μm) to more easily attainable finer semolina (<355 μm) have led to relative simplification of what tended to be extensive processing practices. Finer semolina granulations with wider acceptance over time have somewhat reasonable levels of processing system with improved yields. Such plants are also easier from operation and supervision point of view. There are still some markets where the demand for coarse semolina exists, along with fine semolina, and in such situations, a portion of the coarse semolina is separately ground into a finer semolina particle size, as specified [21].

The expectations of the milling-process outcome are in the production of semolina quality of required specifications consistently. The quality expectations may often be challenged for unavailability of appropriate durum wheat due to crop failure, environmental factors, logistics, or even trade disruptions.

Recent advances in the milling equipment and in the process solutions have greatly improved the situation mitigating the quality shortcoming of wheat by compensating with improved milling technology to a reasonable extent.

4. Wheat Preparation

The primary functions in wheat preparation include the following:

- Removal of all non-wheat material, including damaged and diseased wheat and thin, immature, shrunken, and broken kernels;
- Bringing the wheat to its optimum condition for milling with appropriate tempering or conditioning;
- Removal of surface contaminants, crease-dirt, and the loosened outer layer of the bran following conditioning.

The equipment and process involved in this area have been extensively covered by Bizzarri and Morelli (1988) [21], Kuentzli (2001) [22], Sarkar (2003) [23], and Posner and Hibbs (2005) [24]. Although there have been ongoing continued improvement and development, the basic working principles and functions have remained essentially the same. On the other hand, there have been some noteworthy major developments that have benefitted the semolina milling operations significantly and therefore deserve appropriate discussion.

Ensuring durum wheat free from all foreign materials, damage, and diseased kernels, as well as all types of defects before the milling process begins, has always been the top priority for semolina millers. This operation always received necessary attention due to its importance on performance and impact on quality. Millers have always been able to perform this function very well, although with some degree of challenges primarily related to unavailability of cleaning equipment of high-performance as available today. Essentially, what this means is that, in the "Traditional" cleaning plants, the number of equipment used for cleaning was higher than today, adding to the space requirements and energy costs. This transformation is shown in Table 3. Although it is hard to capture all aspects of transformation without being too comprehensive, the table depicts the essential elements of the major developments.

Table 3. Comparative development of cleaning equipment of significance.

<table>
<tr><th>Sequence of Removal of Impurities and Related Advantage</th><th>Traditional Cleaning System</th><th>Contemporary Cleaning System</th><th>Advantages</th></tr>
<tr><td>Larger and finer impurities than wheat are removed first. Bulk of the impurities is removed here, reducing the load on subsequent equipment</td><td>Grain separator, using screens for separation</td><td rowspan="4">Combi-Cleaner/ Vitaris -combines all 4 functions in one</td><td rowspan="4">Both the Combi-Cleaner and Vitaris perform 4 functions in a single machine, helping in reduced space requirements, energy savings, and supervision.
Vitaris provides additional feature of modularity, allowing for the selection of any desired combination of these 4 functions as per the needs of the plant and improved energy efficiency</td></tr>
<tr><td>Lighter impurities than wheat are removed here, improving the cleaning efficiencies of the subsequent equipment</td><td>Aspiration channel with air-recycling</td></tr>
<tr><td>Separation of heavy and mixed wheat streams. Heavy stream requires removal of stones and no further cleaning; only a smaller fraction of mixed stream with lighter seeds require cleaning</td><td>Concentrator, using air and screens for separation</td></tr>
<tr><td>Stones, glass, and metals of similar size as wheat that could not be removed by screens are removed here. Early removal protects subsequent equipment damage and helps in strict adherence to food safety</td><td>Destoner, using density as the basis of separation</td></tr>
<tr><td>Longer seeds than wheat—oats, barley, and shorter seeds such as cockle and wild buckwheat—are removed here. Since most of the impurities are removed already, as mentioned above, this allows precise adjustment to be made here for the removal of these materials</td><td>Indented cylinders, using shape as the basis of separation</td><td rowspan="2">Optical sorters, using optical measurements, color, shape, and size</td><td rowspan="2">Optical sorters help reduce the percentage of screenings due to efficient and effective removal of the rejects with no loss of wheat. This helps increase the yield of cleaned wheat; it is also space saving and energy saving</td></tr>
<tr><td>Heavier and lighter fraction separation/ removal of ergot, oats, and barley</td><td>Gravity tables/ table separators</td></tr>
<tr><td>Surface contaminants</td><td>Scourers, using friction as the basis of cleaning</td><td>Pearlers, using abrasion/friction</td><td>Removes significant percentage of bran, surface contaminants, microbial load, and superficial discoloration</td></tr>
</table>

As we review Table 3, we observe that the sequence of removal of the impurities remains the same through both phases of development—"Traditional" and "Contemporary". The principal differences are in the compactness and comprehensiveness of each new cleaning machine performing multiple functions with a high degree of efficiency and cost savings in terms of building-space and energy requirements. This is seen by observing the subsequent columns of "Contemporary" that appears to be quite lean compared to the "Traditional" plants. The "Contemporary" model is simpler and more powerful in delivering high level of performance.

The "Traditional" cleaning plants are described and discussed at length by Bizzarri and Morelli (1988) [21], Kuentzli (2001) [22], and Sarkar (2003) [23]. Some of the latest developments and innovations that are helping enormously and being embraced by the industry widely are discussed here.

4.1. Developments and Innovations in Wheat Preparation

Among many developments introduced by equipment suppliers, some of the noteworthy ones that deserve mentioning are Vitaris, Optical Sorters, and Pearlers.

4.2. Vitaris (Cleaning of Bulk of the Foreign Material)

The four functions shown in Table 3 required individual machines to perform in the older plants. The Combi-Cleaner, which was introduced in the 1990s, was able to perform all of those functions, saving space and energy [23–25]. Equipment manufacturers have been concentrating on improvement in energy, space, and efficiency as the basis of development of new equipment. These three aspects together help in raising sustainability to higher standards. The introduction of a multifunctional cleaning equipment "Vitaris", available today meets those aspects of functional attributes as in Combi-Cleaner, with an added practical feature of incorporating modularity [26]. Modularity provides flexibility in terms of combining these functions in any combination of choice of the milling company. The discretion of investing in a multifunctional compact cleaning machine during the startup or adding a function at a time as the need evolves is of practical value. The newer machine, apart from offering flexibility, comes with advanced features of energy saving, space, and superior efficiency.

Such a machine removes the bulk of the foreign materials and impurities. However, there are defects and damaged wheat kernels that can only be removed by optical sorters.

4.3. Optical Sorters

Traditionally, the cleaning of durum wheat has been more involved and challenging compared to common wheat cleaning [25]. The development of optical sorters has simplified the cleaning operation, while improving the cleaning efficiency enormously. The visual quality of durum wheat semolina is of the utmost importance for its end users. A bright yellow-colored semolina with minimal discoloring specks is considered to be an integral part of quality. Any black and brown specks would be conspicuous and can be seen as indications of inferior, contaminated, and impure product. With coarser granulation of durum semolina, the potential size of such specks is going to be large, as well. The presence of these specks can be minimized by thoroughly cleaning the wheat free of all types of impurities and foreign materials. Dark seeds, along with removal of wheat kernels with surface discoloration as in blackpoint and smudge, must be removed, because, apart from creating dark specks, they also impact the pasta color (Dexter and D'Egidio 2012) [27]. A durum-wheat-cleaning facility of about 15 years or more (pre-optical sorter in wheat cleaning) could not remove discolored wheat kernels from clean wheat stream, as they could not be removed by any means since they are part of the wheat kernels. The only way to control this problem was to work with top grades of durum wheat that would restrict the presence of such undesirable defects as part of the grade standards. Another problem was the effective removal of oats, barley, and ergot [23] from durum wheat kernels, as their length would be quite comparable. The only difference was that they would be slightly lighter in density relative to durum wheat. This problem was overcome through the installation of more gravity-based cleaning equipment, such as gravity tables and table separators, making the durum cleaning operation more complex than seen in common wheat. Typically, investment in the cleaning house of a durum mill has been higher compared to that for common wheat for flour production.

The optical sorters helped in the removal of these impurities (ergot, oats, and barley), thereby reducing the number of gravity-based cleaning machines, while minimizing the creation of dark specks. This aspect of quality (black and brown specks) is controlled by the implementation of speck count specifications used for benchmarking semolina quality [28]. The number of black and brown specks limits in semolina required by end users varies amongst companies and from region to region, based on specific requirements of the company, grades of semolina, and method used for its determination [24]. This demonstrates the importance of aesthetics in semolina, as it directly impacts the visual appearance of pasta and couscous products. These products, including the semolina (for home use), are sold in transparent packages or packages with windows, and any discolored specks show up clearly, thus rendering the product unattractive at the point of sale.

With the application of an optical sorter, the problem of dark specks has been managed well, along with replacements of the mechanical equipment of disc separators and indented cylinders and at a reduced energy cost [29]. There are a number of manufacturers of optical sorters in the industry, namely Bühler and Satake, among others.

Apart from assisting in quality improvements, optical sorters are very important from a food-safety point of view. There is growing consumer awareness about allergens and demand for product safety. Whole-grain semolina is more susceptible, as there is no further refinement in the milling process for removal of pieces of any allergenic material that goes into the milling process, along with durum wheat.

The following contaminants that are considered a risk to human health are effectively reduced by optical sorters, making it an important machine from a food-safety point of view.

- Allergens such as peanuts, soy splits, and pieces of soy;
- Mycotoxins such as vomitoxin (DON) through the removal of fusarium-damaged kernels [29];
- Ergot bodies;
- Wide range of foreign materials.

Soy pieces or splits, along with whole soy, apart from being allergens, can also result in the light bleaching of pasta dough made from durum semolina during the mixing stage in a slow and longer mixing environment. The bleaching could become extensive in the case of semolina and flour when being mixed for bread dough. This is due to lipoxygenase activity, which is undesirable for the end user.

The effective removal of ergot [30] reduces the generation of dark specks, and, more important, the removal of ergot bodies eliminates the potent alkaloids associated with them that are a risk to human health [27]. The improved detection of subtle color differences is also very helpful in removing fusarium-damaged kernels from durum wheat, helping reduce the mycotoxin levels of deoxynivalenol (DON) [30]. Fowler described the advancements in optical sorters, from monochromatic to bichromatic application enhancing the detection of subtle color differences. Fowler further stated that innovations in optical sorting have effectively removed fusarium-affected wheat from good-quality wheat.

The recent advances in optical sorters have been noteworthy. Technological developments have resulted in the improvement of performance, reliability, flexibility, ease of use, and connectivity. The new optical sorters by Sortex [31] feature improved hardware and software, along with advanced sorting algorithms that help to optimize machine performance. Developments involve improvements in the design of a camera with a low signal-to-noise ratio. The optical sorter aided by full-color cameras with enhanced spectral purity and higher-intensity LED lights helps improve defect detection. Multilayered algorithms, along with precision ejectors, reduce losses of good wheat kernels, with the rejects resulting in higher yields. All the improvements, as mentioned above, have helped improve defect detection quality while reducing losses of good wheat kernels, along with the rejects. Wheat cost is the largest single cost in the production of semolina, saving even a small quantity of wheat would result in substantial savings. Such a precise sorting capability helps in the utilization of lower-quality wheat as input with acceptable output quality.

Connectivity facilitates the monitoring and control of the machine from anywhere, ensuring high performance and product traceability, along with host of other benefits that include improved productivity, quality, downtime, and reduction of operation costs.

4.4. Pearling

Surface Treatment of Durum Wheat

The last step in the preparation of durum wheat for milling involves scouring or surface treatment. This has been traditionally an important step in the preparation of durum wheat for milling as pasta processors require a low microbiological count for semolina. Typically, intensive scouring was used for this purpose until debranners or pearling systems were found to be more beneficial and effective for the following reasons:

- Removal of significant percentage of bran layers of ~8% in one step, thereby removing surface dirt and contaminants and reducing microbial counts;
- Removal of bran layers help reduce bran content to deal with during milling;
- Significantly increasing the semolina yield [32];

- Improving the quality of semolina in terms of low speck count and ash content as dark spots are removed with the bran layers;
- Improvements in pasta brightness and color [32].

Considerable work in the application of the debranning process, developed by Satake [33], was carried out, and its beneficial impact on durum wheat processing was reported [32,34]. Various equipment-manufacturing companies have developed such equipment of their own based on a similar concept.

This process has become a standard feature now in most durum semolina mills around the world. The durum wheat kernels with a larger size and hard kernel texture lend themselves to be easily pearled than common wheat.

The early generation of the pearling system steadily gained popularity amongst the durum millers due to their improved performance. One such system was described by Gruber and Sarkar (2012) [25]. A newer version of this machine (OSIRIS) offers further improvements with more durable, low-energy, and more effective pearlers by replacing stone grinding with diamond-coated grinding wheels, reducing the contamination of stone particles due to wear. This upgrade improved the pearling degree with minimal breakage, while enhancing the product quality and food safety to a very high standard [31,35].

Pearling/debranning has been researched extensively apart from its central theme of improving milling performance, semolina yield, and quality. The benefits of pearling have been further investigated in the reduction of alpha amylase in wheat and its products following pearling [36,37]. The effects of industrial processing on the distributions of deoxynivalenol, cadmium, and lead in durum wheat milling fractions were also investigated by Cheli, 2010 [38]. Cheli et al. concluded that there was more of a reduction of contamination in milled fractions destined for human use from debranned wheat as compared to wheat milled without debranning. They also proposed that debranning wheat would be even more relevant when working with raw materials with a contamination level closer to the legislated levels. It was noted that the pericarp and testa together, being the peripheral part of the grain, are first colonized by the fungi and often contaminated by the microorganisms, heavy metals, and soil. The debranning/pearling process, in addition to improving semolina yield and quality, also helps in reducing any such contamination present in durum wheat.

As noted earlier, with about 8% of bran being removed in the pearler (debranner), there is less bran to deal with in the mill. This allows the break-system to be simplified: since most of the coarse bran is removed, there is less need of coarse break grinding passages [24,25]. For example, instead of fourth break passage coarse (4BC) and fine (4BF), there may be just one fourth break and, likewise, one fifth break instead of fifth break coarse and fine grinding passages. There may also be fewer sizing passages. Generally, similar milling surfaces remain, as the diagram does not change much. Purification passages also remain similar. The main benefit is in higher semolina yields and improved products.

The optical sorter and the pearling system together have been extremely effective in helping improve the visual quality of the semolina. The visual discoloration on account of seed contamination and defects can be reduced to a level with the help of this combination where no dark specks are visible.

5. Advancements in the Milling Process

Good milling performance is a function of the milling process and milling equipment. Landi (1995) [28] stated that the milling process does not improve or add to the inherent qualities of a wheat; however, it can destroy wheat quality if carried out incorrectly.

The collective effort of selection of good-quality wheat, its preparation for milling, and superior milling performance is vital for product quality, output, and semolina yield.

The milling process is key to ensuring that the following properties are maintained in the semolina produced:

- Low black and brown speck count;
- Appropriate particle size distribution (granulation);

- Low ash content;
- Low Starch damage.

5.1. Black and Brown Speck Count

The sources of specks, especially dark specks, are attributable to foreign material and discolored durum wheat kernels on account of heat damage with black-tipped germ, blackpoint, and smudge [24,25]. A good cleaning plant with an optical sorter and a debranner/pearler can minimize the presence of such specks, as explained in the earlier section under cleaning.

Apart from dark/black specks, brown specks are also evaluated carefully to ensure that the visual quality of semolina meets the requirements of the processor. Brown specks are created due to premature shredding of bran during milling and/or incorrect setting of machines, such as purifiers. This is coupled with the fact that coarse semolina generation would require lighter grinding, elaborate grading and purification, and a comprehensive sizing system to extract a large amount of semolina with minimal production of flour. Even then the semolina yield of good quality is in a range of around 66–68%.

5.2. Semolina Particle Size Distribution (Granulation)

The production of high-quality speck-free yellow color coarse semolina posed a major challenge to durum millers to meet the quality requirement [22]. As mentioned above, despite the employing elaborate break system, grading, purification, and sizing system, the semolina yield remains very limited [24,25].

The traditional semolina granulation from 35 to 40 years ago typically consisted of coarse particles with a very restricted percentage of flour (<2%) [22,25]. Traditional pasta processors showed a preference for coarse particle size with minimal flour. It has been well documented that coarse particles take longer to hydrate, and if they are not fully hydrated, this results in white spots [24,25,28]. Pasta processors' requirements of a very low percentage of flour in the coarse semolina was likely to control the differences in granulation. Kuenzli suggested that the tight tolerance was helpful in preventing any potential adulteration of durum flour with common wheat flour, as, around that time, there were no accurate tests available to determine that [22].

Coarse semolina was also preferred in North Africa, a major durum-consuming region, for the production of couscous. The production of couscous 30 years ago was primarily carried out by hand, requiring the use of coarse semolina to facilitate the easier production of large granules of couscous (1300–500 μm) [19].

Coarse semolina particle size requires a gradual processing system to ensure that a large particle size is maintained while detaching the adhering bran particles from the large chunks of endosperm in the break system. This is followed by a long-extended grading and purification system for the removal of these detached bran pieces. Following the removal of the large pieces of semolina in the purifier, there is a good portion of material with bran pieces attached to the endosperm that still requires detaching. This is sent to the sizing system for further size reduction, carefully keeping the particle size as large as possible while detaching the bran pieces. This is followed by comprehensive purification in sizing purifiers. This entire process can be elaborate in order to maintain a larger semolina particle size; therefore, when a coarser semolina particle is required, the processing system must be more gradual in order to ensure that a large particle size is achieved. This means there is more of a need for equipment to enable the gradual processing. Vertical integration in the pasta industry has been common practice (Industry & Trade Summary 2003) [39]. As suggested by Kuenzli (2001) [22], those durum milling facilities that are part of the pasta processing plants have been benefitting from the production of fine semolina granulation containing a portion of flour. From the literature, it appears that Manser (1985) [40] presented the benefits of finer semolina granulation from the point of view of pasta processors [22,24]. Finer granulation helped with quick and uniform hydration, due to a narrow range of particle size. The emergence of a fast mixing process worked very well with finer granulations of semolina, reducing mixing time, improving the homogeneity of the extruded dough [41], and resulting in pasta products with improved color [22].

Reduced dough-mixing time and high-temperature drying works well with finer semolina granulation tripling the pasta production [28], while helping durum millers to work with simplified fine semolina production diagram [25]. Table 4 below shows the grinding passages of a mill using coarse semolina production as compared to a plant that uses pearlers and production of finer semolina. The number of corresponding passages for finer semolina over coarse semolina in each processing passages is lower, except for the "conversion", where the finer material gets ground into flour.

Table 4. Milling passages for coarse and fine semolina production.

Semolina Granulation	Break Passages	Grading	Purification	Sizing	Reduction	Conversion
Coarse semolina	B1–B7F	Div 1–Div 4	S1–S26	D1–D7	RED 1–RED 3	
Fine semolina	B1–B5	Div 1–Div 2	S1–S12	D1–D3	RED 1–RED 2	C1–C4

Even though the vast majority of durum semolina is being produced with finer granulations, there are markets that still use coarse semolina of 630–200 μm for traditional pasta in parts of Europe. There is still a portion of coarse semolina of 1000–600 μm produced by plants in North Africa for handmade couscous.

Table 5 shows milling surface allocation for durum semolina mills with fine granulation in comparison with traditional coarse granulation. There are appreciable differences between the two practices with respect to roll and purification surfaces. Part of the reason for this may be rooted in the fact that the traditional coarse semolina was being generated in the older, less efficient equipment. It is, however, mainly due to more elaborate grading, purification, and sizing system, which are all required to achieve the desirable quality of the coarse semolina.

Table 5. Milling surface allocation of durum semolina mills for fine and coarse semolina [1].

	Fine Semolina	Coarse Semolina
Granulation, μm	355–0	630–125
Roll, mm/100 kg/24 h	11.3–12.5	15–18
Purifier, mm/100 kg/24 h	3.7–4.5	5–7
Sifter, m^2/100 kg/24 h	0.060–0.064	0.062–0.068

[1] Based on expected average commercial data.

Durum semolina mills in North America commonly produce semolina of granulation <425 μm, which is coarser than the fine semolina granulation produced, as well <355 μm. If milling surface allocations were to be compared between these two granulations, they would be similar. For example, instead of six sizing passages, there may be just one or two passages used for fine semolina, and in place of two reduction passages, there may be four or five passages for fine semolina, as there is more flour produced. Purification passages remain similar. While dry pasta production works well with finer semolina, there is a preference for coarse semolina for fresh pasta products such as ravioli and tortellini.

The most recent developments in durum wheat milling primarily have been in the increased utilization of durum flour (<180 μm) for end products. This helps the durum miller further increase the yield to up to 78% or higher, as compared to 72% to 74% of fine semolina with some flour. There is a growing demand for the increased utilization of durum flour (<180 μm), either alone or in combination with fine semolina. It is common to see a good number of pasta packages sold in North America showing durum flour as the second major item on the ingredient list. This is paving the way in favor of increased durum flour production. Durum flour is also used for the production of artisan breads and hearth breads. Increased yield has certainly helped the profitability for the durum milling business, as a higher yield of 78% plus is very attractive compared to the lower yield of semolina of 72% to 74% and especially when compared to the traditional yields of coarse semolina of 68% that was previously obtained around 30 to 35 years ago. The value of yield is more critical in durum milling, as the return on mill-feed (by-products) both from durum wheat and common wheat is the same, while durum wheat generally tends to be more expensive.

5.3. Ash Content

Ash is mineral-matter residue that is left behind upon the incineration of a sample in a muffle furnace. Ash content progressively increases from the center of the wheat kernel to the bran layers. Since ash is much higher in bran than in pure endosperm, it is a good indicator of the level of refinement and therefore is often used as a measure of milling performance. A mill providing a higher semolina yield at a given ash level is considered as superior in performance over another plant that yields lower semolina at the same ash content but milling wheat from the same source. More efficient milling produces lower ash [42]. A higher semolina yield has associated elevated levels of ash content, as the endosperm is extracted from progressively closer to the bran [43]. Usually, higher extraction also results in an increased speck count; however, with good milling practice, an increase in specks can be controlled to some extent. Studies have shown higher semolina ash content affecting pasta color negatively [44,45]. Although protein content increases with higher ash content due to an increase in semolina yield, semolina quality starts to go down due to poorer color, increased speck count, and finer particle size with potentially increased starch damage levels affecting functional properties. In their study, Joubert M. et al. (2018) [45] concluded that an increase in outer layers in semolina, with increased ash content, reduced pasta brightness and yellowness due to increased brown spots and likely enzymatic activities to pasta, while the increase in arabinoxylans, as a source of reducing sugars after extrusion, led to higher red index possibly due to Maillard reaction. Therefore, they suggested optimizing pasta color by reducing the inclusion of the grain outer layers.

Baking with semolina with increased yields may result in higher water absorption and inferior dough-handling properties on account of trying to extract the remaining endosperm from bran layers. Due to its importance, there are countries that have legislated the maximum allowable ash content, as shown in the Table 6:

Table 6. Semolina ash content regulation.

	Max	**Max (Dry Moisture Basis)**
Country	Moisture, %	Ash, %
Italy (semola) [1]	14.5	0.90
France (SSSE) [2]	14.5	0.80
USA [3]	15.0	0.92

[1] PRESIDENTIAL DECREE N° 187, dated 9 February 2001 (Official Journal n. 117, of 22 May 2001). Article 2. Durum wheat milling products. More details provided in Article 2 [46]. [2] Fixing the characteristics of durum wheat semolina and pasta as modified by the by-laws of 22 July 1959, 13 August 1974, and 6 December 1974. More details provided in Article 3 Superior durum wheat semolina SSSE UNAFPA July 2001 [47]. [3] Code of Federal Regulation. Title 21. Sec. 137.320 Semolina. Food and drugs. [48].

The ash content in wheat is variable among varieties and is also influenced by environmental factors [43]. The legal limits allow the semolina yield advantage for wheat with a low ash content.

5.4. Starch Damage

Due to the kernel hardness in durum wheat, starch damage is easily caused if grinding during milling is not carefully carried out. One of the reasons for coarse semolina particles produced in the traditional semolina is to avoid the risk of inflicting physical starch damage during milling. A coarse semolina particle size, on the other hand, might be difficult to fully hydrate [49]. Starch damage is an intrinsic parameter affecting dough/pasta quality during dough mixing and kneading [49]. Higher starch damage results in surface stickiness and cooking loss under low-temperature drying.

Desirable starch-damage levels can be achieved through several measures; some of them include appropriate tempering to ensure that durum wheat is not excessively hard as it enters the grinding process. Grinding pressure needs to be gradual, as too much compression of grinding rolls generates heat, resulting in a higher starch-damage level. Roll corrugations are dispositioned to cut rather than compress. Smooth rolls should not be used for semolina size reduction. The humidity level in the milling environment should not be too low. Overall, the processing should be gradual.

6. Advances in the Milling Equipment

The three principal pieces of milling equipment—the roller mill, plansifter, and purifier—collectively have gone through major enhancements with respect to performance, low energy consumption, minimal maintenance requirements, and greatly improved design for hygiene and sanitary standards. The focal point of design is easy accessibility of the equipment, allowing cleaning and complete emptying out of any residual material preserving superior level of sanitary standards. Other measures include insulation of walls and doors of sifter compartments for the prevention of condensation that could promote mold development. An important aspect of design is to ensure all the surfaces that the product encounters are made of stainless steel or food-grade material for avoiding contamination.

Although these improvements added functional advantages, the essential principles of operation and function very much remained the same. Two noteworthy developments that significantly contributed to enhancing technology of grinding and became commercially successful are the eight-roller mill and automated roll gap adjustment system. Automated roll-gap adjustment served as an important tool in remote operation, enabling roll gap adjustments while switching wheat mixes. The eight-roller mill, on the other hand, opened opportunities of constructing compact mills with lower investments, increasing the capacity of an existing mill retrofitting under space restrictions and lowering energy costs. Both innovations became available commercially around 1990. Fistes and Rakic (2014) [50] noted numerous advantages of using the eight-roller mill over conventional mills with respect to investment, operating, and maintenance costs. Their study demonstrated that the flour yield and ash content improved by making an appropriate increase in the aperture of the sieve. In a previous study, Fistes et al. (2008) [51] investigated the use of the eight-roller mill on the head reduction passages of a mill. With appropriate adjustment in the roll gap and sieving conditions they obtained similar results as in a conventional process, with investment costs and energy requirements being much lower in favor of the eight-roller mill. The eight-roller mill may be used in durum mills for first and second break grinding passages and for the reduction of semolina [25] or for the regrinding of semolina [22].

All three pieces of equipment are being continually improved for food safety, energy efficiency, and improved performance. As it would be a tedious task to cover most of the improvements in a table, Table 7 offers a quick snapshot of some of these improvements and the related advantages.

The advantages reported in Table 7 have food safety and sustainability efforts as a common element for all three pieces of equipment.

The roller mills of today are equipped with modern sensor technology that enables users to measure the grinding force, which, along with the flow-rate data, ensures that the grinding performance remains stable throughout ensuring the production of a consistent high-quality end product [52]. Dübendorrfer [52] stated that semolina for pasta requires consistently low starch damage to arrive at a desired dough consistency (viscosity) requiring less water absorption for energy saving during drying, as there is less water to be removed. Consistent particle size distribution at the lowest temperature would help achieve the consistent low starch damage required. With the help of built-in grinding-force sensor and temperature-monitoring option, the grinding-gap status and temperature distribution along the rolls are available.

Table 7. Advances in three principal pieces of milling equipment used in a durum milling operation.

Modern Equipment	Features	Advantages
Roller Mills	Advanced use of sensors and automation. Improved design with high food safety standards and energy efficient drives. Quick installation and space saving. Roller mills supplied with a touch screen panel for operational control which can also be remotely accessed using a plant computer or through a wireless connection with a tablet or mobile.	Stable and uniform control on grinding. Higher sanitary standards and improved food safety. Improved standards of sustainable operation with energy- and space-saving features. Ease of operation and accessibility.
Plansifters	large sieving area with optimized space utilization. Top sanitation and Hygiene. Sturdy construction with energy efficient operation. Stable design with lightweight energy efficient motor.	Increased sifting capacity due to increased sifting area Improved food safety. Low maintenance. Sustainable operation with lower energy requirement and improved sifting capacity
Purifiers	High specific capacity at same space. High sanitary standard completely enclosed. Energy efficient.	Improved semolina purification. Improved food safety design. Increased output. Sustainable operation with energy efficiency and more output on smaller footprint.

7. Developments in Quality Measurements

The routine quality testing of durum wheat is carried out to ensure that the quality of incoming raw materials is within the expected range for generating the semolina quality desired. Quality testing on semolina and flour is carried out to ensure that the milling process is well adjusted to generate semolina and flour within the specifications of customers' needs. The frequency of testing samples is considered to be a gauge of the consistency of quality. There needs to be a balance between the frequency of quality testing limited to the lab's capacity and what can be considered as practical.

A study carried out by Cecchini et al. (2020) [53] demonstrated the advantage of a low-cost pocket-size sensor providing a short wavelength NIR range for easier measurements at the sample source over laboratory-based instruments and other expensive portable devices. Such options are good for enabling quality testing at inconveniently located sample source and helping in reducing the laboratory workload.

However, the assurance of consistency in quality can only be truly achieved when the quality is being monitored online, along with the production, as durum milling operation is a 24/7 operation. Davies and Grant (1987) [54], in a review of near infra-red analysis of food, reported the online application of NIR analysis being in the process of development and that this is expected to be one of the most important applications for the future utilization of NIR in the food industry.

The true consistency in the finished products can only be achieved through interventions as and when required, based on faults detected through alarms or notifications. This can only be achieved with the application of online monitoring real-time with the appropriate control systems activating alarm or notification in case of "out of specification" scenarios. Furthermore, automated machine settings drive the process toward the optimized targets as real-time quality-monitoring facilitates enabling of "Automatic control loops" to achieve that (Bühler Inc., Reference [31]).

The key quality attributes measured online in durum wheat, semolina, and flour are shown in Table 8.

These properties along with Falling number and Gluten index are commonly tested and reported in the laboratory of the plant. The quality of semolina is much more tolerant to low falling number in durum wheat especially for the application in pasta products [55]. This eliminates the need for its testing less of a priority in online monitoring.

Table 8. Quality attributes measured online.

Property	Durum Wheat	Semolina and Flour
Moisture	✓	✓
Protein/Gluten	✓	✓
Ash	✓	✓
Color L*, a*, b*		✓
Specks Black and Brown		✓
Particle size distribution		✓

7.1. *Quality Monitoring Online*

7.1.1. Particle Size Distribution

One of the key quality parameters of durum semolina is the appropriate particle-size range for the desired semolina type consistently whether measuring traditional or fine semolina. Modern technology has enabled the online monitoring of this key attribute.

This system, which was introduced by Bühler Inc. [31], covers the measurement of particle size, ranging from as low as 10 μm to as high as 5000 μm. This range is wide enough to cover the requirements of all types of semolina for application in pasta, couscous, and bakery products (Table 9). There are other suppliers of similar products, but this system is advanced in terms of its customization for application in the durum-semolina process control. It applies laser diffraction and image-processing technology in combination to achieve the determination of particle size distribution continuously in the ongoing process. Any deviation to the particle size is detectable by the operating software serving as the basis for a monitored and traceable product quality.

This system has the ability for connecting with an appropriate roller mill, where the grinding gap gets adjusted upon the detection of any deviation between measured values and targeted values.

Table 9. Specifications related to particle size distribution.

Semolina Type	European Semolina (Special)	Common Semolina	Handmade Couscous	Industrial Pasta and Couscous	Extra Fine Semolina Special Bread
	μm	μm	μm	μm	μm
Granulation	630–200	425–125	1000–600	400–212	300–160

7.1.2. Multi Online Analyzer Using NIR and Camera

This system is referred to as NIR Multi Online Analyzer [31]. The unit provides assurance of consistent quality, documentation, and traceability through real-time quality monitoring and recording.

The unit measures moisture, protein, and ash in wheat, semolina and flour using NIR probe. There are up to six measurement points that can be connected to one NIR spectrometer.

The modular system allows for the combining of the camera probe for measuring the visual quality of semolina and flour color (L*, a*, and b*) in the CIE color space and the detection of black and brown specks. System flexibility facilitates the combining of different products and probe positions.

The unit can detect even a small change in color and contamination of the product due to leakages, such as ruptures in the sifter or purifier sieve. An early detection in this scenario is very helpful. The software offers current readings in addition to the trending charts. The unit needs calibration only once for color measurement with a reference method. It has the flexibility of being integrated as part of the process control system or could be used as a stand-alone. Residual starch content in bran can also be measured if so desired as an option. Changes in starch content would be indicative of yield fluctuations.

When connected to a comprehensive plant automation system, the full potential of online quality monitoring system is further realized in the optimization of production process control enhancing semolina yield while maintaining the desired quality in semolina consistently. Consistency in quality is key to millers and end-users alike, as these processes, for the most part, are automated. Inconsistencies in quality, therefore, end up impacting downstream processing and product quality.

Since there are no improving agents or additives that are added to compensate for quality shortcomings in pasta making, it is therefore very important to deliver the required quality consistently through reliable processing.

8. Innovations in Plant Operation, Monitoring, Control, and Digitalization

8.1. Plant Operation Using Programable Logic Controllers (PLCs) with Computer Interface

Going back 35–40 years ago, when operating a milling plant, process monitoring and performing control by replacing hardwired electro-mechanical relays with programable logic controllers (PLCs) and with computer interface (the aid of PCs) was considered a major advancement in plant automation. During this period, some plants were being touted as "lights out" plants, where no personnel were present during the night shift. Although such possibilities were very advanced for the time, the system was only able to shut the equipment down in a failsafe manner, at best, and the status was communicated to the operator by alarm.

The use of digitalization with artificial intelligence, machine learning, cloud computing, and internet of things (IoT) has brought about changes in almost every aspect of the industrialized world. Among the most significant advancements in durum milling are innovations in digital technologies and its application in developing a very powerful and comprehensive plant automation system. Plant monitoring in real time provides enhanced productivity with optimized quality assurance. The leading equipment manufacturing and process solution company Bühler Inc. of Switzerland [31] offers such a system, along with comprehensive support in providing guidance to the milling companies in their pursuit of digital transformation.

Digitalization using a holistic approach of integrating business processes such as Enterprise Resource Planning (ERP), maintenance, quality monitoring, yield management system, and others has made it possible to have enhanced capabilities. These capabilities include increased efficiency, transparency, and traceability, along with controlling the entire production in real time from anywhere. The system integrates data in a central data base system that are used for the optimization of the production process on an ongoing basis. In situations where problems are detected, the system facilitates prompt intervention.

The availability of the automation system, along with offerings of digital services by the equipment manufacturers, has emerged as an area of development of great significance.

8.2. Self-Adjusting or Smart Mill

The operation of a newly developed milling plant capable of using its own process parameters in a closed loop to optimize its production has been widely reported in trade publications [56–58]. The most advanced blend of engineering with the application of digitalization is being harmonized in a highly technically advanced way in the development of the most innovative milling technology of today (Buhler Group) [31]. We are now in a digitalized world, with the application of machine learning and AI enabling us to navigate through complex process optimization. This development of the self-adjusting mill will be the precursor to the smart mill [56].

As was described in the preceding sections, the wheat-milling operation has evolved into a very highly sophisticated technology with the advancements in process and equipment development. Milling companies are expected to generate finished product quality, as specified, consistently in food-safe conditions. The present-day expectations of the above would also require production in a sustainable way. This would mean the use of less energy and the reduction of waste. The use of innovation in advanced automation, sensors' application, and digitalization would potentially help in improving the reduction of waste and energy use with a certain strategic and novel approach in design development.

In this self-adjusting plant, as reported, the system processes a large volume of data points, exceeding 15,000, covering all aspects of the production process for optimization on an ongoing basis.

8.2.1. Building Design

The modular design of the mill is kept in mind with the "plug and play" concept of the equipment, allowing for the reduction in installation time by 30%. A more innovative building design helped reduce the building volume further by cutting the building cost by an additional 30%. The application of an energy-efficient fully integrated grinding system with the resource of full digitalization further helped in the reduction of energy costs, as targeted. The mill is designed for optimum performance, while also allowing for easy accessibility of the equipment for maintenance. The pneumatic conveying of products throughout the plant is carried out by blower units which are preassembled

8.2.2. Consistent Process Optimization

The key parameters of the incoming wheat are checked by the sensors as the wheat comes into the mill, and the sensors in the grinding system check it again and recalibrate the settings based on the changing characteristics of the wheat as appropriate. The ability of the process to optimize its setting on its own is the key feature of the self-adjusting mill [56].

8.2.3. Operation of the System

The sensors send the data to an advanced plant automation system that performs the routine process monitoring and control operation. Sensors also send the data to an IoT hub where algorithms are performed, comparing present production and process parameters with the past. This enables the milling plant to perform optimally, thus achieving the most consistent product.

Additional service modules include Temperature and Vibration Management Service (TVM), Yield Management System (YMS), Error and Downtime Analysis (EDA), and Overall Equipment Effectiveness (OEE). These modules feed data on an ongoing basis on machine and process trends, potential maintenance matters, and how machine performance relates to quality and efficiency [56].

8.2.4. Application of Block Chain Technology

With the planned use of blockchain in the future, the customers of the milling company can benefit from accessing/viewing the production parameters in real time as part of the product certification process. This is achieved through a seamless interface from laboratory systems to plant automation system and IoT hub and then to the milling company's customer through blockchain. The secure data handling of blockchain provides the transparency desired by customers, as they can verify the production parameters used for the production of their product. This would likely result in the reduction and simplification of product sampling and testing. The end result is that this process, while simplifying product testing, will enable a consistent, retraceable food-safe product [56].

8.2.5. Future Development

The development reported with the self-adjusting mill has been phenomenal. Despite the achievement, it is still referred to as a precursor to a fully developed "Smart mill" of the future. Based on the reports, there is a need to further understand the high volume of data being generated by sensors relating to the process for a better understanding of all the machine parameters required for the manufacturing of high-quality end products.

This appears to be very futuristic; nevertheless, such a milling plant is operational. Although the concept and the tools are currently being applied in a flour milling unit, the technology will be transferrable to a durum milling facility in the future.

9. Food Safety

Food safety is a very comprehensive subject with a large volume of documented procedures that continue to evolve in step with developments and changing environment. Hufford (2018) [59] stated that it is a living and breathing document [59]. The reference to food safety here is made within the context of how recent technological advances in durum wheat milling are helping to address food safety issues.

The downstream processing of durum wheat semolina helps in eliminating or reducing the risk of any microbiological contamination that may be present. As most pasta is processed by using an HT (high-temperature) drying process, there is little risk of contamination in the pasta [28].

Since the main source of exposure is through the contamination of incoming grain with soil, dirt, and plant material, as well as the presence of damaged and diseased kernels, molds, fungi, surface contaminants and infested kernels, a thorough inspection of the incoming wheat is an essential part of the routine before the wheat gets precleaned and sent to bins for safe storage. If the wheat received shortly after the harvest has an elevated moisture level that is unsafe for storage, a proper aeration system may be necessary to dry it down to the moisture level that is safe for storage.

Out of the three typical hazards identified [59]—biological, chemical, and physical—for wheat milling plants, all of them are related to wheat to a greater or lesser extent.

The biological hazards of durum wheat could include the cadmium level; contamination with mycotoxins, such as DON and ochratoxin (OTA); and *E. coli* and Salmonella contamination [60].

Chemical hazards include the overtreatment of crop with fungicide, pesticides, and insecticides to control weeds, mold, pests, and insects.

Physical hazards include all types of foreign material, such as chunks of wood, stones, metals, glass, and rubble, which, if they remain present in the wheat stream, they may break down or disintegrate in smaller pieces and pose a serious safety hazard.

In the description of wheat preparation and milling equipment, the design and function with reference to food safety has been adequately covered. Additionally, it is worth noting that magnets are installed at various critical points, especially before any cleaning machine that uses friction as a principle for its operation, such as scourers and pearlers. It has also been noted that pearlers significantly lower the microbial load before wheat is milled and contamination extends to milled products. A study noted that a substantial reduction of bacterial and mold load from >1 log to >5 log was achieved by treating tempering water with antimicrobial agents [61]

A big part of food safety procedures involves the documentation and maintenance of records and traceability. In the present environment of automation and digitalization, this part of the food safety function is easily performed.

The demand for food safety has always been growing for a long time, becoming increasingly important, since cases of foodborne illnesses began being linked to wheat flour. Following the reporting of the detection of pathogens in flour [62] and several reports of a similar nature, there was a raised heightened level of awareness on the issue of food safety. Raw wheat flour may be exposed to Salmonella or Pathogenic *Escherichia coli* (*E. coli*); therefore, raw flour, dough, or batter should not be eaten or tasted [60,63,64].

Food safety has become an important part of quality assurance and protection. The associated functions of documenting, maintaining a required paper-trail, and preparation for audits are time-consuming and a drain on resources, and there are operating expenses when carried out manually. Since production-process- and quality-related data are recorded in the central database, they are all available and traceable, as required, enabling the preservation of related proof-of-product traceability and quality information.

Such a possibility has never been conceived of before. The recent developments have been primarily concentrated in this area of digitalization of all the food processes, enabling the operations' savings on energy and resources while furnishing valuable production data.

10. Recent Trends in the Milling Industry

10.1. Health and Nutrition

The trends related to low-carb, gluten-free, high-fiber, whole-grain, organic, and enriched pasta with a variety of nutritional components have been in the market for some time, and such products have a secure shelf space. While the pasta produced from 100% durum semolina is affected in terms of market share because of gluten-free and other varieties of pasta, it does provide an opportunity to durum millers in diversifying their production involving these other components

10.2. Environmental Sustainability

- Major milling-equipment manufacturers have taken steps to follow sustainable practices; for example, Bühler Inc. set up their corporate target of cutting water, energy, and food waste by 50% in their customer-value chains by 2025 [56]. Ocrim, a major milling equipment manufacturer based in Italy is using an intelligent energy-saving system with high-efficiency motors and other measures to build energy-efficient mills [65].
- Barilla, the largest pasta maker, requires wheat produced under their established farming code to ensure that quality, economic, environmental, and social sustainability are met [66]. Durum millers have established their own goals and have aligned their sustainable practices accordingly by cutting back on energy, using renewable energy [67], and reducing waste. Millers are also partnering with the farm community in support of farm practices that can improve the environment through regenerative agriculture.

10.3. Larger Production Units

There is a growing trend of construction of larger-capacity milling units. Milling companies, particularly those located in a highly competitive environment such as in North America, are leveraging on economies of scale to improve their margins by concentrating their production in large units. Wheat milling, whether common wheat or durum wheat milling, is a low-margin business. The scale of economy has allowed us to take advantage of rationalizing the managing of production, marketing, distribution, and related administrative costs. There is benefit in combining production capacities through mergers [68] to reduce operating costs and improve margins.

11. Conclusions

Advances in quality improvement and milling innovation have led to improvements in the functional attributes, food safety, and sustainable production of durum wheat products. General developments in regard to processing equipment and digital application with IoT platforms have transformed the processing ecosystems, radically enabling greater efficiency and the saving of resources. Quick fault detection and timely intervention help in saving production time, cutting down on interruptions, and eliminating wastage, all of which contribute to more efficient and sustainable production. Sustainability has also been enhanced by efforts toward process simplification and improving building design, which have helped to reduce energy needs. Finally, the application of digitalization has created benefits of consistency in production, quality, transparency, documentation, and traceability, and this has supported the efforts of processors to improve their productivity, ensure food safety, and enhance sustainability efforts.

Author Contributions: Conceptualization, A.S. and B.X.F; writing—original draft preparation, A.S. and B.X.F; writing—review and editing, A.S. and B.X.F.; funding acquisition, B.X.F. All authors have read and agreed to the published version of the manuscript.

Funding: This research received no external funding.

Institutional Review Board Statement: Not applicable.

Informed Consent Statement: Not applicable.

Data Availability Statement: The data presented in this study are available on request from the corresponding author.

Acknowledgments: Many developments and innovations related to milling technology cited in this paper are from various sources published by Buhler Inc., Uzwil, Switzerland. The authors are grateful for the assistance received from Ian Steele and Dimitri Stuetzberger, Bühler Inc., Markham, Ontario, Canada; and Mike Soder, Bühler Inc., Minneapolis, USA.

Conflicts of Interest: The authors declare no conflict of interest.

References

1. Mandato, S.; de Vries, H.; Mayer, C. Potential innovations for the traditional durum wheat food sector. In Proceedings of the 29th EFFOST International Conference, Athens, Greece, 10–12 November 2015. Available online: https://www.archives-ouvertes.fr/hal-02741292/ (accessed on 20 April 2022).
2. Fu, B.; Wang, K.; Dupuis, B.; Taylor, D.; Nam, S. Kernel vitreousness and protein content: Relationship, interaction and synergistic effects on durum wheat quality. *J. Cereal Sci.* **2017**, *78*, 2–9. [CrossRef]
3. Sissons, M.; Cutillo, S.; Marcotuli, I.; Gadaleta, A. Impact of durum wheat protein content on spaghetti in vitro starch digestion and technological properties. *J. Cereal Sci.* **2021**, *98*, 103156. [CrossRef]
4. Clarke, J.M.; Clarke, F.R.; Pozniak, C.J. Forty-six years of genetic improvement in *Canadian durum* wheat cultivars. *Can. J. Plant Sci.* **2010**, *90*, 791–801. [CrossRef]
5. Sissons, M. Role of durum wheat composition on the quality of pasta and bread. *Foods* **2008**, *2*, 75–90.
6. D'Egidio, M.G.; Mariani, B.M.; Novaro, P. Influence of raw material characteristics and drying technologies on pasta cooking quality: A review of our results. *Ital. Food Beverage Technol.* **1993**, *1*, 29–32.
7. Zweifel, C.; Handschin, S.; Escher, F.; Conde-Petit, B. Influence of high-temperature drying on structural and textural properties of durum wheat pasta. *Cereal Chem.* **2003**, *80*, 159–167. [CrossRef]
8. Abecassis, J.; Cuq, B.; Boggini, G.; Namoune, H. Other traditional durum-derived products. In *Durum Wheat Chemistry and Technology*, 2nd ed.; Sissons, M., Abecassis, J., Marchylo, B., Carcea, M., Eds.; AACC International Inc.: St. Paul, MN, USA, 2012; Chapter 10; pp. 177–199.
9. Borrelli, G.M.; Troccoli, A.; Di Fonzo, N.; Fares, C. Durum Wheat Lipoxygenase Activity and Other Quality Parameters that Affect Pasta Color. *Cereal Chem.* **1999**, *76*, 335–340. [CrossRef]
10. Troccoli, A.; Borrelli, G.M.; De Vita, P.; Fares, C.; Di Fonzo, N. Durum wheat quality: A multidisciplinary concept (review). *J. Cereal Sci.* **2000**, *32*, 99–113. [CrossRef]
11. Ficco, D.B.M.; Mastrangelo, A.M.; Trono, D.; Borrelli, G.M.; De Vita, P.; Fares, C.; Beleggia, R.; Platani, C.; Papa, R. The colours of durum wheat: A review. *Crop Pasture Sci.* **2014**, *65*, 1–15. [CrossRef]
12. Borrelli, G.M.; Trono, D. Molecular approaches to genetically improve the accumulation of health-promoting secondary metabolites in staple crops-A case study: The Lipoxygenase-B1 genes and regulation of the carotenoid content in pasta products. *Int. J. Mol. Sci.* **2016**, *17*, 1177. [CrossRef]
13. Dexter, J.E.; Edwards, N.M. *The Implications of Frequently Encountered Grading Factors on the Processing Quality of Durum Wheat*; Association of Operative Millers—Bulletin: Lenexa, KS, USA, 1998; pp. 7165–7171.
14. Clarke, J.M.; Leisle, D.; DePauw, R.M.; Thiessen, L.L. Inheritance of cadmium concentration in five durum wheat crosses. *Crop Sci.* **1997**, *37*, 1722–1726. [CrossRef]
15. Knox, R.E.; Pozniak, C.J.; Clarke, F.R.; Clarke, J.M.; Houshmand, S.; Singh, A.K. Chromosomal location of the cadmium uptake gene (Cdu1) in durum wheat. *Genome* **2009**, *52*, 741–747. [CrossRef] [PubMed]
16. EU Standard for Cadmium August 2021. Available online: https://eur-lex.europa.eu/legal-content/EN/TXT/PDF/?uri=CELEX:32021R1323&from=EN (accessed on 1 February 2022).
17. Haile, J.K.; N'Diaye, A.; Walkowiak, S.; Nilsen, K.T.; MClarke, J.M.; Kutcher, H.R.; Steiner, B.; Buerstmayr, H.; Pozniak, C.J. Fusarium Head Blight in Durum Wheat: Recent Status, Breeding Directions, and Future Research Prospects. *Phytopathology* **2019**, *109*, 1664–1675. [CrossRef] [PubMed]
18. Wang, K.; Taylor, D.; Chen, Y.; Suchy, J.; Fu, B.X. Effect of Kernel Size and Its Potential Interaction with Genotype on Key Quality Traits of Durum Wheat. *Foods* **2021**, *10*, 2992. [CrossRef]
19. Jiménez González, A.T. Milling process of durum wheat. In *Durum Wheat Quality in the Mediterranean Region*; Di Fonzo, N., Kaan, F., Nachit, M., Eds.; CIHEAM: Zaragoza, Spain, 1995; pp. 43–51.

20. Dziki, D.; Laskowski, J. Wheat kernel physical properties and milling process. *Acta Agrophysica* **2005**, *6*, 59–71.
21. Bizzarri, O.; Morelli, A. Milling of durum wheat. In *Durum: Chemistry and Technology*; Fabrianai, G., Lintas, C., Eds.; American Association of Cereal Chemists: St. Paul, MN, USA, 1988; pp. 161–189.
22. Kuenzli, T. Particle size requirements for pasta production. In *Pasta and Semolina Technology*; Kill, R.C., Turnbull, K., Eds.; Blackwell Science Ltd.: Oxford, UK, 2001; pp. 45–64.
23. Sarkar, A. Grain Milling Operation. In *Handbook of Post Harvest Technology*; Chakraverty, A., Majumdar, A.S., Raghavan Vijay, G.S., Ramaswamy, H.S., Eds.; Marcel Dekker Inc.: New York, NY, USA, 2003.
24. Posner, E.S.; Hibbs, A.N. Milling of durum wheat. In *Wheat Flour Milling*, 2nd ed.; American Association of Cereal Chemists: St. Paul, MN, USA, 2005; pp. 329–354.
25. Gruber, W.; Sarkar, A. Durum wheat milling. In *Durum Wheat Chemistry and Technology*; Sissons, M., Abecassis, J., Marchylo, B., Carcea, M., Eds.; AACC International: St. Paul, MN, USA, 2012; pp. 139–159.
26. Würsch, S. *Technical Profile: Grain Cleaning Breakthrough*; World Grain: Kansas City, MO, USA, 2020. Available online: https://www.world-grain.com/articles/13969-technical-profile-grain-cleaning-breakthrough (accessed on 20 April 2022).
27. Dexter, J.E.; D'Egidio, M.G. Grading factors impacting durum wheat processing quality. In *Durum Wheat Chemistry and Technology*; Sissions, M., Abecassis, J., Marchylo, B., Carcea, M., Eds.; AACC International: St. Paul, MN, USA, 2012; pp. 235–250.
28. Landi, A. Durum Wheat, semolina and pasta quality characteristics for an Italian food company. In *Durum Wheat Quality in the Mediterranean Region*; Di Fonzo, N., Kaan, F., Nachit, M., Eds.; CIHEAM: Zaragoza, Spain, 1995; pp. 33–42.
29. Inamdar, A.A.; Suresh, D.S. Application of color sorter in wheat milling. *Int. Food Res. J.* **2014**, *21*, 2083–2089.
30. Fowler, M. *Innovations in Wheat Cleaning: Improving Process Efficiencies*; Kansa State University: Manhattan, KS, USA, 2013.
31. Bühler Group. Durum Wheat Milling. In *Grain Processing. How to Produce High-Quality Semolina*; Bühler Group: Uzwil, Switzerland. Available online: https://www.buhlergroup.com/content/buhlergroup/global/en/industries/Wheat-and-grain/Durum-Wheat.html (accessed on 20 April 2022).
32. Dexter, J.E.; Wood, P.J. Recent applications of debranning before milling. *Trends Food Sci. Technol.* **1996**, *7*, 35–41. [CrossRef]
33. Willis, M.; Giles, J. The application of a debranning process to durum wheat milling. In *Pasta and Semolina Technology*; Kill, R.C., Turnbull, K., Eds.; Blackwell Science Ltd.: Oxford, UK, 2001; pp. 64–85.
34. McGee, B.C. *PeriTec Process and Its Application to Durum Wheat Milling*; Association of Operatives Millers Technical Bulletin, International Association of Operative Millers: Lenexa, KS, USA, 1994.
35. Reidy, S. *Bühler Introduces New Vertical Pearler*; World Grain: Kansas City, MO, USA, 2018. Available online: https://www.world-grain.com/articles/10333-buhler-introduces-new-vertical-pearler (accessed on 20 April 2022).
36. Henry, R.J.; Martin, D.J.; Blakeney, A.B. Reduction of the α-amylase content of sprouted wheat by pearling and milling. *J. Cereal Sci.* **1987**, *5*, 155–166. [CrossRef]
37. Hareland, G.A. Effects of Pearling on Falling Number and α-Amylase Activity of Preharvest Sprouted Spring Wheat. *Cereal Chem.* **2003**, *80*, 232–237. [CrossRef]
38. Cheli, F.; Campagnoli, A.; Ventura, V.; Brera, C.; Berdini, C.; Palmaccio, E.; Dell'Orto, V. Effects of industrial processing on the distributions of deoxynivalenol, cadmium and lead in durum wheat milling fractions. *LWT* **2010**, *43*, 1050–1057. [CrossRef]
39. Office of Industries U.S. International Trade Commission. *Industry & Trade Summary Pasta*; Office of Industries U.S. International Trade Commission: Washington, DC, USA, 2003.
40. Manser, J. Feinheitsgrad von Durum-Mahlerzeugnissen aus der sicht der Teigwarenindustrie. *Getreide Mehl Brot* **1985**, *39*, 117–123.
41. Dexter, J.E.; Marchyllo, B.A. Recent Trends in Durum Wheat Milling and Pasta Processing: Impact on durum wheat quality requirements. In Proceedings of the Second International Workshop on Durum Wheat, Semolina and Pasta Quality: Recent Achievements and New Trends, Montpellier, France, 27 November 2000; Pierre, F., Ed.; Institute National de la Recherche: Montpellier, France, 2001; pp. 139–164.
42. Turnbull, K. Basic Semolina Requirements. In *Pasta and Semolina Technology*; Kill, R.C., Turnbull, K., Eds.; Blackwell Science Ltd.: Oxford, UK, 2001; pp. 43–44.
43. Sissons, M.; Abecassis, J.; Marchylo, B.; Cubadda, R. Methods used to assess and predict quality of durum wheat, semolina, and pasta. In *Durum Wheat Chemistry and Technology*; Sissons, M., Abecassis, J., Marchylo, B., Carcea, M., Eds.; AACC International: St. Paul, MN, USA, 2012; p. 215.
44. Feillet, P.; Autran, J.-C.; Verniere, C.I. MINI REVIEW Pasta brownness: An Assessment. *J. Cereal Sci.* **2000**, *32*, 215–233. [CrossRef]
45. Joubert, M.; Morel, M.-H.; Lullien-Pellerin, V. Pasta color and viscoelasticity: Revisiting the role of particle size, ash, and protein content. *Cereal Chem.* **2018**, *95*, 386–398. [CrossRef]
46. Italian Law (DPR 187/2001) Imposes Limits on Ash Content in Durum Wheat Products Used for Human Food (i.e., 0.9% dmb for Semolina and Pasta). 2020. Available online: https://www.pasta-unafpa.org/public/unafpa/pdf/ITALIA.pdf (accessed on 3 May 2022).
47. Fixing the Characteristics of Durum Wheat Semolina and Pasta as Modified by the by-Laws of 22 July 1959, 13 August 1974 and 6 December 1974 Article 3 Superior Durum Wheat Semolina SSSE UNAFPA July 2001. Available online: https://www.pasta-unafpa.org/public/unafpa/pdf/FRANCIA.pdf.47 (accessed on 20 April 2022).
48. Code of Federal Regulation. *Food and Drugs Standard of Identity*; Parts 100–169, Section 137.320; Code of Federal Regulation, 2003. Available online: https://www.accessdata.fda.gov/scripts/cdrh/cfdocs/cfcfr/CFRSearch.cfm?CFRPart=137&showFR=1 (accessed on 21 April 2022).

49. Bresciani, A.; Pagani, M.A.; Marti, A. Review Pasta-Making Process: A Narrative Review on the Relation between Process Variables and Pasta Quality. *Foods* **2022**, *11*, 256. [CrossRef] [PubMed]
50. Fistes, A.; Rakić, D. Using the eight-roller mill in the purifier-less mill flow. *J. Food Sci. Technol.* **2014**, *52*, 4661–4668. [CrossRef]
51. Fistes, A.; Tanovic, G.; Mastilovic, J. Using the eight-rollermill on the front passages of the reduction system. *J. Food Eng.* **2008**, *85*, 296–302. [CrossRef]
52. Dübendorfer, U. *Technical Profile: Ensuring Optimal Flour Quality*; World Grain: Kansas City, MO, USA, 2021. Available online: https://www.world-grain.com/articles/15594-technical-profile-ensuring-optimal-flour-quality (accessed on 12 April 2022).
53. Cecchini, C.; Antonucci, F.; Costa, C.; Marti, A.; Menesatti, P. Application of near-infrared handheld spectrometers to predict semolina quality. *J. Sci. Food Agric.* **2021**, *101*, 151–157. [CrossRef]
54. Davies, A.M.C.; Grant, A. Review: Near infra-red analysis of food. *Int. J. Food Sci. Technol.* **1987**, *22*, 191–207. [CrossRef]
55. Fu, B.X.; Hatcher, D.W.; Schlichting, L. Effects of sprout damage on durum wheat milling and pasta processing quality. *Can. J. Plant Sci.* **2014**, *94*, 545–553. [CrossRef]
56. Reidy, S. *Whitworth Bros. First to Open Mill E3*; World Grain: Kansas City, MO, USA, 2021. Available online: https://www.world-grain.com/articles/15878-whitworth-bros-first-to-open-with-b%C3%BChlers-mill-e3 (accessed on 10 October 2021).
57. Nayak, T. Buhler to Open Digitally Smart Mill E3. Far Eastern Agriculture. 2021. Available online: https://www.fareasternagriculture.com/example-pages?view=article&id=9188:buhler-to-open-digitally-smart-mill-e3&catid=1089 (accessed on 5 April 2022).
58. Lyddon, C. *A New Era in Milling*; World Grain: Kansas City, MO, USA, 2021. Available online: https://www.world-grain.com/articles/16263-a-new-era-in-milling (accessed on 28 April 2022).
59. Hufford, S. Food Safety in Flour Mills. Miller Refereed. 2018. Available online: https://millermagazine.com/blog/food-safety-in-flour-mill-2658 (accessed on 25 April 2022).
60. U.S. Food and Drug Administration. Handling Flour Safely: What You Need to Know. Available online: https://www.fda.gov/food/buy-store-serve-safe-food/handling-flour-safely-what-you-need-know (accessed on 7 April 2022).
61. López, A.M.M.; Simsek, S. Pathogens control on wheat and wheat flour: A review. *Cereal Chem.* **2020**, *98*, 17–30. [CrossRef]
62. Outbreak Investigation on *E. coli*: Flour. 2019. Available online: https://www.fda.gov/food/outbreaks-foodborne-illness/outbreak-investigation-e-coli-flour-may-2019 (accessed on 4 January 2022).
63. Centers for Disease Control and Prevention, Say No to Raw Dough. Available online: https://www.cdc.gov/foodsafety/communication/no-raw-dough.html?CDC_AA_refVal=https%3A%2F%2Fwww.cdc.gov%2Ffeatures%2Fno-raw-dough%2Findex.html (accessed on 8 April 2022).
64. Neil, K.P.; Biggerstaff, J.K.G.; MacDonald, E.T.; Medus, C.; Musser, K.A.; Stoika, S.G.; Zink, D.; Sotir, M.J. A Novel Vehicle for Transmission of Escherichia coli O157:H7 Infections Associated with Consumption of Ready-to-Bake commercial Prepackaged Cookie Dough—United States, 2009. *Clin. Infect. Dis.* **2011**, *54*, 511–518. [CrossRef] [PubMed]
65. Reidy, S. *Ocrim Completes Energy-Efficient Mill in Italy*; World Grain: Kansas City, MO, USA, 2021. Available online: https://www.world-grain.com/articles/15254-ocrim-completes-energy-efficient-mill-in-italy (accessed on 12 October 2021).
66. Barilla Group. We Use Only Quality Wheats. Available online: https://www.barillagroup.com/en/who-we-are/position/we-use-only-quality-wheats/ (accessed on 21 December 2021).
67. Donley, A. *Milling Operations: Using Solar Power to Make Flour*; World Grain: Kansas City, MO, USA, 2018. Available online: https://www.world-grain.com/articles/10813-milling-operations-using-solar-power-to-make-flour (accessed on 14 December 2021).
68. McKee, D. *Ranking the World's Biggest Flour Millers*; World Grain: Kansas City, MO, USA, 2020. Available online: https://digital.world-grain.com/sosland/wg/2020_01_01/index.php?startid=30#/p/30 (accessed on 15 December 2021).

foods

MDPI

Review

Durum Wheat Couscous Grains: An Ethnic Mediterranean Food at the Interface of Traditional Domestic Preparation and Industrial Manufacturing

Rifka Hammami [1], Reine Barbar [2], Marie Laurent [3] and Bernard Cuq [2,*]

1 Field Crop Laboratory, National Institute of Agronomic Research of Tunisia, Carthage University, Ariana 2049, Tunisia; rifkahammami82@gmail.com

2 L'Institut Agro Montpellier, UMR IATE, Université de Montpellier, 34060 Montpellier, France; reine.barbar@supagro.fr

3 INRAe, UMR IATE, Université de Montpellier, 34060 Montpellier, France; marie.laurent@inrae.fr

* Correspondence: bernard.cuq@supagro.fr; Tel.: +33-(0)4-9961-2860

Abstract: Couscous is the product prepared from durum wheat semolina that agglomerates by adding water and undergoes physical and thermal treatment. Couscous is a traditional food from Mediterranean countries consumed for many centuries. Between ancestral domestic practices and industrial performance, the diversity of methods for couscous processing meets the needs of different consumers, whether they are concerned about preserving family culinary traditions or discovering innovative foods that respond to changing consumption patterns. In this work, we present the story of durum wheat couscous through several complementary visions and approaches: a "historical and societal" approach to discover the origins of couscous, its migrations and its unifying role in Mediterranean societies; a "physicochemical" approach to describe the role of wheat components at the heart of couscous grains; a "technological" approach to compare domestic and industrial production of couscous; a "food science" approach to understand organoleptic characteristics of couscous grains; and a "consumer" approach to understand the motivations associated with the consumption of couscous.

Keywords: couscous; durum wheat; manufacturing processes; domestic preparation; history

Citation: Hammami, R.; Barbar, R.; Laurent, M.; Cuq, B. Durum Wheat Couscous Grains: An Ethnic Mediterranean Food at the Interface of Traditional Domestic Preparation and Industrial Manufacturing. *Foods* **2022**, *11*, 902. https://doi.org/10.3390/foods11070902

Academic Editor: Laura Gazza

Received: 27 February 2022
Accepted: 18 March 2022
Published: 22 March 2022

The *Codex Alimentarius* [1] defined couscous as the product prepared from durum wheat semolina that agglomerates by adding water and undergoes physical and thermal treatment. Couscous is an inexpensive food staple with a long shelf life, and it can be simply prepared in different recipes: salads known as "taboulé" or traditional couscous dish. Knowledge, ancestral know-how and practices related to the production and consumption of couscous have been inscribed in 2020 on UNESCO's list of Intangible Cultural Heritage [2]. Although considered as a traditional foodstuff, couscous still remains mysterious in terms of structuring mechanisms, elaboration processes and qualities of use. The present review addresses the manufacture of durum wheat couscous on traditional and industrial scales and the consumption of couscous in traditional and modern ways. Technical and scientific descriptions of couscous grains still remain patchy, with three book chapters [3–5]. The review was built on the basis of only about fifty publications and patents, a couple of students' reports and PhD theses available to date and expertise from the authors' laboratories.

1. The History of Couscous

1.1. Origins

Couscous is a very ancestral food product, nearly 2000 years old [3,5–7]. The combined origins between Berber food habits and those coming from Andalusia following the wave of mass migration of Muslims and Jews after Andalusia's fall to the hands of Christians are among the many historical characteristics of couscous. Part of the origin of couscous is related to Numidians, the Berber population of Numidia. The culinary historian Lucie Bolens describes primitive pots that closely resemble the main cooking utensil of couscous, which is the couscoussier, found in Kabylia in tombs coming from the period of Berber king Massinissa, who ruled Algeria between 238 and 149 BC [7]. Archaeological evidence found in North Africa dating back to the 10th century covers the kitchen utensils needed to prepare couscous. Steaming grains over a broth in a special pot was first found in the west part of North Africa. The Berbers invented an original way to slightly moisten and roll the semolina of durum wheat into small spherical and succulent grains. These are light, fragrant and nourishing, giving a fluffy mass. Couscous was the basic cereal preparation of Berbers even before the Arabic conquest. Neither in the ancient world nor in the oriental Arab world are we aware of this way of treating grains. The first references issued about couscous were written in the 13th century in the North African cookbook.

1.2. Etymology

The worldwide known etymology of the word couscous may be derived from the Arabic word "*kaskasa*" meaning "to pound small" and also relating to the sound "*keskes*" arising from grains sieving, or also from the Berber "*seksu*", meaning well rolled or well formed [6]. Couscous or "*seksu*" is pronounced "*koos-koos*" in the Berber language.

1.3. Migration

Couscous was spread by Arabs from the Mediterranean basin throughout Europe in the 17th century and moved to the Americas with Portuguese cargoes from Morocco. The Mediterranean basin sees extensive migratory flows: memories, senses, images, tastes and aromas travel with moving groups or individuals [5,8,9]. In the 16th century, couscous arrived in Turkey from Syria. In 1699, a letter mentioned couscous spreading in France and Brittany. Mediterranean migrants from Spain, Italy and Malta were part of the European population in the north of Africa. Since the end of the last century, the Maghreb couscous has become widespread in many countries of the world. Today, couscous is produced and eaten around the world.

1.4. French Context

Couscous had been present in France since the 19th century as the staple food of Kabyle people [9]. Couscous became a significant part of modern French cuisine on the way of the 19th-century colonial route. The first French people to settle in Maghreb were typically colonists of modest means who lived in rural areas [10]. During the 1970s, North African immigration intensified in order to support French economic growth. French law allowed for wives and children of immigrants to join them in France. This shift from the migration of male workers to the migration of families was a key factor in the enduring presence of couscous in France. The return of colonial populations to their native countries, following North African immigration, intensified Maghreb's decolonization [5]. French nationals who colonized Algeria until the end of the Algerian War in 1962 adopted local produce until they could establish means of producing food [11]. During this period, the preparation of grains shifted from being a manual craft to an industrial operation, with the introduction of flour mills in Algeria (Maison Ricci in 1853 or Ferrero in 1970). Following independences between 1956 and 1962, the families of millers from North Africa typically settled in Marseille and developed the unexplored couscous manufacturing industry there [9]. In less than 50 years, couscous would become one of France's three favorite savory dishes.

1.5. Durum Wheat Semolina

Couscous has not been developed at random and responded to a necessity. In North Africa, couscous is made from durum wheat semolina. Beyond agronomic and technological aspects, the durum wheat semolina carries a strong anchorage in the Mediterranean diet. Durum wheat semolina is the "traditional" raw material for the manufacture of couscous grains in Maghreb countries and the Mediterranean region because of the ideally suited color and cooking quality [12]. Sown in autumn to germinate in the rain, it is therefore called "winter wheat". Durum wheat contains a high level of proteins [13]. Crushed, it becomes semolina, simila for the Latins or smilla in Arabic.

2. State Diagram of Durum Wheat Components

Durum wheat semolina is the traditional raw material for making couscous grains [3,4]. The semolina is extracted by milling kernels and corresponds mainly to the starchy endosperm.

2.1. Physical Characteristics

Durum wheat semolina is a powder with a low water content (10–14%) formed of heterogeneous and nonporous particles (Figure 1). It has a strong dispersion of diameter (between 100 and 400 µm) with a median diameter (d_{50}) of nearly 300 µm.

Figure 1. Observation of the microstructure of durum wheat semolina particles by scanning electron microscopy (reproduced from [14] with permission from Elsevier, 2022).

2.2. Composition and Reactivity of Components

The two main components of durum wheat semolina are starch and proteins (Table 1), with low quantities of lipids, fibers and minerals [3,4]. Physicochemical properties of starch and proteins can be described using a state diagram displaying their physicochemical reactivity as a function of temperature and water content (Figure 2) [3,4,15].

Table 1. Order of magnitude of protein and starch contents (g/100 g dry matter) of durum wheat semolina, coproducts generated during the manufacturing process and dry couscous grains (adapted from [3] with permission from Elsevier, 2022, adapted from [4] with permission from authors).

Composition	Durum Wheat Semolina	Wet Recyclates	Cooked and Dry Recyclates	Dried Couscous Grains
Water content	13–15	30–35	8–11	10–13
Gelatinized starch content	84–88	84–88	84–88	84–88
Gelatinized starch content	4–6	15–25	80–90	80–90
Total protein content	11–15	11–15	11–15	11–15
Soluble protein content	11–13	9–11	2–4	2–4
Total pentosan content	1–2	1–2	1–2	1–2
Soluble pentosan content	0–0.1	0–0.1	0–0.1	0–0.1

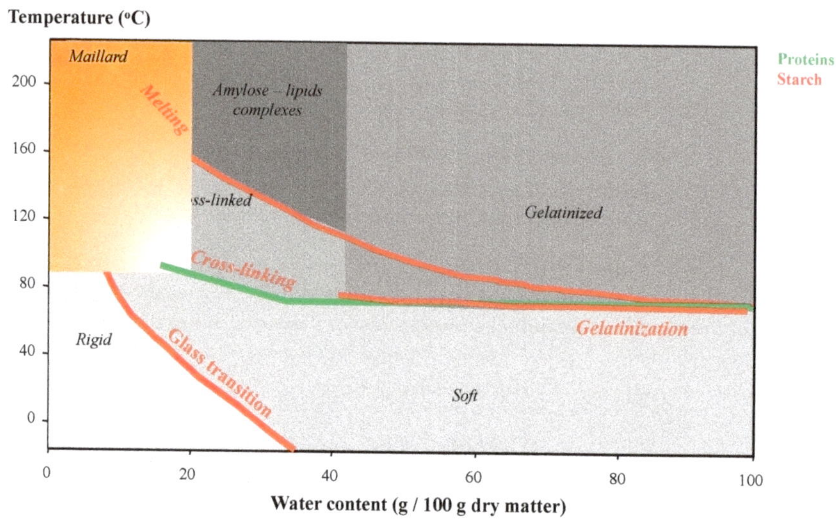

Figure 2. Schematic representation of transitions and physicochemical reactivity zones of wheat components (starch and proteins) according to temperature and water content conditions (adapted from [3,15] with permission from Elsevier, 2022, adapted from [4] with permission from authors).

Starch is the major component (70–75%) of durum wheat semolina. Starch is an assembly of linear amylose chains and branched amylopectin chains that interact through H-bonds. In native state, starch molecules are grouped together in individualized spherical starch granules (diameters between 1 and 20 μm). In starch granules, macromolecules are assembled in concentric layers alternately crystalline and amorphous. For couscous manufacture, starch is mainly involved through its functional properties of water absorption and gelatinization, which are expressed during processes at temperatures above 50–60 °C and water contents above 40% (Figure 2) [3,4]. During couscous processing, gelatinization of starch is described as the disappearance of crystalline structures and the partial release of some amylose chains. The application of heat treatments above 100 °C could induce the formation of noncovalent interactions between released amylose chains and lipids present that lead to the formation of amylose–lipid complexes (Figure 2).

Wheat proteins represent 10–14% of semolina and are structured in amorphous individualized fibrillar form located around the starch granules. Proteins are stabilized by the presence of a high density of low-energy bonds (mainly H-bonds and hydrophobic interactions) and some covalent disulfide bonds. At temperatures above 70 °C, thiol groups present on the protein chains can participate in crosslinking reactions via the formation of disulfide bridges (Figure 2) [3,4]. Some proteins (e.g., alpha amylases, lipoxygenase, polyphenol oxidases) display enzymatic activity which can have technological consequences on couscous.

The Maillard reaction can occur between the free amine function of a protein and the carbonyl group of a reducing function of sugars. The Maillard reaction is mainly observed under conditions of high temperature (>80 °C) and low water content (<18%) (Figure 2) and leads to volatile and/or colored compounds.

2.3. Plasticization and State Diagram of Components

Changes in structure and reactivity of the main wheat components have been described as a function of temperature and water content on their state diagram (Figure 2) [3,4,15]. In native semolina at 20 °C and 12% water content, the structure of macromolecules is stabilized by a high density of H-bonds, in amorphous (starch and proteins) and/or crystalline (starch) states. The high density of H-bonds contributes to their low mobility and low availability to participate in reactions.

The plasticization is associated with the sensitivity of H-bonds to temperature or water content changes [16]. Thermal plasticization describes the decrease in the density of H-bonds between macromolecules due to an increase in temperature. Molecular plasticization describes the ability of added water molecules to establish H-bonds with hydrophilic groups of macromolecules, which induces an overall decrease in interaction density between macromolecules. During the processing of native semolina, small increases in temperature and/or water content thus reduce the density of H-bonds, allowing local movements of small amplitude, involving localized groups of macromolecule chains [3,4,16]. These moderate changes are observed up to a sharp transition zone, which corresponds to the expression of cooperative phenomena generating large amplitude movements involving the entire chains of macromolecules. This abrupt transition zone is called "glass transition" for amorphous structures and "melting" for crystalline structures. The construction of the state diagram of durum wheat components as a function of temperature and water content helps to locate their transition zones and different reactivity areas.

The glass transition of amorphous structures separates two states [3,4,15,16]. Below the glass transition (i.e., at low temperature and/or water content), the high density of H-bonds describes the "glassy" state: macromolecules are poorly mobile and not available to participate in reactions. Above the glass transition (i.e., at high temperature and/or water content), the low density of H-bonds describes the "rubbery" state: macromolecules are mobile and available to participate in reactions. The glass transition zone is classically described by a temperature range at a given water content. The decrease in the glass transition temperature under an increase in the water content reflects the equivalence of thermal and molecular plasticization. In durum wheat, the glass transition affects the amorphous structures of proteins and starch chains within granules. Above the glass transition at temperatures above 80 °C, proteins are available to participate in crosslinking reactions. Above the glass transition at room temperatures, an increase in water content can activate enzyme activities.

For starch, melting of crystalline structures occurs at high temperatures or water contents, above the glass transition of amorphous structures (Figure 2) [3,4,16]. Melting is classically described by a temperature range, called the melting temperature. As for the glass transition, the decrease in the melting temperature under the effect of an increase in water content reflects the equivalence of thermal and molecular plasticization. In the presence of high amounts of water, starch melting is classically associated with gelatinization. Gelatinization occurs at temperatures as low as 60 °C in the presence of high amounts of water.

3. Process Diagram and Structuring Mechanisms

Couscous grains are made from durum wheat semolina according to successive unit operations. A structural model to describe the transformation of durum wheat semolina particles into couscous grains is proposed by considering four phases (Figure 3) [3,4].

- *Phase 1*: Native semolina particles are agglomerated by water addition and mixing to generate the granular structure of couscous grains. The agglomeration stage is followed by a size classification stage to select grains that meet granulometric specifications and to isolate too small or too large grains, which will be recycled.
- *Phase 2*: Wet grains are consolidated by steam treatment to strengthen the internal structure through starch gelatinization, crosslinking of proteins and formation of amylose–lipid complexes. The components form a glue between semolina particles.
- *Phase 3*: Cooked grains are dried to eliminate a large part of the water, in order to ensure physicochemical and microbiological stability of the couscous grains, by reducing the water activity to about 0.5.
- *Phase 4*: Dried couscous grains have to be rehydrated before consumption. Rehydration can be completed by cold water addition, by steaming or by immersion in hot water. This step is essential to give couscous grains the firm, melt-in-the-mouth texture.

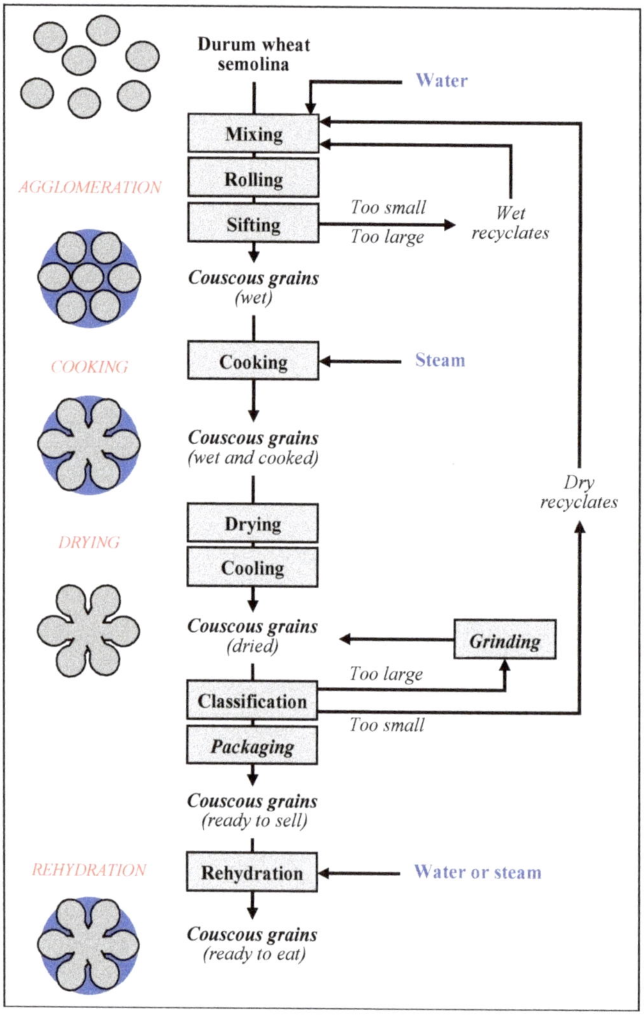

Figure 3. Production diagram and structural model for the processing of couscous grains from durum wheat semolina (adapted from [3] with permission from Elsevier, 2022, adapted from [4] with permission from authors).

The description of the couscous grain process on the state diagram of durum wheat components makes it possible to link changes in the process parameters for each unit operation with induced mechanisms (Figure 4) [3,4].

3.1. Agglomeration

The agglomeration of semolina is induced by the simultaneous addition of water (up to a water content of about 45%) and mixing, at constant temperature around 20–25 °C. Agglomeration mechanisms correspond to the structuring by the assembly of small semolina particles to form larger agglomerates. A necessary and sufficient amount of water must be added to induce the formation of cohesive contacts between native particles, generate a granular structure and ensure the internal cohesion of agglomerates [3,4]. Mixing ensures the homogeneous dispersion of water, promotes contacts between hydrated particles and generates growth mechanisms (Figure 5). Hydration properties of semolina and mechanisms of wet agglomeration by wetting and mixing have been studied in several works [6,14,17–24]. The multiplicity of mechanisms identified during the agglomeration of the semolina controls the characteristics of couscous grains [3,4].

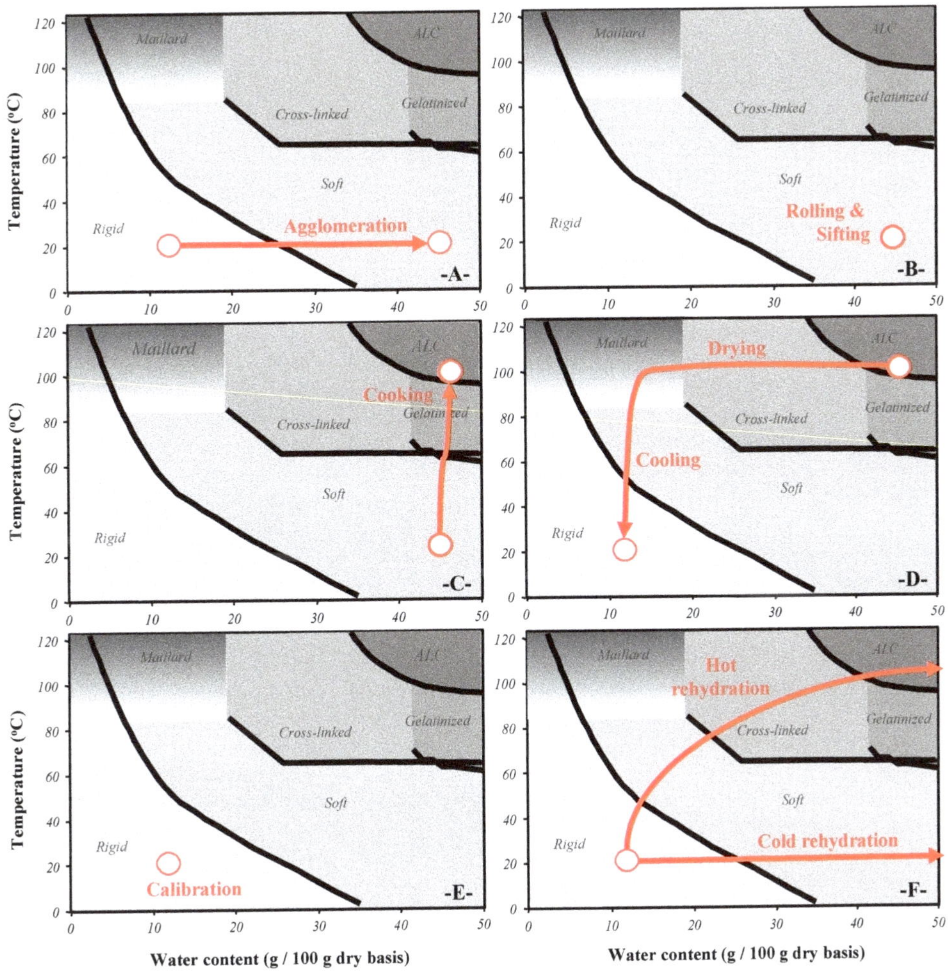

Figure 4. Diagram of hydrothermal paths for each unit operation of the couscous grain process on the state diagram of durum wheat components (ALC represents amylose–lipid complexes): (**A**) agglomeration stage; (**B**) rolling and sifting stage; (**C**) cooking stage; (**D**) drying and cooling stage; (**E**) calibration stag; (**F**) rehydration stage (adapted from [3] with permission from Elsevier, 2022, adapted from [4] with permission from authors).

The hydration and mixing stage plays an essential role in agglomeration mechanisms of semolina particles, generating different heterogeneous granular structures [3,4,18]:

- Fine particles (diameter < 0.5 mm) are residual particles of native semolina, or small wet particles generated by erosion mechanisms of larger agglomerates.
- Wet nuclei (0.5 < diameter < 0.8 mm).
- Wet agglomerates (0.8 < diameter < 2 μm) are the desired structures and will give the couscous grains.
- Large pieces of "dough" (2 mm < diameter).

Figure 5. Schematic representation of wet agglomeration mechanisms of durum wheat semolina for the generation of agglomerates (adapted from [4] with permission from authors).

Physicochemical mechanisms: The addition of water induces the passage of the glass transition of amorphous zones of starch and proteins (Figure 4A), which allows enzymatic reactions involving peroxidases and polyphenol oxidases in semolina [25]. During couscous processing, a significant decrease in carotenoid pigment content was observed during kneading and rolling stages [26]. Even if starch and proteins are not directly involved in agglomeration mechanisms, the passage of the glass transition allows macromolecules to participate in adhesion and sticking mechanisms [27–30]. The agglomeration stage results in a decrease in the insoluble glutenin content and an increase in the soluble glutenin and SDS-soluble protein contents. These mechanisms were attributed to the dissociation of large SDS-insoluble glutenin polymers due to the shearing effects of large dough pieces during mixing. Based on electrophoresis, Lefkir [30] indicated that the agglomeration stage of semolina does not allow the development of continuous protein networks. The amount of added water and mechanical energy inputs are not sufficient to induce the formation of a gluten network.

Agglomeration mechanisms are impacted by the particle size distribution of semolina [30,31]. The decrease in the median diameter of semolina leads to an increase in the proportion of agglomerates and a decrease in the proportion of dough pieces and small particles. Using semolina with a low diameter span reduces the proportion of small particles after agglomeration. Conversely, the protein content of semolina does not impact the proportion of wet agglomerates after mixing [31]. Lefkir [30] showed a slight positive correlation ($r = 0.82$) between the protein content of semolina and the proportion of large dough pieces after mixing. The protein content of semolina does not seem to control the agglomeration yield. Lefkir [30] found a positive correlation between the content of soluble glutenins in wet agglomerates and agglomeration yield.

The amount of added water is a key parameter as it controls the size, density and shape characteristics of wet agglomerates and couscous grains [3,4]. A sufficient amount of water is required to ensure agglomeration yields and to promote mechanisms during the subsequent cooking stage [32]. Increasing the amount of added water results in an increase in the proportion of wet agglomerates and dough pieces and in a decrease in the proportion of fine particles [18,19,21,25,33–35]. An increase in hydration could favor the solubilization mechanisms of glutenins [30]. A decrease in water temperature results in an increase in the proportion of agglomerates at the end of mixing, associated with a decrease in the proportion of dough pieces [34]. An increase in the water temperature is unfavorable for the solubilization of glutenins [27,30]. The mixing time has a significant impact on agglomeration mechanisms [25]. Water homogenization between different fractions was also observed during the mixing stage [36]. Long mixing times increase the proportion of fine fractions and decrease the proportions of medium and coarse fractions due to specific breakage mechanisms.

3.2. Rolling and Sifting

After mixing, the agglomerated powder is submitted to a combined rolling and size classification stage [3,4]. This stage is conducted at room temperature and without changes in water content (Figure 4B). The rolling operation corresponds to the movement of granular objects on the metal surface of a succession of sieves with different opening diameters. The displacement of the wet powder on sieves contributes to structuring mechanisms of the grains, with slight densification and some erosion effects on the agglomerates [37]. Rolling conditions (type of roller, agitation forces, layer thickness, duration) control the intensity of these changes. The displacement of the wet powder on the surface of sieves ensures the classification by the diameter of the wet granular objects. The size classification allows removing too small particles (diameter < 0.8 mm), recovering wet agglomerates with the target size (0.8 < diameter < 2 mm) and removing too large dough pieces (2 mm < diameter). Too small particles are recovered and reintroduced at the mixing stage to again participate in agglomeration mechanisms. Large dough pieces are shredded before being reintroduced into the mixer. Wet agglomerates with a size within the target diameter are wet couscous grains that will undergo the next cooking stage.

3.3. Steam Cooking

The internal structure of grains is consolidated by a steam cooking stage [3,4]. Wet grains are exposed to a stream of steam at 100 °C for a period of about 10–20 min. Steaming results in a rapid increase in the temperature of couscous grains up to 100 °C and induces a slight increase in their water content due to steam condensation and water absorption (Figure 4C). Due to their small diameter (1–2 mm) and circular shape, heat transfers in grains are not limiting factors because they are quick enough (about 1 min) to allow grains to rapidly equilibrate with the steam temperature. During steaming, the water content of grains rapidly increases from 0.48 to 0.53 g/g dry matter [32]. Water absorption by couscous grains during steaming is more important when using fine semolina as raw material [38]. During steaming, the diameter of grains increases from 1.6 to 2.15 mm due to swelling mechanisms, but this does not impact the spherical shape of the grains. The steaming induces different mechanisms involving wheat components that contribute to strengthening the structure of couscous grains [3,4].

(i) Steaming induces the gelatinization of starch granules with loss of crystalline structures and the partial release of amylose chains [3,4]. These changes participate in the formation of a sticky cement between the semolina particles. The extent of starch gelatinization increases rapidly as a function of the steaming time and reaches 80–100% [32,39]. The homogeneity of gelatinization mechanisms controls the water absorption properties of couscous grains [25]. The sticky behavior of couscous grains after rehydration could be due to the presence of amylose chains on the surface of grains or heterogeneous steaming [40].

(ii) Steaming also induces the formation of amylose-lipid complexes [3,4]. After gelatinization of starch granules and at high temperatures, released amylose chains can be available to participate in complexation reactions with monoglyceride molecules, which are present in native semolina at low contents. The extent of the complexation of amylose with lipids depends on the steaming duration [40]. The initial content of lipids would be the limiting factor for improving the culinary quality of couscous. The formation of amylose–lipid complexes strengthens grains, contributes to limiting the sticky behavior of rehydrated couscous grains and reduces the retrogradation phenomena during storage [25].

(iii) Steaming induces insolubilization of wheat proteins through the formation of covalent disulfide bonds [3,4,25,41]. Insolubilization of glutenins occurs rapidly during steaming [30,41]. Conversely, gliadins are little affected by steaming. An increase in the hydration level or a decrease in the water temperature at the mixing stage favors the insolubilization of glutenins during steaming [30]. Crosslinking reactions could contribute to decreasing the stickiness of couscous grains [3,4].

Before steaming, wet couscous grains are formed by the brittle assembly of semolina particles that adhere to each other. The steaming induces a partial melting of semolina particles which then irreversibly bind to each other. After steaming, couscous grains display a homogeneous structure with a "melted" appearance, in which semolina particles are hardly identifiable [32]. The extent of steaming greatly contributes to the water absorption, stickiness and texture properties of rehydrated couscous grains. Steaming makes the starch ingestible due to the gelatinization mechanisms. Couscous grains can be consumed directly, after a simple rehydration using cold water.

It should be noted that the steaming stage is classically called "precooking" and is followed by a "cooking" stage when preparing the couscous grains before consumption.

3.4. Drying and Cooling

Cooked couscous grains are dried in hot dry air to reduce their water content to a value of less than 13.5% in order to comply with the legislation [3]. The small diameter and porosity of couscous grains facilitate water removal mechanisms. A thin layer of couscous grains is exposed to dry air and/or subjected to agitation to favor water extraction. During drying, water transfers are the limiting phenomena [32]. Yüksel et al. [42] calculated effective moisture diffusivities of couscous grains (between 1×10^{-8} and 1.7×10^{-8} $m^2{\cdot}s^{-1}$) according to Fick's second law for sphere geometry in one dimension. The drying curves of a packed couscous bed were linear with a constant drying rate. Increasing air temperature (from 60 to 80 °C) led to an increase in effective diffusivity and drying rates. The extraction of water during drying induces shrinkage of grain structure: the volume of water extracted from the product is almost compensated by the volume contraction of grains and the reduction in diameter [32]. The shrinkage of grains during the drying stage is possible due to the plasticization of wheat components above the glass transition (Figure 4D). When the drying stage is conducted at high temperatures (90–120 °C), complementary mechanisms of glutenin insolubilization and amylose–lipid complex formation can be observed [30]. The drying stage contributes to the strengthening of couscous grains. Drying at high temperatures favors the Maillard reactions, especially at the end when the water activity is low. These reactions can lead to the formation of brown-colored compounds and specific volatile compounds.

After drying, the cooling phase until ambient temperature allows the rigidification of couscous grains thanks to the passage under the glass transition curve of amorphous wheat components which become rigid [3].

3.5. Calibration by Size

After drying, it is necessary to size-grade dried grains on a vibrating sieve column to recover couscous grains and to separate too fine or too large particles [3]. The classification stage is carried out at room temperature and does not significantly change the water content of couscous grains (Figure 4E). Thanks to the size classification stage, the granulometric dispersion of dried couscous grains is relatively low.

3.6. Storage

Dried couscous grains are packed in the appropriate packaging. Only little work describes the behavior of couscous grains during storage. The lipid fraction is critical during couscous storage, through oxidation mechanisms and the appearance of rancidity off-flavors [43]. These mechanisms can be reduced by using a high-temperature drying cycle which can be more effective to inactivate lipase. Guezlane et al. [40] showed that gelatinized starch in couscous grains is not sensitive to retrogradation phenomena during storage, because structures of the amylose–lipid complexes remain present.

3.7. Rehydration before Consumption

Before being consumed, couscous grains must undergo a final rehydration stage [4]. Depending on the intended use, dry couscous grains can be rehydrated by mixing with water at room temperature, by mixing with hot water or by exposure to steam (Figure 4F). For conventional uses, it is recommended to hydrate the couscous by mixing similar volumes of water and couscous to reach a final water content close to 60%. The increase in water content ensures the glassy transition of the wheat components into the rubbery domain, which contributes fully to the smooth texture of the ready-to-eat couscous grains.

4. Domestic Production of Couscous

The manufacture of artisanal or domestic couscous is carried out in summer (from May to September) at home in a clean and well-ventilated room [5]. Experienced women are dedicated to manufacturing couscous following successive steps (Figure 6). Traditional couscous production requires a large workforce. The process involves mixing water and durum wheat semolina in a large wooden dish and then rubbing the mixture between the palms of the hands to form agglomerates or small irregularly shaped granules. Granules are then separated by a set of appropriate sieves, and the desired portion is retained. The control of the agglomeration and hydration processes is very important to produce couscous with desired quality. In Maghreb countries, processes of making artisanal couscous differ from one region to another or even from one person to another. Details concerning the ethnic preparation of couscous in Tunisia are presented in the following sections.

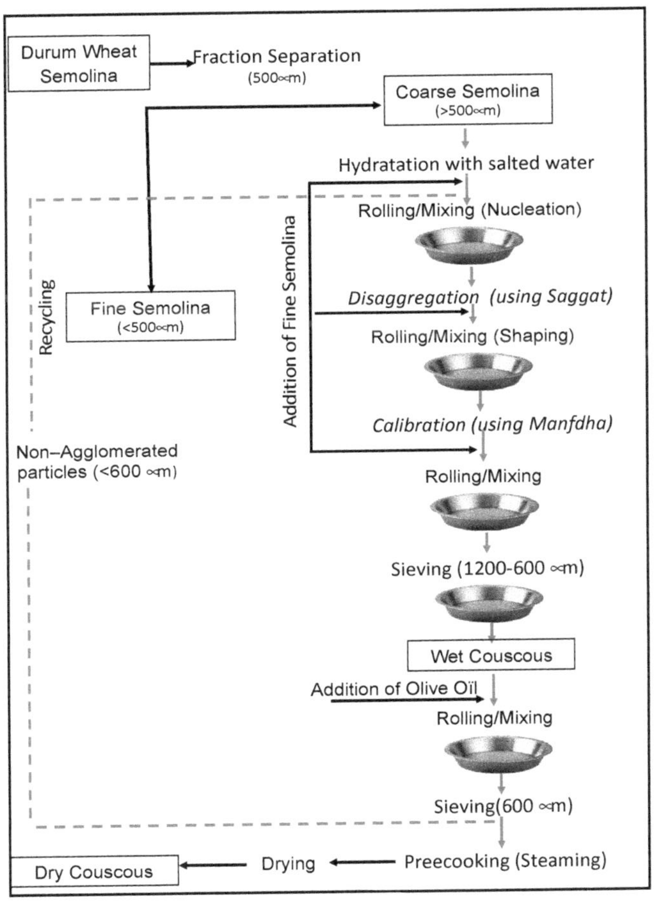

Figure 6. Traditional couscous making diagram identified from the handmade process used in Northern Tunisia.

4.1. Utensils

In Maghreb countries, couscous is still prepared manually at home using different utensils [5] (Figure 7). The humidification and rolling of the semolina are carried out on the *"guassâa"*, a wide bowl and hollow plate in wood or aluminum clay (Figure 8). Sizing and calibration of wet grains are performed on different sieves named *"saggat"*, *"manfdha"*, *"thannaya"* and *"tallâa"* (Figure 9), depending on mesh opening (2.3, 1.2, 1 and 0.6 mm, respectively). The traditional double-chambered food steamer is used to cook couscous by North Africans and now worldwide. This utensil is called "*couscoussier*" in French ("*taseksut*" in Berber language). It is made from ceramic or metal and consists of an upper smaller pot ("*kaskes*" in Arabic) containing holes that allow the passage of steam (Figure 10). The lower part is a large pot ("*borma*" in Arabic) that holds the meat and vegetables to be cooked as a stew in water or soup and produces steam. Once the couscous is steamed, the lower pot is kept at a simmer until cooking is complete [6].

Figure 7. The utensils used for the domestic manufacture of the couscous in Tunisia.

Figure 8. "*Guassâa*" made from aluminum (**a**) and tierra clay (**b**).

4.2. Preparation and Classification of Semolina

Homemade couscous is classically prepared from coarse semolina. Semolina classification is performed to separate two fractions by using a 0.5 mm mesh opening sieve (named "*ghorbel chaâr*"): coarse semolina (named "*fetla*") and fine semolina (named "*dkak*") (Figure 11). Classification improves the agglomeration yield of semolina by allowing the formation of agglomerates rather than clumps of dough. In Tunisia, making artisanal couscous obeys the classification step, but coarse semolina is used directly without sieving. The survey conducted by Chemache et al. [6] indicated that 20% of interviewed women state that the classification operation is not necessary if there is homogeneity in the particle size distribution of the semolina.

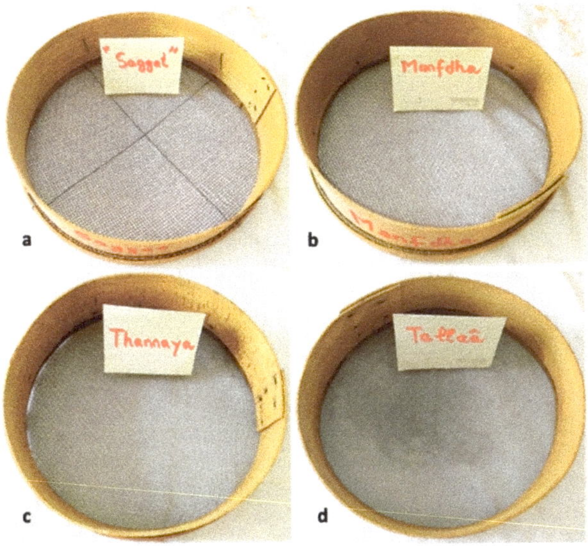

Figure 9. Different sieves used for semolina calibration and couscous shaping: (**a**) "*saggat*" used for lump disaggregation; (**b**) "*manfdha*" used for couscous calibration; (**c**) "*thannaya*" used 2 or 3 times for couscous sieving; (**d**) "*tallâa*" used for semolina sieving and finishing.

Figure 10. Examples of "*couscoussiers*", traditional steamers for preparing couscous, that are made of copper (**a**) or metal (aluminum) (**b**).

Figure 11. Semolina classification: (**a**) sieve used for classifying semolina known as "*ghorbel*" in Arabic and "*agherval*" in Berber language; (**b**) semolina classification; (**c**) semolina separated into 2 fractions: coarse semolina (CS) and fine semolina (FS).

4.3. Hydration and Mixing

The wet agglomeration is usually carried out in the *"guassâa"* (Figure 12a,b). This requires the double effect of mixing and adding salted cold water on coarse semolina which allows particle sticking. This critical step leads to the agglomeration of semolina to form couscous grains [35]. It is very important to obtain a homogeneous wetting of the semolina and to ensure the wetting of the semolina formation of dough pieces that will make the rolling operation difficult. Cold water helps to avoid the formation of large agglomerates. According to a survey [6], salt is added (salt content is 1.6%) to enhance the flavor of the final product. The characteristics of obtained wet agglomerates contribute greatly to the final couscous grain quality [35].

4.4. Rolling and Calibration

Rolling is the operation of shaping couscous by agglomeration of the hydrated semolina particles. This operation is conducted in four main substeps known as nucleation, shaping, sieving and finishing.

(a) Nucleation: The rolling process begins with simultaneous watering and mixing of both semolina fractions. First, the watering is performed gradually with small volumes of salted water using a ladle or by hand (Figure 12b). Second, the whole is mixed in circular movements by hand fingers half bent to distribute the wetting liquid in the bed powder in a homogeneous way (Figure 12c). The addition of small quantities of fine semolina allows the initiation of particle nucleation (Figure 12d). The wetting liquid is absorbed by the fine particles, which serve as nuclei around which coarse particles adhere [6,33,34]. The most influential parameter on the rolling yield is the semolina hydration rate [25]. The rolling operation is easier with coarse semolina [34].

(b) Shaping: Primary grains formed during the hydration step are grown by the addition of fine semolina [6]. At this stage, rolling is carried out by applying energetic and circular movements with the palm on the particle bed (Figure 12e). The fine semolina aggregates onto primary grains (nuclei). This step allows the formation of larger agglomerates through a snowball effect and coalescence. The fine semolina adheres against the voids of grains and gives spherical and smooth agglomerates. According to a survey [6], the rolling operation of semolina is carried out two or three times to ensure that it has absorbed all the added water. This stage allows good cohesion between semolina particles. In this step, lumps are broken down through a mesh sieve (called *"saggat"*) (Figure 12f,g). If there is an exaggerated agglomeration, a small quantity of fine semolina could be added, and lumps are broken down using the *"saggat"* two or three times.

(c) Sieving: The sieving operation corresponds to setting in motion the agglomerates on the surface of a succession of sieves of decreasing mesh to ensure a classification by size. The sieving step is important to obtain the desired homogeneity and size of the couscous. Agglomerates formed during the previous step are broken down through successive sieves (called *"manfdha"*, *"thannaya"* and *"tallâa"*) (Figure 12i,l,m). The under-size fraction will undergo rolling several times before calibration or sieving (Figure 12n). The recycling process is repeated until maximum depletion of semolina, but it is impossible to obtain a rolling yield of 100%. The grains with a size greater than 0.6 mm undergo a finishing step [5,6].

(d) Finishing: This step consists of a rolling operation of wet couscous grains using a small quantity of olive oil. Women add a small quantity of olive oil and perform a circular movement several times (Figure 12o). Olive oil is used to homogenize and improve the texture of couscous grains by giving them a more spherical shape and a smooth surface and producing well-individualized grains (Figure 12p). Chemache et al. [6] reported that instead of olive oil, Algerian women also use wheat flour or corn starch and obtain similar results.

Figure 12. Overview of the traditional handmade couscous according to the method used in Northern Tunisia: (**a**,**b**) semolina hydration with salted water; (**c**) rolling mixing of coarse semolina and salted water; (**d**) adding a small quantity of fine semolina to avoid over-agglomeration; (**e**) rolling mixing; (**f**,**g**) breaking down lumps with "saggat"; (**h**) rolling mixing; (**i**–**m**) sieving using different kinds of opening mesh sieves; (**n**) wet couscous; (**o**) adding a small quantity of olive oil; (**p**) rolling mixing; (**q**) steaming; (**r**) sieving using "*thannaya*"; (**s**) drying.

4.5. Steaming

The finished wet couscous of the desired particle size is put in the upper part of the *"couscoussier"* containing boiled water. Couscous is steamed for 15 min (Figure 12q). Steaming time depends on the thickness of the couscous layer and on the couscous granulometry: large couscous grains require a shorter cooking time as the water vapor circulates more rapidly between coarser grains [44]. The precooking time is determined when steam is on the surface of couscous. Cooked couscous grains break apart between fingers in the form of dough and have a yellow color. Immediately after steaming, the manual lump breaking (Figure 12r) is carried out using the *"thannaya"* sieve to obtain separated cooked grains, ready for drying. In some regions of Tunisia, couscous does not undergo drying. However, before serving, it has to be steamed two or three times. The "moist couscous" is prepared in the same way as the dry couscous (except the drying stage) and is prepared and consumed the same day [45].

During the traditional steaming, the couscous can be subjected to two successive cooking stages interrupted by the addition of fats (butter or olive oil) [40,46]. Amylose chains released by the starch gelatinization during the first steaming stage can form complexes with added lipids, in addition to those present in the native semolina. Enhancing the formation of amylose–lipid complexes reduces stickiness, delamination and caking index and increases the firmness of couscous grains upon rehydration [25,46–50]. The impact on organoleptic characteristics of couscous depends on the type of lipids used.

4.6. Drying

The drying stage of couscous is conducted in two phases. The couscous is first spread out on a clean sheet in the shade at ambient temperature (Figure 12s) for a duration depending on the air temperature and relative humidity. This first phase allows the preservation of couscous qualities. When the couscous is "sufficiently" dried, the couscous is then dried in the sun to ensure optimum water elimination. Couscous is occasionally stirred for a good drying process. The drying step is strictly related to climatic conditions that account for the production of home-made couscous during the sunny summer months [12]. Sun-dried couscous has a long shelf life.

4.7. Grading

The couscous is separated into fine, medium and coarse. The final product is classified in three different sizes: small couscous (diameters < 1.5 mm) is recommended for desert preparation; medium couscous (1.7 < diameter < 2.0 mm) is the most appreciated for traditional dishes; coarse couscous (2.5 mm < diameter) is used to prepare couscous with vegetable sauce [12].

4.8. Storage

The couscous is stored until use in cloth bags or in a large jar named a *"khabiya"* (Figure 13) and kept in a dry place at room temperature. To enhance the shelf life or to improve the organoleptic qualities of couscous, homemakers can add ingredients such as black or red dried pepper and bay leaf [6].

Figure 13. Picture of a jar named *"khabiya"* in Arabic language used to store couscous.

4.9. Couscous Dish Preparation

The final step of rehydration before consumption can be carried out using cold or warm water. The dry couscous is soaked in warm water for a few minutes, followed by draining in a couscous pot (Figure 14). Afterward, the rehydrated couscous is immediately drained, allowed to stand for about 8–10 min, stirred and dispersed from time to time before the rehydrated couscous is added with the fat. Several types of fat can be used, such as olive oil. The choice of added fats is based on their availability and consumption at family events. Melted butter ("*dehane*") is the most preferred when preparing couscous to be served during celebrations. The hydrated couscous is put in the couscous pot "*kaskas*" that is placed over a pot containing the sauce being cooked. Several criteria have been listed to stop the final cooking: the rise of the vapor, the development of the bright yellow color and an increase in the volume of cooked couscous grains [6]. Subsequently, the couscous is crumbled and watered with a small amount of water. Couscous can be served in many different ways and with a variety of foods.

Figure 14. Picture showing the typical dish of couscous prepared with meat and vegetables.

5. Industrial Production of Couscous

The first industrial production of couscous was initiated in North Africa in the 1960s and later in France, Italy, Greece and the United States [3–5,39,43]. Although the consumption of couscous is worldwide, the manufacturing industry is still mainly located around the Mediterranean. The first pieces of industrial equipment were simple transpositions of the lines used for the production of short pasta. From the 1970s, fully automated couscous production lines were developed. The design of equipment sought to reproduce the gestures mastered for the artisanal manufacture, particularly the agglomeration and rolling stages, in order to obtain qualities similar to domestic couscous. For the past 20 years, equipment manufacturers have been offering industrial lines that are specifically adapted to optimize industrial performance and product quality, with flow rates reaching 500 to 1500 $kg{\cdot}h^{-1}$. The specificity of the couscous manufacturing process is the management of the homogeneous treatment of granular materials, from semolina to couscous grains.

5.1. Specifications for Durum Wheat Semolina

Only a few scientific works have investigated the contribution of the characteristics of semolina to the process behavior and qualities of couscous grains. Originally, specifications for semolina for couscous were similar to those for pasta [3–5,12,38,43,51,52]. However, the use of high-quality semolina is not a requirement for the production of couscous. It is classically recommended to use durum wheat semolina with coarse diameter, as the size of semolina plays a role in defining the process settings [3,5]. Because of its higher water absorption, coarse semolina requires less water during mixing and results in a higher couscous yield than fine semolina [3].

The characteristics and content of wheat components (proteins, ash, gluten index, damaged starch, etc.) of semolina only play secondary roles in structuring mechanisms and qualities of couscous [3,5,43,53]. The role of semolina protein content remains unclear [5]. On one hand, Boudreau et al. [51] indicated that the couscous value of semolina depends on its protein content, close to 13.5% being preferred. Debbouz et al. [33] observed that wheat varieties with strong gluten expressed a better yield of couscous than cultivars with weak gluten. They referred to a decrease in stickiness as protein content increased. On the other hand, Ounane et al. [49] demonstrated that the semolina protein content, dry gluten contents and gluten index were poorly related to couscous characteristics. Concerning lipids, no correlation was found between the semolina total lipid content and cooked couscous quality [49]. Conversely, contents of apolar lipids, polar lipids and polar bound lipids of semolina could affect couscous qualities.

5.2. Hydration and Mixing

On an industrial scale, two types of equipment can be used for the agglomeration stage with wetting and mixing unit operations [3,4].

(i) Equipment based on the simultaneous water addition and mixing in a horizontal mechanical mixer with two rotating axes [3,4]. This process mimics traditional gestures and practices while intensifying the technique. Water is supplied by flowing directly onto the semolina during mixing. The high rotation speed of mixing shafts is required to ensure the homogeneous distribution of the water within the semolina and to generate agglomeration mechanisms.

(ii) Equipment based on individual hydration of the particles before mixing [3,4]. The semolina particles are individually hydrated by spraying water in a mechanically intensive system with a high speed of rotation of a mixing shaft to individualize the particles. This system generates water droplets to be evenly distributed over each particle of semolina. Intensive wetting is immediately followed by intense mechanical mixing in a double-axis horizontal mixer to promote agglomeration mechanisms.

The management of the agglomeration stage is critical as it determines the performance of the production lines. The agglomeration stage can generate large amounts of by-products after the rolling stage (i.e., too small or too large particles), which can represent a mass flow up to 2.5 times greater than the flow of the native semolina [3,4]. Minimizing these flows has obvious implications because of the unnecessary energy expended to reincorporate these products and the oversized equipment for mixing and classification.

5.3. Rolling and Sifting

Two types of equipment are available for carrying out the rolling and sifting operations [3,4].

(i) The plansichters consist of a series of superimposed flat vibrating sieves with openings of decreasing diameter. They are used to replicate the manual gestures. The too large particles are retained on the first sieve. Wet agglomerates with a size within the target diameter are retained and collected on the second sieve. Fine particles pass through the second sieve. The rolling operation on vibrating sieves significantly impacts the density and shape of wet agglomerates [54].

(ii) The rotary drum rollers consist of a succession of sections within a slightly inclined cylindrical drum. The first section is formed by unperforated metal plates. The following sections consist of a succession of perforated plates with holes of increasing diameter. The wet granular material is introduced at the inlet of the drum. The drum rotation helps to advance the granular material. Some mechanical effects are generated by the flow of the granular bed induced by the rotation of the drum. Too fine particles are removed at the first screens. Couscous grains are collected at the next grids. Coarse particles flow to the end of the drum. Couscous grains rolled in a rotating drum are more spherical and denser than grains rolled on plansichters [3,4,54]. The rotating screen drum parameters (angle of inclination, rotating speed and product flow rate) do not impact the sieving efficiency and characteristics of the agglomerates (diameter, water content and porosity), as no secondary agglomeration phenomena significantly occur [37].

5.4. Steam Cooking

Wet couscous grains are cooked by steam injection in a continuous tunnel cooker [3,4]. Grains are deposited to form a thick layer of about 20 cm on a perforated metal belt that passes through the tunnel. Steaming is carried out at 100 °C by injecting steam at atmospheric pressure through the product layer. A residence time of 15–20 min is required to ensure steam flow and heat transfer within the product layer. Cooking a thick layer of couscous can result in heterogeneities in the distribution of cooking values, especially between the surface and the core of the couscous bed, or in the case of heterogeneous steam circulation within the product layer. These heterogeneities can result in undercooked or overcooked couscous grains [3,4]. Some industrial lines use steam injectors to spray steam over and under the product, assuring a more homogeneous cooking [43]. At the cooked exit, the layer of cooked couscous grains forms a sort of cohesive "cake" that must be separated mechanically using a specific mixer combined with a calibration sieve to individualize cooked couscous grains before the drying stage.

5.5. Drying

The drying stage of couscous grains is conducted on pods circulating in a hot air drying tunnel, with controlled flow rate, temperature and relative humidity of air [3,4]. Industrial drying of couscous is carried out at high temperatures (90–120 °C) over short periods of time (15–20 min). The movement of the pods allows a "soft" mixing of products to favor mass exchanges with the hot air stream, limiting the formation of a static layer barrier to transfers. The movement of the pods can lead to erosion of grains, resulting in the formation of "fine dried particles". After the drying stage, dry and hot couscous grains are cooled to room temperature in a cooler with a cold air stream [3,4].

5.6. Sifting

Dry couscous grains are graded according to size criteria depending on the target diameter [3,4]. Products collected at the exit of the cooler are deposited at the top of a column of vibrating sieves, with decreasing mesh. The products are separated into three categories.

- Too fine particles with a diameter below the target diameter mainly come from breakage or erosion mechanisms of couscous grains during the drying stage. The flow of fine particles can represent 5–7% of the throughput of dry couscous grains.
- Several dry couscous grains in the diameter target can be produced: fine ($0.63 <$ diameter < 1.25 mm), medium ($1.25 <$ diameter < 1.85 mm) or coarse ($1.85 <$ diameter < 2 mm) couscous grains.
- Too large particles with a size greater than the target diameter are usually clusters of several couscous grains that have stuck together during the cooking or drying stages. These particles are sent to a roller mill for size reduction and then sifted again.

5.7. Recycling Discarded Products

The classification (after rolling) and calibration (after drying) stages discard significant flows of too small or too large products [3,4]. These products are characterized by a composition similar to native semolina, but with specific values of the extent of starch gelatinization and the solubility of proteins (Table 1).

- The dried products discarded after the drying stage display biochemical characteristics equivalent to dry couscous grains, with a high extent of starch gelatinization and a low solubility of the proteins. They are sent to a specific hydration stage before mixing with native durum wheat semolina [3,4].
- The products discarded after the rolling stage display physicochemical characteristics similar to native semolina. The main difference is a higher water content. These products are characterized by a slightly higher gelatinization extent (about 10%) and lower protein solubility than native semolina [3,4]. These differences are not due to the

agglomeration and rolling stages. They are mainly due to the flow of dry discarded products which are reincorporated at the agglomeration stage with the native semolina. The extent of starch gelatinization is consistent with the reincorporation ratio (5–7%) of the fine dried particles [3,4].

Discarded products are reintroduced at the agglomeration stage (Figure 3). The differences in the physicochemical state of the components of the discarded products (Table 1) result in differences in water absorption capacity and physicochemical reactivity compared to the native semolina [3,4]. The contribution of discarded products to the structuring mechanisms of couscous grains is still little known. The reincorporation of discarded products requires specific know-how to adjust parameters, namely the amount of added water and the reincorporation ratio, at the agglomeration stage.

5.8. Rehydration before Consumption

The diversity of marketing channels for couscous grains generates great diversity in rehydration methods, whether by companies preparing tabbouleh-type dishes, by catering companies using couscous grains for cold or hot preparations or by individuals. Several methods of rehydrating couscous grains can be described [4]:

- To prepare the tabbouleh, rehydration by adding an equivalent volume of tap water, mixing and resting for 30 to 60 min.
- For the traditional preparation of couscous, rehydration by contact with a steam flow for a defined period of time.
- For rapid hot preparation, there are several possibilities for rehydration.
- By adding an equivalent volume of boiling water, mixing and resting.
- By immersing a "cooking" perforated plastic bag in excess water for a defined period of time and draining.
- By mixing with an equivalent volume of cold water and heating in a microwave oven for a defined time.

6. Characteristics of Couscous

Over the past 30 years, as research has progressed and needs have arisen, a set of specific analytical methods has been developed to evaluate the qualities of dried couscous grains and of couscous grains after rehydration [3,4,12,22,25,33,39,43,49,55]. The evaluation is based on visual, usage and organoleptic criteria using instrumental or sensory methods. Couscous grain quality depends on the characteristics of the semolina and on process parameters [3,4]. High-quality dry couscous grains are amber in color, uniform in size and lack a particular odor. They must have a high capacity to absorb water. After rehydration, couscous grains must be easy to remove with a fork, not sticky and not bulky. When chewed, they should remain cohesive, distinct and firm, with good taste and neutral flavor.

6.1. Biochemical Composition

Similar methods are used to measure the composition of dried couscous grains and native semolina. As couscous is made exclusively from semolina, the content of proteins, starch, lipids and fibers of dry couscous grains depends on the composition of the native semolina (Table 1) [3,4]. The water content of dry couscous depends on drying conditions. Thermal treatments during processing induce significant differences in the solubility of proteins and the extent of starch gelatinization. Although not a general rule, lower contents in gelatinized starch were found in some homemade types of couscous compared to one industrial type of couscous [36].

6.2. Size Distribution

The diameter distribution of couscous grains is classically measured using a vertical vibrating sifter with sieves of decreasing mesh [3–5]. The *Codex Alimentarius* [1] specifies that the particle size distribution of dry couscous should be between 0.63 and 2 mm, with a tolerance of 6%. The particle size distribution of couscous follows a monomodal distribution (Figure 15). The relative position of the particle size distribution curves is specific to fine, medium or coarse couscous. Although not a general rule, some forms of artisanal couscous display finer grain size compared to medium industrial couscous, close to fine industrial couscous [12,39]. The diameter dispersion of couscous is relatively large and increases with the size (Figure 15). An increase in the extraction rate of semolina could cause a decrease in the median diameter of dry couscous [56]. At the mixing stage, an increase in the hydration level, a decrease in the water temperature or an increase in the duration of the mixing stage allows obtaining dry couscous with a greater diameter and lower dispersion [25,30,34,35]. The diameter distribution of the dried couscous mainly depends on the calibration stages after rolling and after drying.

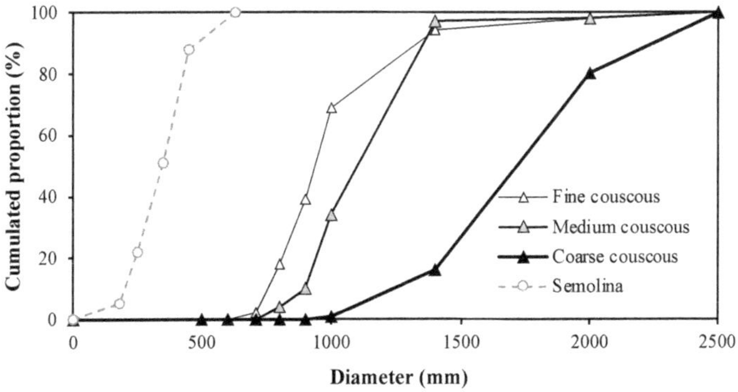

Figure 15. Typical examples of distribution curves of grain diameters for durum wheat semolina and different forms of couscous (adapted using data from Mezroua [50] and Lefkir [30], with permission from authors).

6.3. Grain Shape and Microstructure

The description of the shape of dry couscous grains is based on analysis of images obtained by optical or scanning electron microscopy [3,4,39,57]. A couscous grain appears as an approximately spherical granular object. The shape of grains was described using shape factors, such as circularity (0.68–0.73) and elongation (0.70–0.74) [25]. A couscous grain is formed by the assembly of durum wheat semolina particles that remain visible and partially melted to each other [54]. The melted bridges between particles contribute to the internal cohesion of grains. The quasispherical shape of couscous grains is the result of the mechanical stresses imposed during the process. Although not a general rule, grains of home-made couscous seem smoother and more uniform with rounded and oval shapes (Figure 16), unlike grains of industrial couscous which present more angular and heterogeneous shapes [12,26]. Dried couscous grains rolled in rotating drums seem more spherical than those rolled using plansichters.

6.4. Grain Porosity and Density

The density of a dry couscous grain (1.39–1.41 $g \cdot cm^{-3}$) is slightly lower than the density of native semolina particles (1.46–1.48 $g \cdot cm^{-3}$) (unpublished data). The porosity of couscous grains has been determined by measuring the real density of the grains by X-ray microtomography (XMT) methods [22,58]. Couscous grains are slightly porous objects, with a compactness between 0.68 and 0.88. The porosity is due to the presence of entrapped air between the more or less melted semolina particles (Figure 17). The XMT closed porosity values (0.005–0.011) are 10 times lower than the internal porosity values (0.121–0.206) that were calculated from measured compactness values.

6.5. Bulk Density

The bulk density of couscous has been measured by filling a graduated cylinder [3,25,39]. Although not a general rule, values of bulk density range were found lower for some homemade types of couscous (0.60 g/cm^3) than for one industrial type of couscous (0.79 g/cm^3). The bulk density of couscous depends both on the compactness of the grains (true density) and on the air volume entrapped between the grains.

Figure 16. Scanning electron microscopy observation of the microstructure of homemade couscous (**A**) and commercial couscous (**B**) (adapted from Debbouz and Donnely [39]) and of couscous grains of industrial origin from rotary drum rolling (**C**) or plansichter rolling (**D**) (adapted from [4] with permission from authors).

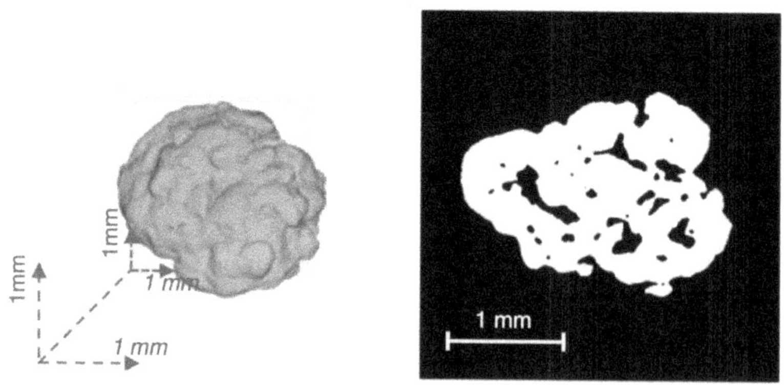

Figure 17. Typical examples of images calculated during the processing steps of XMT image analysis (adapted from [22], with permission from Elsevier, 2022).

6.6. Color

The color of dry couscous grains can be quantified by instrumental methods using colorimeters and the color space parameters: L^* (lightness), a^* (red hue) and b^* (yellow hue) [3]. Couscous grains are characterized by a high lightness L^* (30–75), marked yellow hue b^* (25–45) and low red hue a^* (0–4) [25,26,33,39,47]. Although not a general rule, some artisanal types of couscous have been characterized by slightly higher yellow hue

b^* and lightness L^* than one industrial type of couscous. The color of dry couscous grains greatly depends on the characteristics of the native semolina, in particular on the content of carotenoid and flavonoid pigments and on enzyme activities [3,5]. An increase in the extraction rate of the semolina leads to a decrease in the lightness of the dry couscous [58]. During processing, the hydration stage of the semolina initiates enzymatic reactions. Increasing the hydration level or the water temperature during mixing favors enzymatic reactions and reduces the yellow hue (b^*) and lightness (L^*) [25,30,35]. The degradation of the color can be slowed down by reducing the duration of the mixing stage and the number of recycles in the rolling stage [25,47]. The cooking stage also contributes to the degradation of carotenoid pigments, reduces the lightness (L^*) and increases the yellow hue (b^*) [25,47,59]. The drying stage conducted at high temperature (3 h at 95 °C) has a more marked effect on the color of the couscous compared to drying at low temperature (17 h at 55 °C) [25]. Granulometry of the dry couscous also greatly affects its color.

6.7. Hygienic Characteristics

According to the *Codex Alimentarius* [1], dried couscous must be free from microorganisms that may grow under normal storage conditions and must not contain any substance originating from microorganisms in quantities that may present a risk to health. When the homemade couscous is sun-dried, it can support the growth of *A. parasiticus* with production of aflatoxins during storage in the wet season [60,61]. Mold growth is possible if the product is not well dried or is poorly stored.

6.8. Rehydration Properties

Rehydration properties are important criteria of couscous grains. Determining rehydration properties of couscous grains is not an easy task, as many ways to rehydrate couscous exist. The methods used to determine rehydration properties of couscous grains are based on three complementary criteria: a swelling index which describes the water absorption capacity, a hydration time which describes the kinetics of water absorption and a solubility index which describes the loss of dry matter in water.

The swelling index is the relative increase in volume occupied by couscous grains immersed in an excess of water (at 25 or 100 °C), measured in a graduated tube, compared to the volume initially occupied at the time of immersion [3]. The test tube is placed in a water bath at a controlled temperature and the changes in couscous volume are recorded as a function of time [55]. Water absorption index can also be determined by introducing couscous and water in a centrifuge tube, shaking for 30 min and centrifuging at 2200 g for 10 min. The supernatant liquid is drained and the material remaining in the centrifuge tube is weighed to calculate the water absorption index. A great diversity in measured values of the swelling index for couscous exists, between 130% and 415% [35,39,49,62–64]. It is accepted that high swelling values are indicative of high-quality couscous [25,55]. Starch fraction plays a determinant role in the swelling index of the dry couscous [43]. The content of damaged starch of semolina has been supposed to impact the water absorption index [33,65]. The total lipid content of durum wheat semolina was correlated with the swelling index of couscous [49]. The protein content of the semolina has been partially negatively correlated ($r = -0.571$) with the swelling index of dry couscous [50]. A slight positive correlation was found between the swelling index and the bulk density of the couscous [47].

It remains difficult to discriminate between homemade couscous and industrial couscous by their swelling index values [39,47]. During mixing, an increase in the water temperature has negative effects on the dry couscous swelling [30,35]. The swelling index of couscous increases significantly with increasing hydration level during mixing, due to positive effects on the gelatinization mechanisms during the cooking stage [25]. The swelling capacity of couscous is proportional to the duration of the cooking stage [66]. A high correlation ($r = 0.90$) was found between the swelling index and the extent of starch gelatinization of couscous grains [67]. The drying temperature seems to have a negative effect on the swelling of couscous [25,68]. The insoluble glutenin content of dry couscous after the cooking and drying steps has been positively correlated ($r = 0.73$) with the swelling capacity [30].

The rate of water absorption by the dry couscous is often associated with the empirical term "optimal time" or "cooking time", in comparison with the method of assessing the quality of pasta [3]. Rehydration time is assessed by measuring the time required for maximum water absorption by the couscous during the rehydration test [33,39]. A decrease in the particle diameter of native semolina results in a decrease in the rehydration time of dry couscous [33]. The time required for couscous to swell decreases with an increase in the water temperature during mixing [25,46]. Some homemade types of couscous were characterized by lower rehydration times due to the smaller particle size compared to one industrial type of couscous [39].

Rehydration of dried couscous in excess water can result in the solubilization of some dry matter in the water phase. The solubility index of couscous in water expresses the degree of disintegration of couscous [3]. The terms "cooking loss" and "degree of delitescence" are also used [33,49]. There is a great diversity in the published values of water solubility index for couscous, between 3% and 16% [39,49,63,69]. A low value of the water solubility index is generally associated with good-quality couscous. The water solubility index is correlated with the stickiness of the rehydrated couscous. The swelling index of couscous is inversely proportional to its water solubility index [47]. The use of fine semolina leads to couscous with high water solubility index [56,70]. The protein content of semolina is partly negatively correlated ($r = -0.59$) with the delitescence index of couscous [50]. The apolar lipid content is significantly correlated with the water solubility index [49]. It was difficult to discriminate between some homemade types of couscous and one industrial type of couscous by their water solubility index values [39,47]. During mixing, a decrease in the hydration level or in the water temperature reduces the water solubility index of the couscous [34,35]. The water solubility index of dry couscous has been correlated with the denaturation state of proteins after hydrothermal treatments. The content of insoluble glutenins in dry couscous was negatively correlated ($r = -0.47$) with the degree of delitescence [30,71].

6.9. Stickiness and Caking Index

The caking index is related to the aggregation of couscous grains after rehydration [3]. It can be determined by an instrumental method (proportion of grains with a diameter greater than 3 mm formed after rehydration and drying) or by sensory analysis. A low value of the caking index is an indicator of a high-quality couscous. There is a great diversity of values for the caking index (between 5% and 80%) in the literature [25,47,49,50,53,68,72]. The caking index is inversely proportional to the granulometry of the couscous: fine couscous is perceived as stickier than medium couscous. The caking index was positively correlated ($r = 0.91$) with the water solubility index. The solubilized dry matter at the time of rehydration contributes to the sticky character of the grains and favors their caking. The sticky character of couscous was associated with the extent of starch gelatinization and the possible diffusion of amylose chains on the surface of the couscous grains. An increase in cooking time increases the stickiness of couscous.

The use of fine semolina (instead of coarse semolina) results in dry couscous that is stickier and easier to disperse [33,56]. The stickiness of couscous decreases with increasing protein content and gluten index value of the semolina [33,70]. The extraction rate of the semolina has no significant effect on the caking index [56]. No correlation was found between the total lipid content of semolina and the caking index [49]. During mixing, increases in the water temperature, hydration level or mixing time result in increases in the stickiness and in the caking index values [25,30,34]. The drying temperature also affects the stickiness of the couscous [68,73]. Although not a general rule, some homemade types of couscous showed lower stickiness than one industrial type of couscous [39,47].

Belaïd et al. [46] showed that the incorporation of 1% monoglycerides during processing reduces the stickiness of couscous.

6.10. Texture Properties

Texture properties of couscous are described during its preparation and consumption by sensory analysis [3]. Texture qualities relate to the couscous ability to be fragmented with a fork and to its texture during chewing, with terms such as firmness, consistency, elasticity, smoothness, chewiness or stickiness [12,25,33,39,49,50,72]. Couscous of good culinary quality can be forked after rehydration; it should maintain its firmness it should have a not too firm consistency and a soft appearance and it should be easy to chew. An instrumental compression method was developed to evaluate the viscoelastic characteristics of a rehydrated couscous bed by transposing a method classically used for pasta [49,55,72]. However, the instrumental firmness is not correlated with sensory analysis [53]. The use of coarse (instead of fine) semolina results in firmer couscous grains [56]. Protein and total lipid contents of semolina were not correlated with the texture of the couscous [49]. Although not a general rule, a higher elasticity was found in some homemade types of couscous compared to one industrial type of couscous [26]. The incorporation of 1% monoglycerides during processing increases the firmness of couscous [46]. The firmness of the couscous is proportional to its granulometry: fine couscous is less firm than medium couscous.

6.11. Nutritional Characteristics

The nutritional qualities of rehydrated couscous are typical of a food based only on durum wheat semolina. Couscous is a source of proteins with "good" nutritional quality (except the low lysine content) and is a source of energy (350 kcal/100 g of dry matter) due to its high starch content. During chewing, couscous grains are not easily broken down and have slow rates of softening [74]. Couscous grains display a slightly higher absorption rate (glycemic index between 60 and 65) than pasta (glycemic index between 50 and 55). The porous granular structure, the high specific surface area and the lack of protein network around the starch granules in couscous grains are favorable for enzymatic attacks during digestion. The vitamin content of couscous is influenced by the thermal treatment. The content of vitamins (thiamin and riboflavin) was found to decrease with increasing steaming time [61]. As riboflavin is very sensitive to processing conditions, especially heat and light, its content in the traditional sun-dried samples tended to be lower than that in industrial samples.

7. Uses and Consumers of Couscous

In North Africa, couscous is an iconic food. It permits the expression of national identities and ways of life. It has religious and symbolic meanings. According to Habib Bouguiba, ex-leader of Tunisia, the border of Maghreb is marked by an imaginary line corresponding to a cultural boundary: to the east, the staple food is rice; to the west, the staple food is couscous [5,43]. Couscous is the dish that united the history and geography of Maghreb. The couscous dish allowed Morocco, Algeria, Tunisia and Mauritania to submit a joint file to UNESCO in order to obtain international recognition of the couscous dish as an intangible world heritage. Only the difficult political conditions that Libya is experiencing can explain Libya's absence.

7.1. A Traditional Ethnic Food

In rural regions of Tunisia, women make the couscous alone at home or sometimes they ask their cousins or neighbors for help. Women choose a sunny day during summer and dedicate it to making a large quantity of couscous which covers the needs of their family throughout the year. This special day is named "Al Oula for one year", which refers to joy and happiness [5]. Habitually, couscous-preparation knowledge was passed from mother to daughter and played a crucial role in North Africa's patriarchal society. The know-how was an important "intangible" element of a young woman's dowry.

Couscous is a staple in the Arab Maghreb region: it is the most popular in all the countries of the Maghreb. It symbolizes comfort, warmth and tradition. Women usually prepare couscous dish during a family celebration, and it is eaten during a family feast, thereby associating both the product and the dish with solidarity. Couscous accompanies the traditional Arabic weekend (Friday and Saturday); the end of Ramadan celebrations; and Muslim year, birth and wedding feasts. The association of couscous with these festivities also attaches it to the concepts of abundance, fertility, fidelity and Barakah (God's blessing). While preparing couscous, women are used to making a kind of invocation and converse about religious facts, prosperity and positive feelings [5]. Couscous for interviewees with North African connections is first and foremost a dish that never needs to be paid for as it is a family dish and thus is not eaten in restaurants. A Mediterranean notion of sharing and valuing home and clan atmospheres is coupled with values that are centered not around money but around exchanges, and which see cultural prowess as having a sense of hospitality: to share a couscous is to be associated with other people and to express one's attachment to the group [9]. The secret of couscous grains makes known the context of the Arab community in the host country, relating different aspects of the integration process, such as family relationships reasoned on solidarity-based and shared identity values [75]. The couscous, in the family, has a sociability function.

Couscous is well known to be consumed with a vegetable sauce. It can be prepared with vegetables, pulses and different types of meat, making the dish of couscous a complete one. When preparing the sauce, up to three vegetables can be included at the same time. The most commonly encountered are carrots, green beans, zucchini, potatoes, turnips, chard, cabbage, tomatoes, etc. Onion and garlic are added to the sauce as spices [6]. Couscous is a simple product that can also be prepared with a simple knob of butter or a little sugar and cinnamon.

7.2. *Diversity of Couscous Market Offer and Consumption Patterns in the World*

Ethnic consumption: In Algeria, there are more than 300 ways to prepare couscous, and spices and seasonings are one of the most important elements that distinguish the flavor of the dish from one country to another. There are as many recipes as there are villages, or even families in the Maghreb countries, each jealously guarding the secret of the recipe passed down from one generation to the other. Among "ethnic" consumers, couscous gives rhythm to daily and religious life. Couscous is the spiritual food of North Africans. In many North African families, the week cannot end without the Friday afternoon bowl of couscous after prayers. During Ramadan, mesfouf—couscous sweetened with cinnamon and raisins—is served before sunrise just before fasting, to keep everyone going until sunset. The end of Ramadan is celebrated in many homes by a more elaborate couscous than usual [76].

Couscous consumption: In France in 2020, the consumption of couscous was 1.5 kg per inhabitant [77]. Couscous is one of the favorite dishes of the French. In a 2006 survey, it came in second place after "blanquette de veau" and before "moules-frites" [78]. The traditional consumption can be apprehended with large packagings of 5 or 25 kg, purchased in France mainly by restaurant owners and North African consumers. A large part of the purchases in supermarkets of couscous in small packagings (0.5 or 1 kg) are made by consumers. In other European countries, the consumption of couscous is mainly made by the Maghrebian immigrant communities (Moroccan in Belgium and Germany, Tunisian in Italy). English and Polish markets are dominated by a so-called "modern" consumption of couscous (side dishes, tabbouleh, etc.).

From traditional dish to side dish: Today, couscous is available in traditional, canned, frozen and microwaveable forms. It has incorporated the codes of contemporary consumption: organic, fair trade, prepackaged meals, takeaway and so forth. It is served at canteens, restaurants, cafés, markets and catered events. It can even be the single unifying factor behind virtual communities, forums for couscous recipes and so forth. Couscous is a Mediterranean symbol of cultural interpenetration [9]. In the Western market, couscous is prepared due to its taste, quick preparation when presteamed and usage in salads (tabbouleh). In France, we observe a clear evolution in the consumption modes of couscous: 80% of the consumers used the grain in traditional dishes at the end of the 1980s; this proportion is only 60% today. Since the beginning of the 1990s, the consumption followed the rise of the tabboulehs in the delicatessen departments. Couscous grains are widely used in salads or as a vegetable side dish, such as rice or pasta. We have witnessed an important development of couscous called "flavored", becoming widespread in the Anglo-Saxon countries, Great Britain and the United States [76]. Their success tends to prove that the traditional perception of the product is no longer the only one in the mind of the consumer.

International cultural cooperation and fusion: The "knowledge, know-how and practices related to the production and consumption of couscous" testify of a widely confirmed sociocultural importance in related countries. Throughout history, couscous has been able to travel and spread to other regions, such as Sahel and the Mediterranean islands. In the 20th century, it reached Europe, the Americas and Asia. It embodies and reflects successful cultural exchange and sharing. In France, the arrival of North African workers and repatriated French (after independence) in the mid-20th century largely contributed to popularizing the dish. It is in Sicily that since 1998 the "world championship of couscous", the Couscous Fest, "festival of cultural integration", takes place [79]. According to historians, *couscuz*, as it is called in Brazilian Portuguese, is a food that has its origins thousands of years ago among Berber peoples of North Africa, particularly in Morocco. It first crossed the Mediterranean to the Iberian Peninsula and then the Atlantic until it reached Latin America, where it was reinvented, rediscovered. Couscous in Brazil is a clear legacy of the Moroccan (Moorish) presence in Portugal. Today, couscous, which is celebrated every year on 19 March, World Couscous Day and the Feast of St. Joseph, is one of the main components of the intangible capital of the Northeast Region of Brazil [80]. In Brazil, couscous can be made with flour or starch from corn, rice or cassava. Salted and slightly moistened, the dough is marinated to incorporate the seasoning. It is steamed and can be enhanced with other ingredients, as is the custom in the Southeast, or simply accompanied by milk, eggs, butter or dried meat, as preferred in the Northeast.

7.3. Culinary Precisions as Explained by Science

Is a couscoussier necessary? The particularity of the preparation of couscous dish lies in the very particular kitchen utensil that is used: the *"couscoussier"*, a large metal pot in 2 parts: the steam basket above to cook the grains and the "big pot" at the bottom to cook the broth, meat and vegetables. This is the best way to cook the grains, as they are impregnated with the aromas of the broth. The steam will make them swell, making them light and more digestible.

Can couscous be boiled? Couscous is not cooked on the stove in boiling water. It can be steamed for a few minutes or prepared in a container simply by covering it with boiling water and seasoning with a drizzle of oil.

How is light and easy-to-digest couscous made? Couscous grains must be light and digestible. Couscous grains absorb the right amount of water necessary to swell, which will make them soft and not doughy. The grains can be coated with a thin film of fat provided by the oil with which we cover our hands to "roll" it.

8. Conclusions

Couscous, of Berber origin, has been eaten since at least the Middle Ages. If it is difficult to be definitive on its history, but everybody has fallen in agreement on this truth of couscous: "The best couscous, is the one of my mother". Couscous cannot be summarized only by the emblematic dishes which contain it: couscous is much more than a dish; it is a moment, memories, traditions, know-how and gestures which are transmitted from generation to generation. There are as many couscous recipes as there are families and an infinite variety of nuances between regions, making couscous a true mirror dish of the societies where it is cooked. Outside the Maghreb region and outside Europe, most couscous is produced industrially: this may be responsible for the worldwide growth of couscous consumption. The complementary uses of couscous made by industrialists or at home fully meet the diversity of consumer needs, between tradition and innovation. Traditional couscous and industrial couscous are not in competition but ensure the perpetuity and expansion of consumption in the world.

Author Contributions: All authors have contributed writing, reviewing and editing the original draft. All authors have read and agreed to the published version of the manuscript.

Funding: This research received no external funding.

Institutional Review Board Statement: Not applicable.

Informed Consent Statement: Not applicable.

Conflicts of Interest: The authors declare no conflict of interest.

References

1. *Codex Alimentarius, CODEX STAN 202-1995*; Norme Codex pour le Couscous. Codex Alimentarius Commission: Rome, Italy, 1995.
2. UNESCO. Le Couscous Entre au Patrimoine Culturel Immatériel de l'UNESCO. Available online: https://news.un.org/fr/story/2020/12/1084812 (accessed on 15 September 2021).
3. Abecassis, J.; Cuq, B.; Boggini, G.; Namoune, H. Other traditional durum-derived products. In *Durum Wheat: Chemistry and Technology*, 2nd ed.; Sissons, M., Abecassis, J., Marchylo, B., Carcea, M., Eds.; AACC: St. Paul, MN, USA, 2012; pp. 177–199.
4. Ruiz, T.; Rondet, E.; Cuq, B. La graine de couscous: De l'artisanat à la croissance fractale. In *Sciences Culinaires*; Lavelle, C., Ed.; Collection Echelles, Belin: Paris, France, 2014; pp. 28–51.
5. Hammami, R.; Sissons, M. Durum wheat products, couscous. In *Wheat Quality for Improving Processing and Human Health*; Igrejas, G., Ikeda, T.M., Guzman, C., Eds.; Springer Nature: Cham, Switzerland, 2020; pp. 347–367.
6. Chemache, L.; Kehal, F.; Namoune, H.; Chaalal, M.; Gagaoua, M. Couscous: Ethnic making and consumption patterns in the Northeast of Algeria. *J. Ethnic Foods.* **2018**, *5*, 211–219. [CrossRef]
7. Samrakandi, M.H. Arts culinaires en Méditerranée et pratiques culturelles. *Horiz. Maghrébiens Droit Mémoire* **2006**, *55*, 4–5.
8. Caglar, A. McDöner: Döner kebap and the social positioning struggle of german turks. In *Marketing in a Multicultural World: Ethnicity, Nationalism and Cultural Identity*; Costa, J.A., Bamossy, G.J., Eds.; Sage Publications: London, UK; New Delhi, India, 1995; pp. 209–239.
9. Béji-Bécheur, A.; Ourahhmoune, N.; Özçağlar-Toulouse, N. The polysemic meanings of couscous consumption in France. *J. Consum. Behav.* **2014**, *13*, 196–203. [CrossRef]
10. Hubert, A. Destins transculturels. In *Mille et une Bouches: Cuisines et Identités Culturelles*; Bessis, S., Ed.; Autrement: Paris, France, 1995; pp. 114–118.
11. Tabois, S. Cuisiner le passé. Souvenirs et pratiques culinaires des exilés Pieds Noirs. *Diasporas Hist. Sociétés* **2005**, *7*, 81–92.
12. Kaup, S.M.; Walker, C.E. Couscous in North-Africa. *Cereal Foods World* **1986**, *31*, 179–182.
13. Toussaint-Samat, M. Couscous. In *Histoire Gourmande des Grands Plats*; Canta, L., Dechaux, C., Eds.; Casterman: Paris, France, 1994.
14. Saad, M.; Barkouti, A.; Rondet, E.; Ruiz, T.; Cuq, B. Study of agglomeration mechanisms of food powders: Application to durum wheat semolina. *Powder Technol.* **2011**, *208*, 399–408. [CrossRef]
15. Cuq, B.; Icard-Vernière, C. Characterization of glass transition of durum wheat semolina using modulated differential scanning calorimetry. *J. Cereal Sci.* **2001**, *33*, 213–221. [CrossRef]
16. Cuq, B.; Abecassis, J.; Guilbert, S. State diagrams help describe wheat bread processing. *Int. J. Food Sci. Technol.* **2003**, *38*, 759–766. [CrossRef]
17. Hébrard, A.; Oulahna, D.; Samson, M.F.; Galet, L.; Morel, M.H.; Abecassis, J.; Fages, J. Étude morpho-granulométrique et structurale des semoules de blé dur en relation avec leurs propriétés d'hydratation et d'agglomération en couscous. *Récents Progrès Génie Procédés* **2001**, *77*, 489–496.
18. Hébrard, A.; Oulahna, D.; Galet, L.; Cuq, B.; Abecassis, J.; Fages, J. Hydration properties of durum wheat semolina: Influence of particle size and temperature. *Powder Technol.* **2003**, *130*, 211–218. [CrossRef]
19. Barkouti, A.; Rondet, E.; Delalonde, M.; Ruiz, T. Influence of physicochemical binder properties on agglomeration of wheat powder in couscous grain. *J. Food Eng.* **2012**, *111*, 234–240. [CrossRef]
20. Oulahna, D.; Hébrard, A.; Cuq, B.; Abecassis, J.; Fages, J. Agglomeration of durum wheat semolina: Thermodynamic approaches for hydration properties measurements. *J. Food Eng.* **2012**, *109*, 619–626. [CrossRef]
21. Barkouti, A.; Delalonde, M.; Rondet, E.; Ruiz, T. Structuration of wheat powder by wet agglomeration: Case of size association mechanism. *Powder Technol.* **2014**, *252*, 8–13. [CrossRef]

22. Hafsa, I.; Kim, S.J.; Cuq, B.; Le Bail, A.; Ruiz, T.; Chevallier, S. Description of internal microstructure of agglomerated cereal powders using X-ray microtomography to study of process–structure relationships. *Powder Technol.* **2014**, *256*, 512–521. [CrossRef]
23. Murrieta-Pazos, I.; Galet, L.; Patry, S.; Gaiani, C.; Scher, J. Evolution of particle structure during water sorption observed on different size fractions of durum wheat semolina. *Powder Technol.* **2014**, *255*, 66–73. [CrossRef]
24. Deng, L.; Manthey, F.A. Flowability, wet agglomeration, and pasta processing properties of whole-durum flour: Effect of direct singe-pass and multiple-pass reconstituted milling systems. *Cereal Chem.* **2019**, *99*, 708–716. [CrossRef]
25. Guezlane, L. Mise au Point de Méthodes de Caractérisation et Etude des Modifications Physico-Chimiques sous L'effet des Traitements Hydrothermiques en vue D'optimiser la Qualité du Couscous de blé dur. Ph.D. Thesis, INA El-Harrach, Oued Smar, Algeria, 1993.
26. Guezlane, L.; Selselet-Attou, G.; Senator, A. Étude comparée du couscous de fabrication industrielle et artisanale. *Ind. Céréal* **1986**, *43*, 25–29.
27. Icard-Vernière, C.; Feillet, P. Effects of mixing conditions on pasta dough development and biochemical changes. *Cereal Chem.* **1999**, *76*, 558–565. [CrossRef]
28. Kratzer, A. Hydration, Dough Formation and Structure Development in Durum Wheat Pasta Processing. Ph.D. Thesis, ETH Zurich: Swiss Federal Institute of Technology, Zürich, Switzerland, 2007.
29. Joubert, M.; Lullien-Pellerin, V.; Morel, M.H. Fate of SDS-insoluble glutenin polymers from semolina to dry pasta. *Food Chem.* **2018**, *240*, 189–195. [CrossRef]
30. Lefkir, S. Étude et Optimisation des Conditions D'hydratation des Semoules de blé dur sur L'expression de la Qualité Technologique du Couscous. Modifications Biochimiques des Protéines. Ph.D. Thesis, Ecole Nationale Supérieure Agronomique El-Harrach, Oued Smar, Algeria, 2018.
31. Bellocq, B.; Duri, A.; Cuq, B.; Ruiz, T. Impacts of the size distributions and protein contents of the native wheat powders in their structuration behaviour by wet agglomeration. *J. Food Sci. Eng.* **2018**, *219*, 29–37. [CrossRef]
32. Bellocq, B.; Duri, A.; Cuq, B.; Ruiz, T. Contribution of the cooking and drying stages on the structure of the couscous grains of durum wheat semolina. *Cereal Chem.* **2018**, *95*, 646–659. [CrossRef]
33. Debbouz, A.; Dick, J.W.; Donnelly, B.J. Influence of raw material on couscous quality. *Cereal Foods World* **1994**, *39*, 231–236.
34. Lefkir, S. Influence des Conditions de L'hydratation sur la Qualité Technologique du Couscous. Master's Thesis, INA El-Harrach, Oued Smar, Algeria, 2005.
35. Lefkir, S.; Karima, K.; Yesli, A.; Ounane, G. Hydration rate influence on the couscous quality. *J. Food Agric. Environ.* **2017**, *15*, 5–11.
36. Mandato, S.; Cesar Taliania, C.; Aït-Kaddour, A.; Ruiz, T.; Cuq, B. In-line monitoring of durum wheat semolina wet agglomeration by near infrared spectroscopy for different water supply conditions and water addition levels. *J. Food Eng.* **2013**, *119*, 533–543. [CrossRef]
37. Bellocq, B.; Ruiz, T.; Delaplace, G.; Duri, A.; Cuq, B. Impact of rolling conditions on the characteristics of unsaturated wet agglomerates of semolina. *J. Food Eng.* **2016**, *195*, 235–246. [CrossRef]
38. Quaglia, G.B. Other durum wheat products. In *Durum Chemistry and Technology*; Fabriani, G., Lintas, C., Eds.; AACC: St. Paul, MN, USA, 1988; pp. 263–282.
39. Debbouz, A.; Donnelly, B.J. Process effect on couscous quality. *Cereal Chem.* **1996**, *73*, 668–671.
40. Guezlane, L.; Morel, M.H.; Abecassis, J. Effect du traitement hydrothermiques du couscous de blé dur sur les modifications physiques de l'amidon. *Ann. Inst. Nat. Agron.* **1998**, *19*, 62–71.
41. Guezlane, L.; Morel, M.H.; Abecassis, J. Effect du traitement hydrothermiques du couscous de blé dur sur le comportement des proteins. *Ann. Inst. Nat. Agron.* **1998**, *19*, 72–81.
42. Yüksel, N.A.; Oner, M.D.; Bayram, M.; Oner, M.E. Rediscovery of Couscous in the World. *Glob. J. Med. Res. Nutr. Food Sci.* **2018**, *18*, 24–30.
43. D'Egidio, M.G.; Pagani, M.A. Pasta and couscous: Basic foods of Mediterranean tradition. *Tecnica Molit. Int.* **2010**, *6*, 105–115.
44. Angar, O.; Belhouchet, L. Granulométrie du Couscous: Relation avec Quelques Paramètres de Fabrication et la Qualité Culinaire. Ph.D. Thesis, Université de Constantine, Constantine, Algeria, 2002.
45. Derouiche, M. Couscous: Enquête de Consommation a Constantine, Fabrication Artisanale et Qualité. Master's Thesis, Université Constantine 1, Constantine, Algeria, 2003.
46. Belaïd, M.R.; Khendek, D.; Belarbi, A. Rôle des Monoglycérides dans L'expression de la Qualité du Couscous de Blé Dur. Complexation Amylose-Monoglycérides. Ph.D. Thesis, INA El Harrach, Oued Smar, Algeria, 1994.
47. Yousfi, L. Influence des CONDITIONS de Fabrication sur la Qualité du Couscous Industriel et Artisanal. Master's Thesis, Université de Constantine, Constantine, Algeria, 2002.
48. Feliachi, K.; Guerfi, N. Influence de L'incorporation de la Matière Grasse Durant la Cuisson sur la Qualité Culinaire du Couscous. Ph.D. Thesis, Université Mentouri Constantine, Constantine, Algeria, 2003.
49. Ounane, G.; Cuq, B.; Abecassis, J.; Yesli, A.; Ounane, S.M. Effects of physio-chemical characteristics and lipid distribution in Algerian durum wheat semolinas on the technological quality of couscous. *Cereal Chem.* **2006**, *83*, 377–384. [CrossRef]
50. Mezroua, L. Etude de la Qualité Culinaire de Quelques Couscous Industriels et Artisanaux et Effet D'adjonction de la Matiére Grasse Durant la Cuisson. Master's Thesis, Université Mentouri Constantine, Constantine, Algeria, 2011.
51. Boudreau, A.; Matsuo, R.; Laing, W. L'industrie des pâtes alimentaires. In *Le Blé. Éléments Fondamentaux et Transformation*; Boudreau, A., Menard, G., Eds.; Les Presses de l'Université: Laval, QC, Canada, 1992; pp. 193–223.

52. Feillet, P. *Le Grains de Blé, Composition et Utilisation*; INRA: Paris, France, 2000.
53. Leygues, J.P.; Samson, M.F.; Barron, C.; Lullien-Pellerin, V.; Abecassis, J.; Cuq, B.; Crepon, K.; Braun, P. La transformation: Différentes opérations pour amener le grain du silo de collecte à l'assiette du consommateur. In *Blé Dur: Synthèse des Connaissances pour une Filière Durable*; Abécassis, J., Massé, J., Allaoua, A., Eds.; Éditions Quae Arvalis: Versailles, France, 2021; pp. 199–237.
54. Hébrard, A. Agglomération de la Semoule de Blé Dur en Couscous. Ph.D. Thesis, École Nationale Supérieure Agronomique de Montpellier, Montpellier, France, 2002.
55. Guezlane, L.; Abecassis, J. Méthodes d'appréciation de la qualité culinaire du couscous de blé dur. *Ind. Aliment. Agric.* **1991**, *108*, 966–971.
56. Idir, D. Influence du Taux D'extraction et de la Granulométrie de Semoule sur la Qualité Technologique du Couscous de Blé Dur. Master's Thesis, INA El-Harrach, Oued Smar, Algeria, 2000.
57. Pons, M.N.; Vivier, H.; Belaroui, K.; Bernard-Michel, B.; Cordier, F.; Oulhana, D.; Dodds, J.A. Particle morphology: From visualisation to measurement. *Powder Technol.* **1999**, *103*, 44–57. [CrossRef]
58. Hafsa, I.; Mandato, S.; Ruiz, T.; Shuck, P.; Jeantet, R.; Mejean, S.; Chevallier, S.; Cuq, B. Impact of the agglomeration process on structure and functional properties of the agglomerates based on the durum wheat semolina. *J. Food Eng.* **2015**, *145*, 25–36. [CrossRef]
59. Bekradouma, D. Influence des Traitements Hydrothermiques de Précuisson et de Séchage sur la Qualité Nutritionnelle du Couscous de Blé Dur. Master's Thesis, Université de Montpellier, Montpellier, France, 1992.
60. Boutrif, E.; Morse, R.E. Aflatoxin production on couscous. *Can. Inst. Food Sci. Technol.* **1976**, *9*, 186–188. [CrossRef]
61. Rahmani, N.; Muller, H.G. The fate of thiamin and riboflavin during the preparation of couscous. *Food Chem.* **1996**, *55*, 23–27. [CrossRef]
62. Doukani, K. Etude comparative entre le couscous industriel et le couscous à base de glands. *Nat. Technol. B-Sci. Agric. Biol.* **2015**, *13*, 2–11.
63. Messia, M.C.; Oriente, M.; Angelicola, M.; De Arcangelis, E.; Marconi, E. Development of functional couscous enriched in barley β-glucans. *J. Cereal Sci.* **2019**, *85*, 137–142. [CrossRef]
64. Cankurtaran, T.; Bilgiçli, N. Improvement of functional couscous formulation using ancient wheat and pseudocereals. *Int. J. Gastron. Food Sci.* **2021**, *25*, 100400. [CrossRef]
65. Pagani, M.A.; Bottega, G.; Mariotti, M.; Caramanico, R.; Lucisano, M.; Marti, A. Characteristics of couscous samples prepared with different semolina and process parameters. In Proceedings of the AACC International Annual Meeting 2009, Baltimore, MD, USA, 13–16 September 2019.
66. Khendek, D.; Guezlane, L. Rôle des monoglycérides dans l'expression de la qualité technologique du couscous industriel de blé dur. *Céréaliculture* **1994**, *32*, 10–14.
67. Debbouz, A. Influence of Raw Material and Processing on Couscous Quality. Ph.D. Thesis, North Dakota State University, Fargo, ND, USA, 1992.
68. Yettou, N. Les Méthodes Instrumentales D'appréciation de la Qualité Culinaire du Couscous de Blé Dur. Master's Thesis, INA El-Harrach, Oued Smar, Algeria, 1998.
69. Benayad, A.; Taghouti, M.; Benali, A.; Benbrahim, N.; Aboussaleh, Y. Development and nutritional, technological, microbiological, cooking and sensory characterization of durum wheat couscous partially enriched with lentil semolina. *Food Biosci.* **2021**, *42*, 101062. [CrossRef]
70. Tigroudja, F.; Benjouddiouada, A. Influence de la Granulométrie de la Semoule sur la Qualité Technologique du Couscous Artisanal de Blé Dur. Ph.D. Thesis, INA El-Harrach, Oued Smar, Algeria, 1999.
71. Boudouira, N. Mise au Point D'une Méthode D'approche des Interactions Impliquées dans la Constitution D'agglomérats de Couscous. Master's Thesis, Université Constantine 1, Constantine, Algeria, 2014.
72. Yettou, N.; Ait Kaci, M.; Guezlane, L.; Ait Amar, H. Détermination des caractéristiques viscoélastiques du couscous cuit au moyen du viscoélastograghe Chopin. *Ind. Aliment. Agric.* **1997**, *12*, 844–847.
73. Badaoui, D. Influence du Séchage sur la Qualité du Couscous Industriel. Ph.D. Thesis, INA El Harrach, Oued Smar, Algeria, 1984.
74. Drechsler, K.C.; Bornhorst, G.M. Modeling the softening of carbohydrate-based foods during simulated gastric digestion. *J. Food Eng.* **2018**, *222*, 38–48. [CrossRef]
75. Gândido, G.S. The secret of a couscous: Food and identity. *Demetra* **2015**, *10*, 607–621.
76. Franconie, H.; Chastanet, M.; Sigaut, F. *Couscous, Boulgour et Polenta—Transformer et Consommer les Céréales dans le Monde*; Karthala: Paris, France, 2010.
77. Syndicat des Industriels Fabricants de Pâtes Alimentaires de France. Available online: https://www.cfsi-sifpaf.com/sifpaf-chiffres-cles.php (accessed on 15 September 2021).
78. Rambourg, P. Le Couscous. Prenez-en de la Graine! Historia Number 856. 2018. Available online: https://www.historia.fr/gastronomie/le-couscous-prenez-en-de-la-graine (accessed on 15 September 2021).
79. Le Couscous un peu Italien, Espagnol et Français Aussi. Available online: https://www.lexpress.fr/actualites/1/culture/le-couscous-un-peu-italien-espagnol-et-francais-aussi_1984425.html (accessed on 15 September 2021).
80. Attoubata, K. Couscous Marocain, la Surprenante Traversée de l'Atlantique Jusquà Devenir un Symbole Culinaire au Brésil. Available online: https://maptourisme.ma/gastronomie/couscous-marocain-la-surprenante-traversee-de-latlantique-jusqua-devenir-un-symbole-culinaire-au-bresil/ (accessed on 15 September 2021).

foods

Review

Assessing the Rheological Properties of Durum Wheat Semolina: A Review

Cristina Cecchini [1], Andrea Bresciani [2], Paolo Menesatti [3], Maria Ambrogina Pagani [2] and Alessandra Marti [2,*]

1 Consiglio per la Ricerca in Agricoltura e L'analisi Dell'economia Agraria (CREA), Centro di Ricerca Ingegneria e Trasformazioni Agroalimentari, Via Manziana 30, 00189 Rome, Italy; cristina.cecchini@crea.gov.it
2 Department of Food, Environmental and Nutritional Sciences (DeFENS), Università Degli Studi di Milano, Via G. Celoria 2, 20133 Milan, Italy; andrea.bresciani@unimi.it (A.B.); ambrogina.pagani@unimi.it (M.A.P.)
3 Consiglio per la Ricerca in Agricoltura e L'analisi Dell'economia Agraria (CREA), Centro di Ricerca Ingegneria e Trasformazioni Agroalimentari, Via Della Pascolare 16, Monterotondo, 00015 Rome, Italy; paolo.menesatti@crea.gov.it
* Correspondence: alessandra.marti@unimi.it

Abstract: Empiric rheology is considered a useful tool for assessing the technological quality of wheat. Over the decades, several tests have been adapted from common to durum wheat, and new approaches have been proposed to meet the needs of the players of the durum wheat value chain. Breeders are looking for reliable methods to test the functional quality of wheat lines at early stages, where there are limited amounts of sample; millers need fast and reliable methods for checking wheat quality right at the point of the receiving station; and pasta-makers are looking for suitable methods to predict end product quality. This review provides an overview of the strengths and weaknesses of the rheological tests currently used to evaluate the quality of durum wheat semolina, with the emphasis on Europe. Moreover, the relationships among the parameters obtained from different rheological approaches are extrapolated from the literature and integrated with the data obtained from 74 samples of durum wheat semolina. Although numerous efforts have been made to propose rapid and reliable tests for semolina characterization, the ideal test has yet to be proposed, indicating that researchers and pasta companies need to focus on perfecting the way to assess the quality of durum wheat and pasta.

Keywords: durum wheat; semolina; gluten quality; protein network; rheology; pasta

Citation: Cecchini, C.; Bresciani, A.; Menesatti, P.; Pagani, M.A.; Marti, A. Assessing the Rheological Properties of Durum Wheat Semolina: A Review. *Foods* **2021**, *10*, 2947. https://doi.org/10.3390/foods10122947

Academic Editor: Mike Sissons

Received: 28 October 2021
Accepted: 25 November 2021
Published: 30 November 2021

1. Introduction

Durum wheat semolina is considered the ideal raw material to produce dry pasta; this statement is well accepted by all the players of the durum wheat value chain, from breeders to pasta-makers and consumers. This is true not only in Italy, Greece, and France—where only semolina can be used to produce dry pasta legally—but also outside of the Mediterranean area. Specifically, the suitability of semolina for pasta-making is due to the ability of the corresponding dough to withstand the numerous physical stresses occurring during processing [1]. This property is mainly due to the quantity and quality of its protein fractions. Indeed, the combination of protein quantity and quality results in—after cooking—a continuous and coagulated protein network that surrounds the gelatinized starch granules. As described by Resmini and Pagani [2] in the 1980s, on the basis of ultrastructure observations and confirmed in more recent years by several authors [3–5], pasta is considered to be of good quality, i.e., with high firmness (i.e., the degree of resistance to the first bite) and no stickiness (i.e., the adhesion rate of pasta to tongue, teeth, palate, and/or fingers) and no or minimal bulkiness (i.e., the adhesion rate of cooked pasta strands among them), if after cooking, a three-dimensional, continuous, almost non-deformable and elastic protein network surrounds each starch granule. This optimal structure is guaranteed if proteins coagulate before starch swelling (due to the large availability of water and high temperatures), so that the starchy material will remain mostly trapped within the protein network, with negligible amylose release into the cooking water and equally limited quantities of amylopectin on the surface of pasta [3,6].

The quality of raw materials and processing conditions are responsible for the cooking behavior of pasta. The role of drying temperature in enhancing the formation of a regular protein structure around the starch granules has been elsewhere described [5,7–11]. However, the selection of high-quality semolina remains the objective of pasta-making companies. Although several factors contribute to the definition of "high-quality semolina", the focus of this review is the evaluation of the technological quality, i.e., the tendency to form the peculiar structure described above that leads to the typical "al dente" firmness of cooked pasta. Various approaches have been proposed—from molecular to macroscopic—to assess the technological quality of durum wheat semolina. In this context, the present work summarizes the main factors determining the technological quality of durum wheat semolina and reviews the main approaches widely used (especially in Italy and France) for gluten quality evaluation, indicating the strengths and weaknesses of each test. Moreover, the relationships among the parameters of the different rheological tests are presented and discussed.

2. Defining Gluten Quality

The amount of protein in durum wheat is the first parameter that dry pasta producers consider when choosing the raw material. According to the voluntary classification used in Italy [12] in the field of durum wheat, semolina is classified into three classes: for the lowest quality class protein content ranges from 10.5% to 11.9%; the medium class includes samples with 12.0–13.5% protein content, while the excellent semolina quality exhibits at least 13.5% protein. A high amount of protein, in fact, is the prerequisite for a dough in which the gluten matrix is sufficiently thick and well developed even in conditions of non-optimal hydration, like those used in pasta-making (i.e., 30–32% moisture content).

Although the high protein—and consequently gluten (about 30% wet basis; >11% dry basis)—content is an important quality requirement [7,10], this characteristic is not enough to guarantee the good cooking behavior of the corresponding pasta. Indeed, regardless the particle size and protein content, pasta made with common wheat differs in structure and/or firmness from that made with good durum wheat semolina. Moreover, Fuad and Prabhasankar [13] stated that the use of common wheat flour in pasta-making is associated with good cooking quality when additives and optimized technologies are used. The superiority of durum wheat over common wheat is not, in fact, only related to protein content (on average two percentage points higher than common wheat), but to the composition of protein fractions. In this regard it has been shown that the suitability of durum wheat in pasta-making is related to specific combinations of alleles at the storage protein loci: glutenin alleles at low molecular weight (LMW) locus Glu-B3 and at high molecular weight (HMW) locus Glu-B1 [14]. With regard to common wheat, HMW glutenins (HMW-GS) are crucial in guaranteeing the formation of a gluten network suitable for bread-making above all for the presence of Glu-D1 locus that is absent in durum wheat [14,15]. On the other hand, in durum wheat, the formation of a structure suitable for pasta-making is related to the high density of cross links between the shorter chains of LMW glutenins (LMW-GS) [16].

At this point it is necessary to clarify what "suitable for pasta-making" means. Several researchers have used different terms to describe the features of durum wheat gluten that mainly affect pasta quality: strength, tenacity, and elasticity (Table 1). All of them refer to the dough and/or gluten rheological properties, which describe the interactions between the different macromolecules that lead to the formation of the gluten network and, therefore, of the dough. The protein network developed during the mixing and kneading phase of pasta-processing is stabilized by both covalent bonds (disulfide bonds) and bonds of lower energy, such as hydrogen bonds and hydrophobic interactions between non-polar amino acid residues [17–19]. The adjective "strong" is often referred to gluten characterized by high tenacity and/or strength, whereas the adjective "weak" is used to describe gluten with low tenacity and/or strength, and high extensibility.

Table 1. The main attributes used to describe the properties of gluten of durum wheat dough.

Gluten Property	General Definition	Applied to Durum Wheat Dough and Pasta
Viscoelasticity	Ability of solids to have simultaneous viscous and elastic properties	The determinantal characteristic of gluten, necessary for pasta-making process
Viscosity	Resistance of a liquid to flow	It determines in which way the dough flows through the press and the dye
Elasticity	Ability of solids to recover their initial shape after deformation	It allows the mass to withstand strong compression (about 10 MPa) during the extrusion phase and to assure regular shrinkage during drying (shape maintenance)
Extensibility	Maximum degree of deformation reached by solids before breakage	Excessive extensibility doesn't counteract the mechanical stresses during processing
Tenacity	Resistance of dough to deformation	It allows the mass to resist, without breaking, the high/intense mechanical stresses (shear and stretching) occurring during the extrusion phase
Strength	Ability of solids to resist mechanical stress	It allows proteins to form a regular and continuous network that promotes good cooking quality

3. Assessing Gluten Quality

Dough is one of the most difficult materials to characterize from a rheological point of view [20]. In fact, it exhibits viscoelastic behavior defined as plastic by Bushuk [21], or in other words, its behavior ranges between that of an elastic solid and that of a viscous liquid. Moreover, the characteristics of dough change at each stage of the process (especially due to temperature changes occurring during pasta-making) and it is therefore difficult to predict its behavior during processing. This complexity justifies the development of so-called "empirical" rheological tests, which are widely used in the industry. In any case, as pointed out by Dobraszczyk [20], it is essential to define, for each processing variable (humidity, temperature, pressure, etc.), the range of values applied in the step/phase that is under investigation.

Since the 1980s, several rheological tests have been proposed to characterize durum wheat semolina and to objectively describe its pasta-making performance, as summarized in Figure 1.

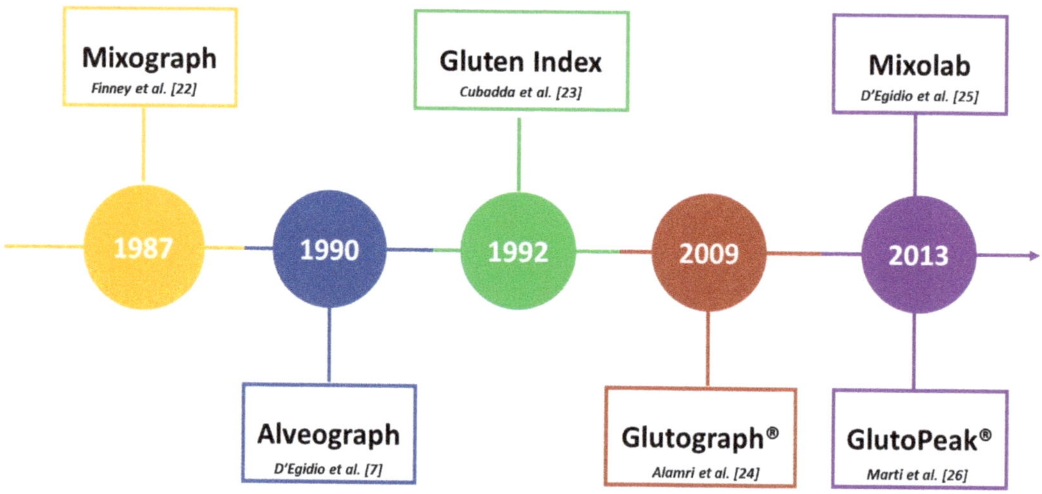

Figure 1. Time sequence of the rheological tests adopted in the durum wheat value chain [7,22–26].

In general terms, the tests used for the rheological characterization of durum wheat can be classified according to various criteria. Some of them (i.e., Gluten Index and Glutograph®tests) directly evaluate the quality of gluten after its extraction from semolina, while others are carried out on dough (i.e., Alveograph, Mixograph, and Mixolab tests) or slurry (i.e., GlutoPeak test) systems. Some of them provide information mainly about strength (i.e., Mixograph, and Mixolab tests), others also provide details on extensibility (e.g., Alveograph) or elasticity (e.g., Glutograph®test). Some of them (i.e., Alveograph, GlutoPeak) test sample breakage, others do not (i.e., Glutograph®, Mixograph, Mixolab). Some of them are used more in Europe than in the United States or Canada, and vice versa, depending on the country of the company that produced the device. For example, the Alveograph is mostly used in European countries, whereas the Mixograph is widely used in North America [27]. As the present review is focusing on the rheological approaches used in Mediterranean countries, the use of the Mixograph for semolina characterization will not be addressed. For further information, readers should see the research of Dick and Youngs [28], Rath et al. [29], Finney [30], Khatkar et al. [31], Kovacs et al. [32], and AbuHammad et al. [33].

The main approaches used for semolina characterization are summarized in Table 3. From the well-known methods to the most recent, the common goal has been to respond to the needs of the operators of the supply chain who are often asked to provide, as quickly as possible, a reliable prediction of the behavior of the raw material during both the pasta-making process and cooking. In particular, breeders need to analyze in a short time a very large number of new breeding lines and released varieties, for which the quantity of material represents a limiting factor [34]. The milling industry needs fast, simple, and reliable methods to control the quality of wheat during the reception phase, in terms of milling yield and semolina characteristics that define its commercial value. Finally, the pasta industry also needs rapid and reliable methods that determine the pasta-making ability of the semolina and predict the cooking quality of the finished product. In this context, near infrared (NIR) spectroscopy, a rapid and non-destructive technique widely used in the industry to determine moisture and protein content [35], is becoming increasingly studied as a technique to predict some of the indices expressed by rheological tests to define the technological quality of semolina [36,37]. Nevertheless, the NIR prediction of qualitative rheological parameters requires robust calibration models to extract information from the spectral data [38].

Most of the tests were proposed for the common wheat sector, so they do not simulate, either by type or intensity, the stresses that arise during pasta extrusion and drying. In some cases, the method has been adapted to measure the quality of durum wheat by making some modest/small changes (e.g., resting time of the dough in the Alveograph). It follows that the information gathered from the current tests is mainly useful for classifying semolina in broad classes (excellent, good, or poor gluten quality), whereas the screening of samples within each class is still challenging. Although the limitations of such tests are well known, most of them are widely used, as a reference in the industry, to predict the pasta-making quality of durum wheat semolina [7,10,34,39–42].

Basically, the procedures have not changed over the years, but additional data integration systems have been developed for directly processing the values of the parameters provided by the instruments. A brief description of each test will be provided in the following sections, whereas the main indices provided by each test are reported in Table 2.

Table 2. Main indices provided by the rheological approaches used for semolina characterization.

Test	Index	Description	Type of Information
Gluten Index	Value from 0 to 100	Percentage of wet gluten retained in the sieve	Gluten strength
Glutograph®	Stretching time	Time to reach deflection or value after time threshold (shear/stretch angle)	Gluten extensibility
	Relaxation	Recovery angle after 10 s of stress removal	Gluten elasticity
Alveograph	P	Maximum pressure (mmH_2O) needed to deform the dough till breakage	Dough tenacity
	L	Length of the curve (mm)	Dough extensibility
	P/L	Ratio between P and L	Balance between dough tenacity and extensibility
	W	Energy (in 10^{-4} J) required for dough deformation till breakage; area under the curve	Dough strength
	Ie	Ratio between P200 (i.e., the pressure 4 cm from the beginning of the curve) and the value of P	Dough elasticity
GlutoPeak®	Maximum consistency (BEM)	Maximum height of the peak	Consistency of gluten upon aggregation
	Peak maximum time (PMT)	Time required to reach the maximum height	Time for gluten aggregation
	Aggregation energy	Area from 15 s before to 15 s after the maximum peak	Gluten strength
	Total energy	Area from 0 s before to 15 s after the maximum peak	Gluten strength
Mixolab	Water absorption	Amount of water to add to semolina to reach an optimal consistency of 1.10 Nm (C1)	The higher the value, the higher protein quantity/quality
	Development time	Time needed to reach C1	The higher the value, the higher protein quantity/quality
	Stability	Time around C1 where the torque is higher or equal to the real value of C1–C1*11%	Dough resistance to mixing
	Torque C2	The lowest point of the curve when the device starts heating the dough	Weakening of protein
	C1–C2	Difference between Torque C1 and C2	Gluten strength
	Torque C3	The maximum torque obtained after C2 during the heating phase.	Starch gelatinization
	Torque C4	The minimum torque after the holding period at 90 °C	Stability during heating and mixing
	Torque C5	Torque at the end of the test	Starch retrogradation tendency

P, maximum pressure; L, maximum length; P/L, pressure:length ratio; W, area under the curve; Ie, P200/P (P200: pressure at 4 cm from the beginning of the curve).

Table 3. Rheological approaches used for semolina characterization.

Test	Principle	Hydration Level	Features	Standard Method for Durum Wheat Semolina
Gluten Index	Gluten ability to pass through a sieve after centrifugation	not required	- Short time for analysis (10 min) - Small amount of sample (10 g) - Need to extract gluten - Overestimation of the value in case of low protein content samples - Low capacity of discriminating semolina of medium quality	Yes [43,44]
Glutograph®	Gluten resistance to stretching	not required	- Short time for analysis (20 min, including extraction and resting time) - Small amount of sample (10 g) - Need to extract gluten - High variability	No
Alveograph	Dough resistance to tridimensional extension	≈52 g water/100 g semolina (14% moisture basis)	- Long time for analysis (50 min) - Large amount of sample (250 g) - High influence of the analyst - Widely used in the field, especially in Europe	Yes [45]
GlutoPeak®	Aggregation kinetics of gluten proteins	≈100 g water/100 g semolina (14% moisture basis)	- Short time for analysis (5–10 min) - Small amount of sample (9 or 10 g) - Low influence of the analyst - Few available studies	No
Mixolab	Dough resistance to both mechanical and thermal stress	≈60 g water/100 g semolina (14% moisture basis)	- Long time for analysis (45 min) - Large amount of sample (50 g) - Low influence of the analyst - Difficulty in following the set temperature profile	No

3.1. Approaches Using Extracted Gluten

Gluten Index and the Glutograph®test have in common the short time required for analysis (about 10–20 min), the small quantity of sample (10 g) and their applicability even to whole grain flours, eliminating the refinement process to obtain semolina. While providing useful information on specific properties of gluten, both tests might underestimate the effect of other wheat constituents and their interactions with proteins because they are measuring extracted gluten. On the other hand, such interactions might affect sample behavior during the pasta-making processing, and thus the quality of the final product. Although some researchers have highlighted high variability related to the extraction phase, both tests have great potential for use in breeding programs.

3.1.1. Gluten Index

The Gluten Index (GI) is a measure of the quality of gluten after mechanical extraction at room temperature: the higher the value, the stronger the gluten. This method is widely used for the screening of durum wheat varieties based on gluten strength [23], as well as in international trade specifications [27]. Briefly, wet gluten is extracted (using the

Glutomatic®instrument, Perten part of Perkin Elmer, Waltham, MA, USA) from semolina by washing both starch and soluble proteins with a sodium chloride solution during mechanical mixing to form the gluten. After which the wet gluten is centrifuged to force it through a specific sieve under standardized conditions: the percent of gluten that remains above the sieve corresponds to the GI. Durum wheat and/or semolina with GI values higher than 80 are the preferred raw materials to produce high quality pasta [33]. A negative correlation was found between GI and both gliadin content and the gliadin to glutenin ratio [46–48]: semolina with high gliadin/glutenin ratios are more extensible, resulting in low GI values. On the other hand, a strong relationship of GI to unextractable polymeric protein was found [47–49].

Conflicting results were found for the relationship between GI values and HMW-GS/LMW-GS ratio: Sissons et al. [47] highlighted a positive correlation, while Edwards et al. [49] found that high proportions of HMW-GS consistently corresponded with low GI values. These findings are in agreement with the statement that in durum wheat the formation of a well-developed network would preferentially involve LMW-GS over HMW-GS: the shorter chain lengths result in greater density of cross links for a given volume and therefore impart greater strength [16].

GI is relatively independent of protein content [10,41]. On the other hand, a negative correlation was found between dry gluten content and GI (average value of $r = -0.506$; $p < 0.01$; [50]) with a possible over-estimation of the index itself, probably attributable to purely mechanical causes. A lower gluten mass encounters a lower centrifugal force as compared to a higher gluten mass (thus a higher percentage of wet gluten remained on the sieve), resulting in a higher GI [51]. Furthermore, some authors report that samples characterized by a very tenacious gluten (such as durum with Glu-D1 HMW-GS 5+10) often fails to form a gluten ball, so this test is able to give no data [52].

Another drawback of this method is its low capacity to differentiate semolina of medium quality. Indeed, for semolina samples with a GI in the 30–65 range, the GI did not show any significant correlation with quality attributes (i.e., firmness, stickiness, bulkiness, and overall quality) of pasta dried using a low temperature drying cycle [42].

Interestingly, the test is not influenced by the extraction rate of the semolina: indeed, although the higher the extraction rate, the higher the protein content, proteins present in bran are not gluten proteins [53]. More recently, the test was successfully used to assess the quality of old cultivars compared to modern ones: the latter showed stronger gluten than the former due to both their genotypic and phenotypic differences [54].

3.1.2. Glutograph®

The Glutograph®(Brabender, Duisburg, Germany) device measures the extensibility and elasticity of gluten quantifying its resistance to stretching and its recovery. Although widely used by industries to evaluate semolina quality, the Glutograph®is rarely mentioned in the literature. The measuring system of the instrument consists of two parallel, round, finely corrugated plates set at a pre-determined distance. The analytical conditions for this test are those indicated in the manufacturer's manual (Brabender, Germany) as there are no official methods. During the test, while the upper plate remains still, the lower plate is turned with a constant moment till a fixed angle is reached (i.e., 800 BU) (stretching phase). This constant force determines the deformation of the dough. After stretching, the force is released for 10 s (relaxing phase) and the sample contracts according to its elasticity. Strong gluten requires prolonged "stretching" times and low relaxation values compared to weak gluten (Figure 2). Based on the results of the present study and in agreement with those obtained by AbuHammad et al. [33], very strong gluten exhibits stretching time >75 s, whereas values between 30 and 74 s are typical of strong gluten. On the other hand, moderately good gluten and weak gluten show stretching time of 12–29 s and <11 s, respectively.

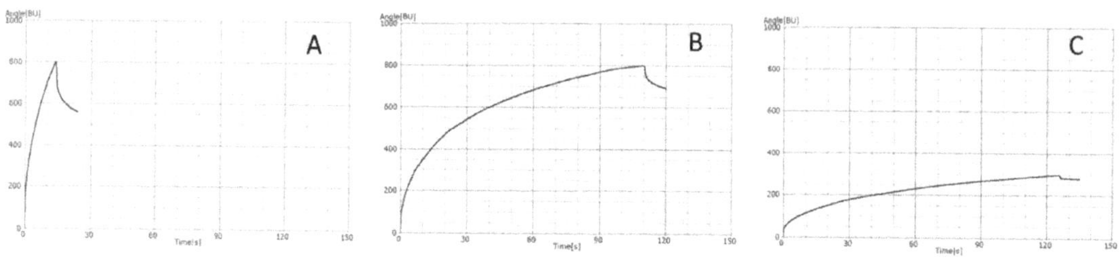

Figure 2. Glutograph®profile of semolina with poor (**A**) and good (**B**,**C**) quality. Strong gluten (**B**): stretching angle expressed in seconds; very strong gluten (**C**): stretching angle expressed in Brabender units (BU).

On the other hand, relaxation values are not a good indicator of gluten strength and cannot be used to differentiate cultivars according to their gluten quality.

Moreover, in case of tenacious gluten, poor repeatability and reproducibility of the results are observed. Furthermore, for very tenacious gluten the stretching angle (800 BU) is not reached during the shear phase (Figure 2C). In this case the result is counted as the stretching angle reached at the stretch abort time (125 s) and is expressed in BU. Therefore, the results of strong gluten expressed with different units (for example, the samples in Figure 2B,C) are difficult to compare. In addition, an unusually high coefficient of variability for the indices was observed compared with other parameters, likely due to either the high level of variability among cultivars or execution of test procedures [33].

3.2. *Approaches on Dough System*

Although using different hydration levels, the Alveograph and the Mixolab provide information on dough behavior during specific stresses (Table 3). The Alveograph, while requiring a large amount of semolina (250 g), is widely used internationally in the rheological characterization of doughs due to its ability to simultaneously define dough strength and extensibility. Both approaches require a long run time for each sample (45–50 min) which makes them unsuitable for rapid evaluation of gluten quality, as required by the industry and breeding programs.

3.2.1. Alveograph

The Alveograph test (Chopin, Villeneuve-la-garenne, France) was developed for the characterization of common wheat flour. Widely used in Europe, it evaluates dough resistance to three-dimensional expansion, thus simulating biological leavening and, therefore, the development dough volume due to the accumulation of carbon dioxide produced by yeasts. Nevertheless, it can be applied to durum wheat semolina by increasing the kneading time (from 8 to 26 min). In this case, the Alveograph test could provide information on the ability of the dough to withstand mechanical stress during pasta processing. The pressure promoted by air insufflation that is necessary for the blowing—until breakage—of a dough disc, is measured and recorded as an alveogram, yielding the indices reported in Table 3.

Figure 3 reports an example of graphs for semolina samples with poor (Figure 3A) and good (Figure 3B) pasta-making performances. In the pasta-making sector, high P/L values (i.e., >1) are associated with strong gluten, while low values (i.e., <0.5) indicate weak gluten, not suitable for pasta production.

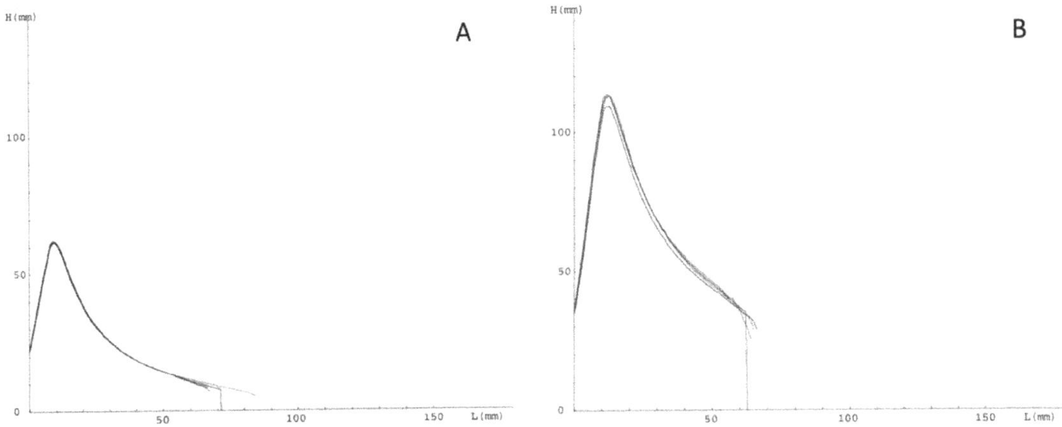

Figure 3. Alveograph profile of semolina with poor (**A**) and good (**B**) quality. L, length; H, height.

The W value can be considered a measure of the gluten network quality: high W values are associated with the formation of a strong network able to retain starch granules during cooking. Therefore, the W value is considered a valid parameter to predict the cooking quality of pasta [7,32,33]. Recently, Phakela et al. [55] found a negative correlation between both the HMW glutenins and α- and ω-gliadins with dough strength (W). On the other hand, W was positively correlated with the γ-gliadins. As regards dough extensibility (L), it was negatively correlated with LMW-GS.

The test is carried out at a constant level of hydration. In the case of durum wheat dough, this aspect might be critical since it does not consider the influence of some characteristics of the raw material (including protein amount and damaged starch content) on dough consistency and its ability to absorb and retain water [56]. Thus, in the case of strong flours, the high P value might be due to the incompletely and insufficiently homogeneous hydrated protein matrix. The hydration level reached in the Alveograph test (about 52% for semolina with 14% moisture content) does not guarantee the complete hydration of durum wheat proteins but is closer to the moisture used in pasta processing (water is added to semolina to obtain a mass of 30–32% moisture).

3.2.2. Mixolab

Mixolab (Chopin, France) is used to measure the rheological properties of a dough subject simultaneously to mechanical kneading and heating with a temperature gradient. This approach is potentially capable of giving information on both protein and starch properties in a single analysis [57]. In addition to the "water absorption" index—which is of particular interest for common wheat—the test provides indications on dough behaviour during mixing, therefore, on the strength of the gluten network, on the effect of amylase activity, as well as on the gelatinization and retrogradation of starch (Figure 4). The method was initially developed for the evaluation of dough from common wheat flour, but it was also adapted to characterize durum wheat semolina [25]. The correlations between other rheological tests and Mixolab were reported by D'Egidio et al. [25], particularly the parameters related to the protein component (stability, C2 and C1–C2) showed a correlation with protein content and Alveograph W. Good quality semolina samples show higher stability during mixing than poor semolina (Figure 4).

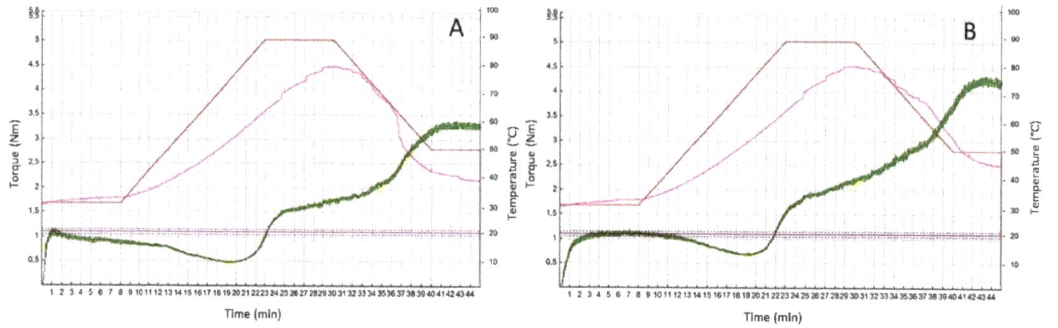

Figure 4. Mixolab profile of semolina with poor (**A**) and good (**B**) quality. Green line: torque; red line: temperature profile; purple line: dough temperature.

Protein and starch are related to each other as the intensity of the gelatinization of starch is inversely related to the protein and gluten content. The swelling rate and gelatinization level both depend on the availability of water in the dough [58], as well as the formation of the gluten network, whereas a higher protein level could signify less water availability for starch gelatinization [25].

Recently, Mixolab®has been successfully used to quickly detect the damage caused by sunn pests in durum wheat [59]. Moreover, Torbica et al. [60] applied the Mixolab to characterize fourteen durum wheat breeding lines grown during two production years with different climate conditions: genotypes greatly affected indices related to protein quality, while the production year influenced indices related to starch.

3.3. Innovative Approach: The GlutoPeak®

GlutoPeak®(Brabender, Germany) has recently been proposed for the evaluation of wheat quality by determining the aggregation properties of gluten. Compared to conventional tests (Table 3), the analysis performed with GlutoPeak®has several advantages, in terms of the quantity of sample required (<10 g), analysis time (<5 min), ease of use and operator influence (very low). It measures the aggregation behaviour of gluten when water (in excess) is added and mixed at a high speed (up to 3000 rpm).

The curve is characterized by an increase in consistency up to a peak (also called BEM, and expressed in GlutoPeak Units, GPU) that corresponds to the maximum gluten aggregation. The time of maximum consistency is called peak maximum time (PMT). After this point, the consistency decreases following the breaking of the gluten network due to intense mechanical action. Generally, low values of BEM and PMT indicate poor aggregation properties, and thus low pasta-making performance [26,61], as reported in Figure 5.

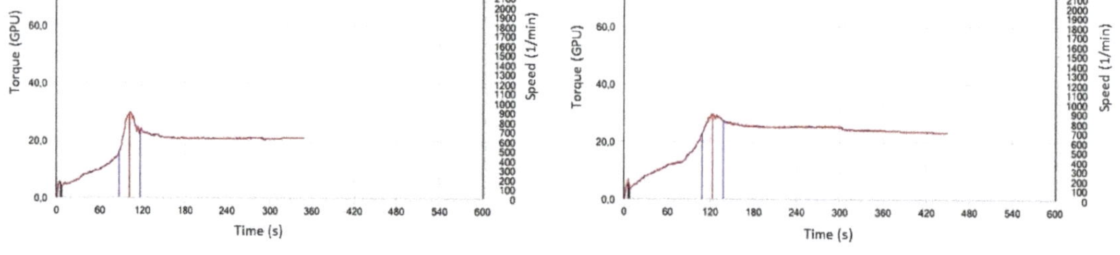

Figure 5. GlutoPeak profile of semolina with poor (**A**) and good (**B**) quality.

Until now there has been no official methodology for this test. From the literature it emerges how each researcher adopts different analysis conditions (i.e., sample:water ratio, paddle speed, temperature, and solvent type) and that the latter can change according to the type of sample. It is therefore difficult to compare the results obtained from different papers and their interpretation. For example, Marti et al. [42], when proposing the use of the energy index for the first time (area underlying the curve, up to 5 min), expressed this index in arbitrary units. This parameter could discriminate semolina on the basis of pasta performance [42]. With updated software, the energy index is automatically calculated and expressed as aggregation energy (i.e., the area under the curve 15 s before and 15 s after the BEM) or as total energy (i.e., the area under the curve from the beginning until 15 s after the BEM). More recently, Sissons and Smit [62] proposed the use of an index not currently provided by the software but interesting to evaluate gluten strength: the gluten strength index (GSI), obtained as a product of BEM and total energy.

As regards applications, Grassi et al. [61] adopted the GlutoPeak test for differentiating durum wheat cultivars based on the glutenin to gliadin ratio. Specifically, high and medium-high quality varieties were differentiated from those of low and medium-low quality based on the aggregation energy and BEM.

Some authors [62,63] have tried to optimize the analysis conditions regarding the paddle rotation speed (1900–2700) and the semolina:water ratio (7/10:10). The best results, in terms of low coefficients of variation for the PMT and total energy indices, were obtained using semolina:water ratio of 9:10 and a rotation speed of 2700 rpm. Lower rotation speeds (for example 1900 rpm) are better to discriminate semolina samples of different strengths, but they result in higher variability [62].

When applied to wholemeal semolina, the GlutoPeak test could be considered as a useful tool in the genetic selection phase of durum wheat lines [62]. Wholemeal semolina showed a shorter PMT and a higher BEM compared to refined semolina; however, significant correlations were found for PMT (r = 0.816), total energy (r = 0.814), and GSI (r = 0.804) indices obtained from refined semolina and those obtained from wholemeal [62]. The analysis on wholemeal would further reduce analysis times related to the preparation of the refined sample in genetic selection studies. In addition, the use of wholemeal would limit the variability linked to semolina particle size, an aspect of great importance in the durum wheat sector.

3.4. Non-Rheological Approach: The Sedimentation Test

Although this review is an overview of rheological tests to evaluate the quality of gluten, we cannot omit the sodium dodecyl sulphate (SDS) sedimentation test, still one of the most popular and used approaches. Unlike the other tests previously described, it is a physical–chemical method that provides indications on the quantity and quality of those protein fractions that define the characteristics of gluten. The test, in fact, is based on the property of gluten proteins to swell and flocculate in an acid medium. Under specific conditions, a suspension of wholemeal in a lactic acid-SDS solution forms a sediment whose volume represents the sedimentation index [63,64]. When the volume (or index) is high, the sedimentation is slow, and the quality of the flour is better. SDS values of 30–40 mL indicate good quality gluten, and values greater than 40 mL indicate excellent quality and, therefore, strong gluten [33].

The SDS test was initially developed for the evaluation of the baking quality of common wheat [64]. Subsequently, Dexter et al. [65] applied the test on durum wheat: by increasing the SDS concentration, the absolute values changed, but the qualitative differences among the samples were maintained. The increase in the SDS concentration therefore promotes greater differences in sedimentation volumes among durum wheat of different quality, allowing better differentiation among samples of similar quality.

The test is commonly used as a rapid method in quality controls and to predict gluten quality in wheat selection programs in early generations when the quantity of seed is a limiting factor. The test, in fact, requires a lower quantity of sample (6 g) and fewer manual skills; a "micro-method" was also developed, applicable on just 1 g of wholemeal flour [66].

4. Relation among the Main Rheological Approaches and Relevance for Cooking Quality

Assessing the quality of gluten is an old but still relevant topic. Indeed, the selection of new lines/varieties as well as the impact of climate change on the quality of wheat crops and, consequently, on gluten quality, account for the number of studies evaluating the pasta-making potential of durum wheat samples. At the same time, several studies aimed at correlating the parameters obtained from different rheological approaches. Such studies are certainly not a pure publication exercise, but are driven by various aspects, for example:

(I) Every time a new device appears on the market, it is necessary to verify its reliability by correlating its indices with those obtained by well-established, conventional approaches;

(II) Each rheological approach provides information on a specific gluten attribute (e.g., elasticity, tenacity, extensibility, and strength); hence, the need to evaluate semolina quality by using all the available rheological tests and/or to find one approach that in the shortest possible time provides information that can be correlated to as many attributes/indices as possible.

Thus, Pearson coefficients and their significance have been extrapolated from the studies on the rheological properties of durum wheat semolina and summarized in Table S1. The bibliographic data were integrated with the results obtained by applying the main tests described in the previous section (i.e., Gluten Index, Glutograph®, Alveograph, GlutoPeak®, and SDS) to a set of 74 samples of durum wheat semolina from Italian experimental trials of varietal comparison conducted in the agricultural year 2016/2017.

SDS and GI are widely used among breeders to select durum wheat varieties [67]. If a positive correlation is found between SDS and protein content, the GI appears to be relatively independent of proteins. Both SDS and GI are significantly correlated to various parameters for evaluating the rheological quality of semolina, but the most interesting correlations were observed with the Alveograph indices [7,10,32,34]. Nevertheless, as genotype x environment interaction significantly affects both Alveograph indices and GI, these tests should be carried out on samples from different environments [33]. Moreover, some authors [48,68] pointed out that Alveograph indices do not seem to distinguish the contribution of the amount of protein from its quality. In other words, a high value for Alveograph strength (W) may be related either to the high percentage of protein or to the high quality of the protein network [69]. This issue seems to be obviated when gluten viscoelasticity is assessed by the Glutograph test. Correlated with SDS, GI, W, and P values, the stretching value is a good indicator of gluten strength (Table S1). However, analyzing extracted gluten instead of semolina dough might 'hide' the potential role of other compounds—as well as their interactions with proteins—in defining the technological potential of semolina samples.

Among the GlutoPeak indices, the area under the curve—which takes into consideration both the peak torque and peak time—seems to be the most indicative index, since it is correlated with W index. As regards the GSI proposed by Sissons and Smit [62], although correlated with dough strength (W), correlations with pasta cooking quality remain to be investigated. In this context, some studies showed a negative correlation between PMT and pasta stickiness and bulkiness [42,70]; although, more samples need to be evaluated.

Although all operators in the durum wheat supply chain support the use of rheological tests for predicting the quality of cooked pasta, only a few studies showed relationships between semolina and pasta quality. As regards the GI, it is positively correlated with cooked firmness [33]. According to Alamri et al. [71] some Glutograph indices are positively correlated with cooking quality (i.e., stretching time versus cooking loss and firmness), whereas others exhibit a negative correlation (i.e., relaxation versus cooking loss). Alveograph indices are the most frequently related to pasta cooking behavior. In particular, the W parameter was significantly correlated with firmness tested by devices for texture analysis [33] as well as the quality judgment expressed by a trained panel [7,32]. Among the latest approaches, GlutoPeak test measurements of maximum torque and energy provided information on firmness, stickiness, and bulkiness of cooked pasta [40,70]. The suitability of this rheological approach in predicting the stickiness of cooked pasta was also confirmed by Sissons [70]. Finally, although the quantity and quality of the proteins in the raw material are universally considered the crucial properties in determining the cooking quality of pasta, the role of starch in cooking behavior should not be forgotten. This is confirmed by the results of D'Egidio et al. [25], which highlight how the C3 parameter (related to starch gelatinization) of Mixolab test is negatively correlated to bulkiness and to the overall judgment of the pasta by a trained panel. The number of analyzed samples (generally low), differences in the characteristics of the raw materials (i.e., gluten content and quality, as well as amylose content) differences in pasta-making conditions (e.g., extrusion pressure, and drying temperature), in cooking procedures (ratio pasta:water, optimal or pre-fixed cooking time, etc.) and in methods used for cooked pasta evaluation (sensory evaluation by trained personnel or devices) among the studies might account for the difficulty in determining relationships between semolina and pasta quality.

5. Conclusions

The assessment of semolina quality continues to be of interest to researchers and pasta companies, suggesting that the ideal test to determine pasta-making potential has not yet been found. However, in the last few years, numerous efforts have been made to propose rapid and reliable tests for semolina quality. For most of them, the lack of a standard method limits their diffusion. Furthermore, some tests are based on the evaluation of the gluten extracted from the dough; this approach is controversial as the extraction of a component can alter and modify real interactions between the different (macro)molecules of the "native" system. Moreover, almost all the rheological tests adopted so far in the pasta-making sector derive from tests developed for bread-making using common wheat flour, simulating the phenomena occurring in that process. In addition to the different particle size between semolina and flour, which is not a secondary parameter in influencing the rheological behavior of a raw material, the conditions adopted in pasta-making and bread-making are very different, both in terms of hydration level during kneading, and of the type and intensity of physical stresses developing during processing. Thus, the relation between the rheological properties of semolina and pasta quality is often weak. Furthermore, although the positive correlations between rheological properties and cooking quality reported by some authors are significant, they remain relatively weak, indicating the considerable variation between measurements to test quality and pasta quality as perceived by consumers. It should be noted that the studies that have been carried out on durum wheat varieties and pasta are generally lab-scale in dimension, whereas industrially, a mixture of varieties is processed into making pasta.

In addition, since each test addresses a specific gluten property (e.g., tenacity, elasticity, etc.), several authors have proposed a correlation among the indices obtained by the various approaches. A multivariate approach might help in identifying which attributes best differentiate pasta samples according to gluten quality. In addition to gluten, starch is also involved in determining pasta quality. Thus, the relation between pasta quality and semolina pasting properties should be taken into consideration in further studies.

Over the years, breeding programs have improved the qualitative characteristics of durum wheat, resulting in varieties that are increasingly rich in proteins, making for very strong dough [72]. Consequently, the rheological tests reported in the literature for semolina may have used raw materials of poor quality. Therefore, a fast, reliable approach to predict the behavior of durum wheat semolina and the cooking quality of its corresponding pasta needs to be elaborated.

Supplementary Materials: The following are available online at https://www.mdpi.com/article/10.3390/foods10122947/s1. Table S1: Pearson's correlation coefficients between the rheological indices used to define semolina quality.

Author Contributions: Conceptualization, C.C., M.A.P. and A.M.; formal analysis, C.C. and P.M.; investigation, C.C. and A.B.; writing—original draft preparation, C.C. and A.B.; writing—review and editing, M.A.P. and A.M.; visualization, A.B. and A.M.; supervision, P.M., M.A.P. and A.M.; Funding acquisition: P.M. All authors have read and agreed to the published version of the manuscript.

Funding: This research received no external funding.

Institutional Review Board Statement: Not applicable.

Informed Consent Statement: Not applicable.

Acknowledgments: The authors thank Guido Arlotti and Alessandro D'Alessandro (Barilla S.p.A., Parma, Italy) for the valuable discussion of the content of this review. This review was part of a collaboration with Barilla S.p.A. (Parma, Italy).

Conflicts of Interest: The authors declare no conflict of interest.

References

1. Liu, C.Y.; Shepherd, K.W.; Rathjen, A.J. Improvement of durum wheat pasta making and breadmaking qualities. *Cereal Chem.* **1996**, *73*, 155–166.
2. Resmini, P.; Pagani, M.A. Ultrastructure Studies of Pasta. A Review. *Food Struct.* **1983**, *2*, 2.
3. Cunin, C.; Handschin, S.; Walther, P.; Escher, F. Structural changes of starch during cooking of durum wheat pasta. *LWT Food Sci. Technol.* **1995**, *28*, 323–328. [CrossRef]
4. Bruneel, C.; Pareyt, B.; Brijs, K.; Delcour, J.A. The impact of the protein network on the pasting and cooking properties of dry pasta products. *Food Chem.* **2010**, *120*, 371–378. [CrossRef]
5. De Noni, I.; Pagani, M.A. Cooking properties and heat damage of dried pasta as influenced by raw material characteristics and processing conditions. *Crit. Rev. Food Sci. Nutr.* **2010**, *50*, 465–472. [CrossRef]
6. Dalbon, G.; Pagani, M.A.; Resmini, P.; Lucisano, M. Einflüsse einer hitzebehandlung der weizenstärke wärhrend des trocknungsprozesses. *Getreide Mehl. Brot.* **1985**, *39*, 183–189.
7. D'Egidio, M.G.; Mariani, B.M.; Nardi, S.; Novaro, P.; Cubadda, R. Chemical and technological variables and their relationships: A predictive equation for pasta cooking quality. *Cereal Chem.* **1990**, *67*, 275–281.
8. Novaro, P.; D'Egidio, M.G.; Mariani, B.M.; Nardi, S. Combined effect of protein-content and high-temperature drying systems on pasta cooking quality. *Cereal Chem.* **1993**, *70*, 716–719.
9. Zweifel, C.; Handschin, S.; Escher, F.; Conde-Petit, B. Influence of high-temperature drying on structural and textural properties of durum wheat pasta. *Cereal Chem.* **2003**, *80*, 159–167. [CrossRef]
10. Cubadda, R.; Carcea, M.; Marconi, E.; Trivisonno, M.C. Influence of gluten proteins and drying temperature on the cooking quality of durum wheat pasta. *Cereal Chem.* **2007**, *84*, 48–55. [CrossRef]
11. Petitot, M.; Brossard, C.; Barron, C.; Larré, C.; Morel, M.H.; Micard, V. Modification of pasta structure induced by high drying temperatures. Effects on the in vitro digestibility of protein and starch fractions and the potential allergenicity of protein hydrolysates. *Food Chem.* **2009**, *116*, 401–412. [CrossRef]
12. UNI 10940:2001. Prodotti Derivati dal Grano duro Destinati alla Pastificazione—Definizione, Caratteristiche e Classi di Qualità. Available online: http://store.uni.com/catalogo/uni-10940-2001?josso_back_to=http://store.uni.com/josso-security-check.php&josso_cmd=login_optional&josso_partnerapp_host=store.uni.com (accessed on 27 October 2021).
13. Fuad, T.; Prabhasankar, P. Role of Ingredients in Pasta Product Quality: A Review on Recent Developments. *Crit. Rev. Food Sci. Nutr.* **2010**, *50*, 787–798. [CrossRef]
14. Mastrangelo, A.M.; Cattivelli, L. What makes bread and durum wheat different? *Trends Plant Sci.* **2021**, *26*, 677–684. [CrossRef] [PubMed]
15. Payne, P.I.; Nightingale, M.A.; Krattiger, A.F.; Holt, L.M. The relationship between HMW glutenin subunit composition and the bread-making quality of british-grown wheat varieties. *J. Sci. Food Agric.* **1987**, *40*, 51–65. [CrossRef]
16. Edwards, N.M.; Mulvaney, S.J.; Scanlon, M.G.; Dexter, J.E. Role of gluten and its components in determining durum semolina dough viscoelastic properties. *Cereal Chem.* **2003**, *80*, 755–763. [CrossRef]
17. Shewry, P.R.; Halford, N.G.; Belton, P.S.; Tatham, A.S. The structure and properties of gluten: An elastic protein from wheat grain. *Phil. Trans. R. Soc. Lond. B* **2002**, *357*, 133–142. [CrossRef]
18. Sissons, M.J. Role of durum wheat composition on the quality of pasta and bread. *Foods* **2008**, *2*, 75–90. [CrossRef]
19. Bonomi, F.; Ferranti, P.; Mamone, G. Wheat Flour: Chemistry and Biochemistry. In *Bakery Products Science and Technology*; Zhou, W., Ed.; Wiley Blackwell: Hoboken, NJ, USA, 2014; pp. 55–74. [CrossRef]
20. Dobraszczyk, B.J. Wheat and Flour. In *Cereals and Cereal Products Chemistry and Technology*; Aspen Publisher Inc.: Gaithersburg, MD, USA, 2001; pp. 100–113.
21. Bushuk, W. Rheology: Theory and application to wheat flour doughs. In *Rheology of Wheat Products*; Faridi, H., Ed.; AACC: St. Paul, MN, USA, 1985; pp. 1–26.

22. Finney, K.F.; Yamazaki, W.T.; Youngs, V.L.; Rubenthaler, G.L. Quality of hard, soft and durum wheats. In *Wheat and Wheat Improvement*, 2nd ed.; Heyne, E.G., Ed.; Agronomy Monograph 13; ASA; CSSA; SSSA: Madison, WI, USA, 1987; pp. 677–748.
23. Cubadda, R.; Carcea, M.; Pasqui, L.A. Suitability of the gluten index method for assessing gluten strength in durum wheat and semolina. *Cereal Foods World* **1992**, *37*, 866–869.
24. Alamri, M.S.; Manthey, F.; Mergoum, M.; Elias, E.; Khan, K. Use of the Glutograph Instrument in durum wheat quality evaluation. *Cereal Food World* **2009**, *54*, 124–131.
25. D'Egidio, M.G.; Moscaritolo, S.; Marti, A.; Pagani, M.A. Adaptation of the Mixolab for durum wheat testing. In *Mixolab a New Approach to Rheology*; Dubat, A., Rosell, C.M., Gallagher, E., Eds.; AACC: St. Paul, MN, USA, 2013; Chapter 13; pp. 89–98.
26. Marti, A.; Seetharaman, K.; Pagani, M.A. Rheological approaches suitable for investigating starch and protein properties related to cooking quality of durum wheat pasta. *J. Food Quality* **2013**, *36*, 133–138. [CrossRef]
27. Marchylo, B.A.; Dexter, J.E.; Clarke, F.R.; Clarke, J.M.; Preston, K.R. Relationships among bread-making quality, gluten strength, physical dough properties, and pasta cooking quality for some Canadian durum wheat genotypes. *Can. J. Plant Sci.* **2001**, *81*, 611–620. [CrossRef]
28. Dick, J.W.; Youngs, V.L. Evaluation of durum wheat semolina and pasta in the United States. In *Durum Wheat: Chemistry and Technology*; Fabriani, G., Lintas, C., Eds.; AACC: St. Paul, MN, USA, 1988; pp. 237–248.
29. Rath, C.R.; Gras, P.W.; Wrigley, C.W.; Walker, C.E. Evaluation of dough properties from 2 g flour using the Mixograph principle. *Cereal Foods World* **1990**, *35*, 572–574.
30. Finney, K.F. A five-gram Mixograph to determine and predict functional properties of wheat flours. *Cereal Chem.* **1989**, *66*, 527–530.
31. Khatkar, B.S.; Bell, A.E.; Schofield, J.D. A comparative study of the inter-relationships between mixograph parameters and bread making qualities of wheat flours and glutens. *J. Sci. Food Agric.* **1996**, *72*, 71–85. [CrossRef]
32. Kovacs, M.I.P.; Poste, L.M.; Butler, G.; Woods, S.M.; Leisle, D.; Noll, J.S.; Dahlk, G. Durum wheat quality: Comparison of chemical and rheological screening tests with sensory analysis. *J. Cereal Sci.* **1997**, *25*, 65–75. [CrossRef]
33. AbuHammad, W.A.; Elias, E.M.; Manthey, F.A.; Alamri, M.S.; Mergoum, M. A comparison of methods for assessing dough and gluten strength of durum wheat and their relationship to pasta cooking quality. *Int. J. Food Sci. Tech.* **2012**, *47*, 2561–2573. [CrossRef]
34. Peña, R.J. Durum wheat for pasta and bread-making. Comparison of methods used in breeding to determine gluten quality-related parameters. In *Durum Wheat Improvement in the Mediterranean Region: New Challenges*; Royo, C., Nachit, M., Di Fonzo, N., Araus, J.L., Eds.; Options Méditerranéennes: Serie A Séminaires Méditerranéennes; CIHEAM: Zaragoza, Spain, 2000; pp. 423–430.
35. EN 15948:2015. Cereals—Determination of Moisture and Protein—Method Using Near-Infrared-Spectroscopy in Whole Kernels. Available online: https://standards.iteh.ai/catalog/standards/cen/f9d75e71-8643-4bb3-8dc4-c34b5b51d890/en-15948-2015 (accessed on 27 October 2021).
36. Sinelli, N.; Pagani, M.A.; Lucisano, M.; D'Egidio, M.G.; Mariotti, M. Prediction of semolina technological quality by FT-NIR spectroscopy. *J. Cereal Sci.* **2011**, *54*, 218–223. [CrossRef]
37. Cecchini, C.; Antonucci, F.; Costa, C.; Marti, A.; Menesatti, P. Application of near-infrared handheld spectrometers to predict semolina quality. *J. Sci. Food Agric.* **2021**, *101*, 151–157. [CrossRef] [PubMed]
38. Porep, J.U.; Kammerer, D.R.; Carle, R. On-line application of near infrared (NIR) spectroscopy in food production. *Trends Food Sci. Technol.* **2015**, *46*, 211–230. [CrossRef]
39. Autran, J.C.; Abecassis, J.; Feillet, P. Statistical evaluation of different technological and biochemical tests for quality assessment in durum wheats. *Cereal Chem.* **1986**, *63*, 390–394.
40. Feillet, P.; Dexter, J.E. Quality requirements of durum wheat for semolina milling and pasta production. In *Pasta and Noodle Technology*; Kruger, J.E., Matsuo, R.B., Dick, J.W., Eds.; AACC: St Paul, MN, USA, 1996; pp. 95–131.
41. Ames, N.P.; Clarke, J.M.; Marchylo, B.A.; Dexter, J.E.; Schlichting, L.M.; Woods, S.M. The effect of extra-strong gluten on quality parameters in durum wheat. *Can. J. Plant Sci.* **2003**, *83*, 525–532. [CrossRef]
42. Marti, A.; Cecchini, C.; D'Egidio, M.G.; Dreisoerner, J.; Pagani, M.A. Characterization of durum wheat semolina by means of a rapid shear-based method. *Cereal Chem.* **2014**, *91*, 542–547. [CrossRef]
43. ICC Standard No. 158:1995. Gluten Index Method for Assessing Gluten Strength in Durum Wheat (Triticum durum). Available online: https://icc.or.at/store/158-gluten-index-method-for-assessing-gluten-strength-in-durum-wheat-triticum-durum-pdf (accessed on 26 October 2021).
44. ISO. *EN ISO 21415-2:2015. Wheat and Wheat Flour—Gluten Content—Part 2: Determination of Wet Gluten and Gluten Index by Mechanical Means*; ISO: Geneva, Switzerland, 2015.
45. UNI 10453:1995. Grano Duro e Semole. Determinazione Delle Caratteristiche Reologiche Mediante Alveografia. Available online: http://store.uni.com/catalogo/uni-10453-1995 (accessed on 27 October 2021).
46. Fois, S.; Schlichting, L.; Marchylo, B.; Dexter, J., Motzo, R.; Giunta, F. Environmental conditions affect semolina quality in durum wheat (*Triticum turgidum* ssp. durum L.) cultivars with different gluten strength and gluten protein composition. *J. Sci. Food Agric.* **2011**, *91*, 2664–2673. [CrossRef]
47. Sissons, M.J.; Ames, N.P.; Hare, R.A.; Clarke, J.M. Relationship between glutenin subunit composition and gluten strength measurements in durum wheat. *J. Sci. Food Agric.* **2005**, *85*, 2445–2452. [CrossRef]

48. Ohm, J.B.; Manthey, F.; Elias, E.M. Variation and correlation of protein molecular weight distribution and semolina quality parameters for durum genotypes grown in North Dakota. *Cereal Chem.* **2017**, *94*, 780–788. [CrossRef]
49. Edwards, N.M.; Gianibelli, M.C.; McCaig, T.N.; Clarke, J.M.; Ames, N.P.; Larroque, O.R.; Dexter, J.E. Relationships between dough strength, polymeric protein quantity and composition for diverse durum wheat genotypes. *J. Cereal Sci.* **2007**, *45*, 140–149. [CrossRef]
50. D'Egidio, M.G.; Cantone, M.T.; Gosparini, E.; Cecchini, C.; Dottori, A.; Gulli, S. Caratteristiche qualitativa delle varietà in prova nel 2004/05. *Molini d'Italia* **2006**, *1*, 87–105.
51. Acar, O.; Sanal, T.; Köksel, H. Effects of wheat kernel size on hardness and various quality characteristics. *Qual. Assur. Saf. Crop* **2019**, *11*, 459–464. [CrossRef]
52. Sissons, M.J.; Pleming, D.; Margiotta, B.; D'Egidio, M.G.; Lafiandra, D. Effect of the introduction of D-genome related gluten proteins on durum wheat pasta and bread making quality. *Crop Pasture Sci.* **2014**, *65*, 27–37. [CrossRef]
53. Dexter, J.E.; Doust, M.A.; Raciti, C.N.; Lombardo, G.M.; Clarke, F.R.; Clarke, J.M.; Marchylo, B.A.; Schlichting, L.M.; Hatcher, D.W. Effect of durum wheat (*Triticum turgidum* L. var. durum) semolina extraction rate on semolina refinement strength indicators and pasta properties. *Can. J. Plant Sci.* **2004**, *84*, 1001–1013. [CrossRef]
54. Giunta, F.; Bassu, S.; Mefleh, M.; Motzo, R. Is the Technological quality of old durum wheat cultivars superior to that of modern ones when exposed to moderately high temperatures during grain filling? *Foods* **2020**, *9*, 778. [CrossRef] [PubMed]
55. Phakela, K.; van Biljon, A.; Wentzel, B.; Guzman, C.; Labuschagne, M.T. Gluten protein response to heat and drought stress in durum wheat as measured by reverse phase-high performance liquid chromatography. *J. Cereal Sci.* **2021**, *100*, 103267. [CrossRef]
56. Peressini, D. Metodi di valutazione delle proprietà reologiche degli impasti. *Tec. Molit.* **2001**, *52*, 337–379.
57. Dubat, A. The Mixolab. In *Mixolab a New Approach to Rheology*; Dubat, A., Rosell, C.M., Gallagher, E., Eds.; AACC: St. Paul, MN, USA, 2013; Chapter 1; pp. 3–13.
58. Lee, Y.E.; Osman, E.M. Correlation of morphological changes of rice starch granules with rheological properties during heating in excess water. *J. Kor. Agric. Chem. Soc.* **1991**, *34*, 379–385.
59. Blandino, M.; Marinaccio, F.; Ingegno, B.L.; Pansa, M.G.; Vaccino, P.; Tavella, L.; Reyneri, A. Evaluation of common and durum wheat rheological quality through Mixolab®analysis after field damage by cereal bugs. *Field Crop Res.* **2015**, *179*, 95–102. [CrossRef]
60. Torbica, A.; Drašković, M.; Tomić, J.; Dodig, D.; Bošković, J.; Zečević, V. Utilization of Mixolab for assessment of durum wheat quality dependent on climatic factors. *J. Cereal Sci.* **2016**, *69*, 344–350. [CrossRef]
61. Grassi, S.; Gullì, M.; Visioli, G.; Marti, A. Gluten aggregation properties as a tool for durum wheat quality assessment: A chemometric approach. *LWT Food Sci. Technol.* **2021**, *142*, 111048. [CrossRef]
62. Sissons, M.; Smit, J. Small-scale methods to assess the gluten properties of durum wheat. *Cereal Chem.* **2018**, *95*, 456–468. [CrossRef]
63. Wang, K.; Dupuis, B.; Fu, B.X. Gluten aggregation behavior in high-shear-based glutopeak test: Impact of flour water absorption and strength. *Cereal Chem.* **2017**, *94*, 909–915. [CrossRef]
64. Axford, D.W.E.; McDermott, E.E.; Redman, D.G. Small-scale test of breadmaking quality. *Mill. Feed Fertil.* **1978**, *16*, 8–20.
65. Dexter, J.E.; Matsuo, R.R.; Kosmolak, F.G.; Leisle, D.; Marchylo, B.A. The suitability of the SDS-sedimentation Test for assessing gluten strength in durum wheat. *Can. J. Plant Sci.* **1980**, *60*, 25–29. [CrossRef]
66. Dick, J.W.; Quick, J.S. A modified screening test for rapid estimation of gluten strength in early-generation durum wheat breeding lines. *Cereal Chem.* **1983**, *60*, 315–318.
67. Clarke, F.R.; Clarke, J.M.; Ames, N.P.; Knox, R.E.; Ross, R.J. Gluten index compared with SDS-sedimentation volume for early generation selection for gluten strength in durum wheat. *Can. J. Plant Sci.* **2010**, *90*, 1–11. [CrossRef]
68. Dexter, J.E.; Preston, K.R.; Martin, D.G.; Gander, E.J. The effect of protein content and starch damage on the physical dough properties and bread-making quality of Canadian durum wheat. *J. Cereal Sci.* **1994**, *20*, 139–151. [CrossRef]
69. Peressini, D. Valutazione della qualità tecnologica del frumento mediante indici reologici di tipo fondamentale. *Boll. Panta Rei* **2017**, *17*, 5–13.
70. Sissons, M. GlutoPeak: A breeding tool for screening dough properties of durum wheat semolina. *Cereal Chem.* **2016**, *93*, 550–556. [CrossRef]
71. Alamri, M.S.; Manthey, F.; Mergoum, M.; Elias, E.; Khan, K. The effects of reconstituted semolina fractions on pasta processing and quality parameters and relationship to Glutograph parameters. *J. Food Technol.* **2010**, *8*, 159–168. [CrossRef]
72. Malalgoda, M.; Ohm, J.B.; Manthey, F.A.; Elias, E.M.; Simsek, S. Quality characteristics and protein composition of durum wheat cultivars released in the last 50 years. *Cereal Chem.* **2019**, *96*, 508–519. [CrossRef]

foods MDPI

Review

Pasta-Making Process: A Narrative Review on the Relation between Process Variables and Pasta Quality

Andrea Bresciani, Maria Ambrogina Pagani and Alessandra Marti *

Department of Food, Environmental and Nutritional Sciences (DeFENS), Università degli Studi di Milano, Via G. Celoria 2, 20133 Milan, Italy; andrea.bresciani@unimi.it (A.B.); ambrogina.pagani@unimi.it (M.A.P.)
* Correspondence: alessandra.marti@unimi.it

Abstract: Pasta is an increasingly popular food worldwide and different formulations have been developed to improve its nutritional profile. Semolina that is high both in protein and gluten content is recognized as the ideal raw material to produce conventional dry pasta. When alternative raw materials are used, an understanding of the relationship between processing variables and pasta quality is crucial in order to optimize the redesign of the production process. This review aims to: (1) investigate the main challenges of the pasta-making process, highlighting the processing variables that most affect pasta quality; and (2) indicate the unknown factors that influence the pasta-making process and which need to be studied. After overviewing the last twenty years of research in the pasta sector, the interplay/relationship between processing variables and pasta quality is examined, together with the main innovations proposed for each step of pasta processing. An analysis of all the variables involved in the process and their influence on each other will elucidate how to optimize certain parameters to ensure the production of pasta with the desired characteristics.

Keywords: pasta making; pasta; hydration; extrusion; drying; cooking quality

Citation: Bresciani, A.; Pagani, M.A.; Marti, A. Pasta-Making Process: A Narrative Review on the Relation between Process Variables and Pasta Quality. *Foods* **2022**, *11*, 256. https://doi.org/10.3390/foods11030256

Academic Editor: Joan M. King

Received: 23 December 2021
Accepted: 17 January 2022
Published: 19 January 2022

1. Introduction

Pasta is one of the most common and popular staple foods thanks to its sensory and nutritional value, convenience, and versatility [1]. It is reported that about 14.3 million tons of pasta are produced annually worldwide. The main producer is Italy, followed by the United States, Brazil, Turkey, and Russia. Italians are the main pasta consumers, with 23.1 kg per capita per year, followed by Tunisians (17 kg), Venezuelans (12 kg) and Greeks (11.4 kg) [2]. According to Italian law, "dried pasta" must be produced with water and durum wheat (*Triticum durum* Desf.) (i.e., semolina, coarse semolina, or wholemeal semolina) [3]. Although in the rest of the world (except for France and Greece) common wheat (*Triticum aestivum* L.) can be used for pasta production, it is well-known that only durum semolina can assure the best product quality, in terms of dough rheological properties, cooking quality and consumer acceptance [4,5]. However, it should be noted that common wheat is approximately 20–25% cheaper than durum wheat, making it an interesting raw material for worldwide production thanks to its high availability and (cost-effectiveness/relatively low cost [6].

Pasta plays a key role in the Mediterranean Diet. WHO (the World Health Organization) and FAO (the Food and Agriculture Organization of the United Nations) described pasta as a healthy, sustainable, and quality food model. Moreover, in 2010, UNESCO (United Nations Educational, Scientific and Cultural Organization) declared pasta an intangible cultural heritage of humanity [7]. One of the main reasons for the success of pasta is its nutritional profile. Indeed, pasta generally is very nutritious, due to its low amount of fats and readily digestible carbohydrates [8]. Moreover, pasta can supply healthy components, such as fibre or prebiotics [9,10]. The low cost and long shelf life of pasta make it popular with many diverse groups of consumers [11].

Despite being considered a traditional product, pasta (and the pasta sector in a broader sense) has been able to evolve over the years to meet the needs of the market that has expanded from Italy throughout the world through the improvement of production efficiency, on the one hand, and the enhancement of product quality from hygienic, sensory and nutritional stand points on the other. The above-mentioned aspects are the driving force behind pasta innovation. The various references present on the market—including wholegrain, multigrain, gluten-free, pulse and vegetable-enriched pasta—are examples of product innovation. Consumers certainly appreciate the taste and cooking behavior of semolina pasta [12] and the healthy features of fiber-enriched pasta [13]. However, what consumers ignore are the challenges of producing these kinds of products, the know-how and processing innovation behind each package of pasta. The change of a single variable—such as the type of raw material (refined vs. wholegrain semolina)—can affect the entire process and product quality. In this context, it is important to single out the current factors (i.e., what process variables are affected by alternative raw materials) in order to adapt the process properly in order to obtain a high-quality end product. This review focuses on individuating the main process variables that influence the quality of the product. Understanding the relationship between processing variables and pasta quality is essential in "redesigning" the process when alternative raw materials (i.e., ingredients other than durum wheat semolina) are used.

The present review is divided into three sections. Firstly, we provide an overview of research on pasta and the pasta-making process carried out in the last twenty years. Secondly, for each step of pasta processing, the interplay between the main variables in affecting and determining the quality of the final product is discussed, together with the main innovations published in research articles. Finally, the last section focuses on the main knowledge gaps of the sector (i.e., how to produce pasta from alternative raw materials), with the hope of stimulating further study in this field.

2. Overview of Research on Pasta

A search using "pasta" or "spaghetti" as keywords (to be searched in the title of documents) was carried out on the Web of Science database. More than 50% of the research articles published in the Food Science and Technology category were published in the last 10 years, with a progressive increase in number over the years and an average of 80 contributions per year over the last five years.

There are numerous reasons that explain this trend which, among other things, coincide with the reasons that accompanied the transition of this food from a "traditional Italian product" to a "product of international success" [14]. Pasta products are popular due to their simplicity in terms of formulation (they can be prepared with only two ingredients: semolina (from durum wheat) or flour (from common wheat) and water), the technological process involved (it is a continuous process, completely automated and consisting of few operations) and methods of preparation by the consumer. Dry pasta is also characterized by long shelf life, up to three years, thanks to its low humidity (generally lower than 12.5%), as well as its great adaptability to different tastes and traditions. In addition, in the presence of a vegetable or meat- or fish-based condiment, it represents a complete and balanced dish from a nutritional point of view, with a medium–low glycemic index [15]. This last characteristic is due to the technological process that leads to the formation of a compact final structure that is slowly accessible to digestive enzymes [16,17].

Most studies focus on pasta formulation, including flours from grains other than durum wheat (or their fractions) or other ingredients (including vegetables) to improve the nutritional profile of the pasta [11,18–23]. Consumer interest in different types of pasta reflects an evolving market trend (see Figure 1) to obtain certain nutritional benefits deriving from the specific alternative raw materials used in pasta production.

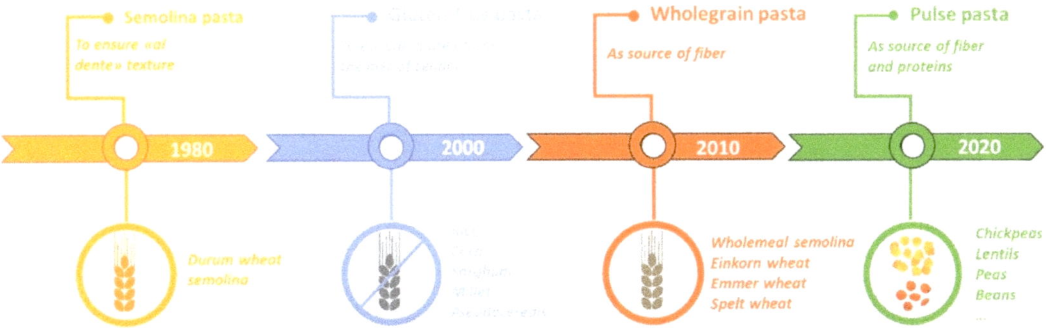

Figure 1. Evolution of types of pasta and the related raw materials.

Strategies and opportunities for producing functional pasta have been widely reviewed in the last ten years [11,18–23]. What these studies have in common is the awareness that pasta can be considered an important and interesting carrier for bioactive compounds, especially dietary fiber. For example, a portion of 80 g of whole wheat pasta provides up to 6 g of the recommended daily 25 g of dietary fiber for those with energy intakes of less than 2000 kcal/day [24]. From the literature, it emerges that the main aim of researchers is to identify the maximum level of fiber (or source of fiber) enrichment possible in order to benefit from a nutritional standpoint, without compromising the quality of the final product in terms of cooking quality and sensory profile. Overall, the quality of enriched pasta is generally similar to that of traditional pasta for enrichment levels of less than or equal to 10% [20]. For higher levels, quality can be significantly lower, suggesting the need for further studies to optimize the pasta-making processes of fiber-enriched pasta.

In a recent study, Cecchini et al. [25] elaborated the results of a literature search through VOSviewer software using the Scopus database and the keywords "quality and pasta" and "quality and durum wheat". Compared to our search, our colleagues have, on the one hand, limited the search to pasta from durum wheat and, on the other, extended it to the quality of the raw materials. Using these criteria, about 2000 studies were published on pasta from 1987 to 2018, dealing with the following topics (Figure 2): (1) varietal and genetic aspects of wheat; (2) agronomic practices and their effect on wheat quality; (3) rheological properties of the raw material, process, and quality of pasta; (4) nutritional aspects. Specifically, the latter seem to have gained more interest in recent years compared to topics related to genetics and breeding. Although the contribution of Cecchini et al. [25] mapped the evolution of durum wheat and pasta quality research topics, it did not provide insights into either the relation between processing conditions and pasta quality or the recent advances in the pasta-making process, which are the objectives of the present review.

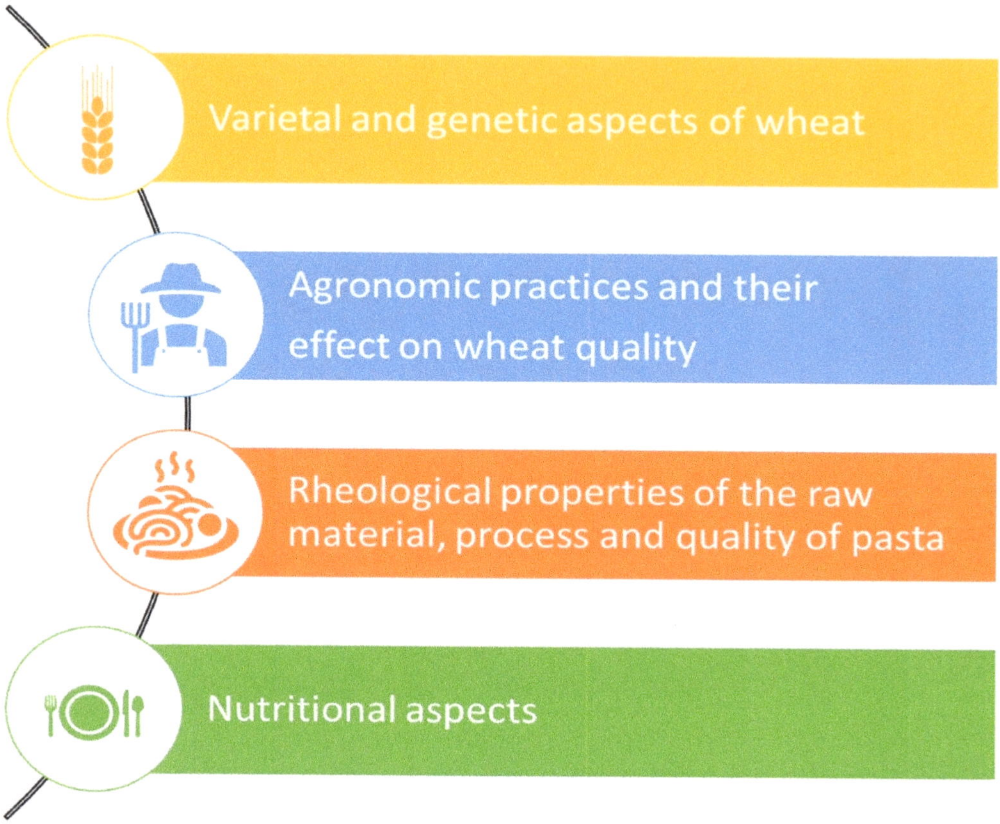

Figure 2. The main topics related to durum wheat and semolina pasta in the last 40 years.

3. Overview of Research on the Pasta-Making Process

As regards processing, pasta-making is a continuous process, consisting of three main steps: dosing and mixing, kneading and shaping (by extrusion or sheeting), and drying. Despite the vast amount of bibliographic information on pasta, the debate over the question "Does the raw material or the pasta-making process matter more?" is still ongoing, also considering that wholegrain pasta is becoming more and more popular.

It is well known that durum wheat semolina characterized by high protein content and strong gluten—able to withstand the physical stress occurring during extrusion, drying, and cooking—is the ideal raw material for high-quality pasta [4]. However, even starting from good-quality semolina, the production of good-quality pasta is not ensured if each step of the continuous pasta-making process is not properly carried out. Table 1 summarizes the aim of each step of the pasta-making process, together with the intrinsic and extrinsic parameters affecting the dough and/or pasta. It is worth noting that the pasta-making process from gluten-free raw materials is reported elsewhere [26–29].

Table 1. Parameters affecting dough/pasta quality.

Operation	Aim	Intrinsic Parameters Affecting the Dough/Pasta	Extrinsic Parameters Affecting the Dough/Pasta
Dosing, mixing and kneading	• To dose in the right proportions both semolina and water (25–27 parts of water/100 parts of semolina) • To hydrate starch and proteins	• Semolina particle size • Semolina protein, ash, fiber, damaged starch content • Enzyme activities • Water temperature and residue	• Presence of a pre-mixer • Vacuum degree
Kneading and shaping by extrusion	• To (partially) form gluten networks • To knead and give a shape to the dough	• Gluten tenacity • Dough humidity • Dough temperature • Dough viscosity	• Mixture feeding into the extruder • Geometrical characteristics of the screw (length, design, etc.) • Extrusion conditions (specific mechanical energy, screw speed, heat regulation system, etc.) • Shape of the extruded product • Die material • Open surface of the die (number and position of the inserts)
Drying	• To remove water • To assure shape integrity • To maintain nutritional quality	• Gluten tenacity • Starch pasting properties	• Air temperature • Air relative humidity • Drying time

So far, the effect of each step of the pasta-making process has been evaluated with respect to its impact on pasta structure and quality [17,30–32]. On the other hand, the effect of processing variables (i.e., hydration level, extrusion pressure/temperature/mechanical energy) on pasta quality has not yet been exhaustively investigated. Based on these considerations, the role of the main variables—involved in each step of the pasta-making process—on pasta quality will be discussed in the following sections.

3.1. From Dosing to Mixing

In the first step of pasta-making, semolina and water are carefully dosed and blended together to form a hydrated mixture with a total moisture content of about 30–32%. The amount of water added to semolina (27–29 g/100 g) is far from the hydration level used in bread-making (50–60% water absorption, namely 45–50% moisture), which is essential for promoting the even water dispersion inside the solid mass. In other words, in the pasta-making process, hydration ensures the correct solvation of proteins while gluten is only partially developed at this stage. Only appropriate protein hydration will assure—in the following steps—the formation of a continuous gluten network capable of restricting and preventing excessive starch swelling during cooking.

Besides the amount of water, other factors may affect semolina hydration and thus the physical properties of pasta and its quality. Among them, protein, ash, fiber, and damaged starch content, as well as particle size (Table 1). Semolina samples with low ash and damaged starch content result in a dried product characterized by an amber-yellow color, low brown specks (due to the presence of bran particles), and low heat damage [30]. Low ash content is assured by a medium–low extraction rate (60–65%), whereas low-damaged starch is assured by milling durum wheat to large particle sizes (>400 μm). However, the choice is not so easy, because a medium extraction rate is synonymous with a low milling yield (and thus productivity), whereas a large particle size might result in low hydration kinetics with inadequate moistening of semolina. A regular and even protein structure is essential to guarantee a product with good cooking behavior, resulting in high firmness and absence of stickiness and bulkiness [33].

Particle size also plays a key role in wholewheat semolina. A positive correlation between the geometric mean diameter of flour particles and the cooking behavior of wholemeal semolina pasta (i.e., high firmness and low cooking loss) was assessed. At the same time, broad particle distribution negatively impacts pasta cooking quality [34]. As regards reconstituted semolina/bran blends, bran particle size doesn't seem to impact on pasta cooking behavior but the semolina/fine bran blend is preferred since the resulting pasta showed higher mechanical strength than pasta from the semolina/coarse bran blend [34].

3.2. The Effect of Hydration on the Extrusion Process and Pasta Quality

The amount of water added to semolina and its uniform dispersion inside the mass are critical parameters because mistakes made in this first operation can hardly be corrected in the following steps of pasta-making. In the case of uneven hydration (often caused by limited amounts of water), the final product may form the characteristic white spots which indicate a potential weak structure and decrease the quality of the product in terms of both appearance and texture. On the contrary, excessive hydration results in a sticky product, with low mechanical resistance and poor cooking quality [35].

De la Peña and Manthey [36] evaluated the effect of different levels of hydration (from 30 to 34%) on the extrusion properties of refined or wholemeal semolina (alone or in combination with flaxseed flour) and on the cooking behavior of the respective pasta samples. The results of this study showed that specific mechanical energy (SME) and extrusion pressure decrease as the level of hydration increases. Specifically, as regards the extrusion pressure, the formulation of pasta seems to have a significant effect: the semolina dough registers a decrease in pressure lower than that observed for wholemeal semolina. The plasticizing action of water facilitates the handling of the mixture inside the extruder, reducing the extrusion parameters. A correlation was highlighted between the viscosity of the dough (measured by a capillary rheometer) and the parameters of extrusion pressure, as well as mechanical energy [37]. Specifically, increased hydration promoted a decrease in the apparent viscosity of the dough without increasing the extrusion rate. Moreover, high levels of hydration (32–34%) are associated with a reduction in brightness/luminosity and an increase in the degree of red (a*) but do not affect the degree of yellow (b*) [37].

As a result of the decrease in extrusion pressure and mechanical energy, the diameter and density of the spaghetti decreases as hydration increases. Since dough at 30% hydration shows high consistency and resistance to flow, it was hypothesized that these systems could bring high pressure to bear on the Teflon coatings of the die inserts, compressing them and thus resulting in an increase in the diameter of extruded spaghetti [37]. On the contrary, formulations hydrated at 34% level, showing lower consistency, exert a lower pressure during extrusion by reducing the diameter of the spaghetti. It is also possible that spaghetti produced with the highest hydration levels (and which is therefore heavier) may have been slightly stretched on the reeds during drying. The smaller diameter in the 34% hydration formulations seems to be responsible for the reduced hardness of the cooked pasta and the greater cooking losses, as a result of the faster migration of water towards the core of the spaghetti [37]. The role of porosity/compactness as a consequence of the decrease in extrusion pressure should also be considered.

3.3. The Effect of Formulations on Hydration Levels

The hydration operation is even more critical when grains other than durum wheat or ingredients are included in the formulation. Re-formulating pasta makes it essential to study and optimize the level of hydration as this affects not only the characteristics of the dough (in particular its processing during the extrusion phase) but also the quality of the final product. Fiber, due to its high hydrophilicity, competes with proteins for water absorption; it could, therefore, reduce the water available for their solvation, compromising the formation of a uniform network. Furthermore, the worsening of pasta quality in the presence of fiber is due to the dilution effect of gluten (as a result of the lower amount of semolina in the formulation), as well as a discontinuity in the protein network caused by the interference of the non-starch polysaccharides [38].

Considering the effect on the formulation, wholemeal semolina doughs showed higher viscosity, even at high levels of hydration (34%), compared to the reference sample (refined semolina). The bran fractions in the wholemeal sample would require more water to show the same rheological behavior as the reference sample [37]. Similar results were found in the presence of flaxseed flour, buckwheat bran or durum wheat bran [39]. However, based on the effect of hydration levels on extrusion parameters, the study of de la Peña et al. [37] states that hydration should not exceed 32% for semolina, whole wheat semolina and their blends. This level should be reduced to 30% in the case of formulations rich in lipids, such as flaxseed flours [37]. Lipids, in fact, can have a plasticising effect on the dough and act as a lubricant by reducing the friction generated inside the extrusion cylinder, thus reducing extrusion pressure, mechanical energy and SME [37].

A possible solution to limit the competition for water between fiber and proteins could be to hydrate the two ingredients separately (for example, semolina and bran) before extrusion, as proposed by La Gatta et al. [40]. From a sensory point of view, separate hydration seems to have a positive effect on color and the resistance to breaking of uncooked pasta and on the elasticity, firmness, adhesiveness and bulkiness of cooked pasta. In addition, a decrease in cooking losses was measured. This approach would allow semolina proteins to solvate and interact optimally by limiting the interference caused by the fiber, forming a structure capable of retaining the swelling and solubilization of starch during cooking. This approach would produce better-quality pasta while maintaining suitable hydration levels for the extrusion process. However, despite the encouraging results obtained on a laboratory scale, the scale-up of the process remains to be investigated.

3.4. New Trends in Hydration Systems

Since hydration is mostly influenced by the physicochemical characteristics of the raw material (Table 1), to ensure correct protein hydration, raw materials with a low extraction rate should be preferred (such as semolina obtained from the innermost part of the endosperm, for its low ash and fibre content) and with low starch damage, therefore medium-to-large sized semolina particles. According to many pasta producers, semolina with particle sizes ranging from 250 and 450 μm seems to guarantee homogeneous hydration. However, a large particle size (more than 450 μm), highly appreciated by some Italian pasta-makers for the low starch damage, makes it difficult to hydrate the semolina particles correctly, promoting the formation of white spots. In this context, besides accurate devices for the dosing step, various hydration systems have been proposed to guarantee a more homogeneous hydration of the raw materials. Indeed, at the end of the mixing step in the conventional extrusion press, dough appears as "lumps" of different sizes. In the innovative devices, the premixing and mixing steps are usually combined in a single operating unit. Among the proposed systems, the Polymatic press (Bhuler, Uzwil, Switzerland) mixes and develops pasta dough in 20 s. A twin-screw extruder forms the dough, which is directly sent to the extruder. The entire system is under vacuum, which assures excellent pasta color. Other advantages of this system include the rapid changeover of dies, which helps when different forms of short pasta are being manufactured, as well as a clean-in-place system for excellent sanitation [41].

Among the other solutions suggested to improve the initial steps of the pasta-making process, the effects of the innovative Premix® and Bakmix® mixing systems (Storci S.p.a., Collecchio, Italy) were compared with a conventional system (V50, Storci Spa) [42]. The centrifugal force applied in the Premix® would promote the rapid (1–2 s) and uniform hydration of the surface of each individual semolina particle, followed by a rest phase (10 min) before extrusion. In the Bakmix® system, hydration is divided into two phases: 2 s in the Premix® system and 18 s in an extruder operating at low pressure (about 10^6 Pa). All the mixing processes result in products (fresh pasta obtained by two shaping approaches: extrusion or lamination) of acceptable quality and characterized by good cooking behavior, with cooking loss values lower than 3 g/100 g pasta.

In general, the new systems facilitate a more uniform distribution of water throughout the flour compared to traditional mixing but in a significantly shorter time; therefore, a well-developed protein matrix may not be formed [42]. This results in a pasta dough that is less extensible and more resistant to deformation, characteristics considered to be negative for fresh pasta. The authors of the work suggest that, due to the short mixing time, it may be necessary to increase the level of hydration to obtain a better-quality product. However, this theoretical solution does not seem to be the best for obtaining a good-quality pasta, as stated by Manthey et al. [43] and already discussed. Moreover, their study focused on understanding the effect of the different hydration systems on the characteristics of fresh pasta; it would therefore be interesting to re-propose this experimental plan completing the pasta-making process with drying; in fact, drying could reduce the differences highlighted in fresh pasta by Carini et al. [42]. Moreover, the effect of non-traditional hydration systems might be particularly successful when applied to wholegrain semolina.

In the new pasta-making plans, the pre-mixer system is connected to a stabilization belt mixer (Beltmix®; Storci S.p.a., Collecchio, Italy), in place of the traditional shaft and blade mixing tank. The belt mixer consists of a slow-moving conveyor belt. Since the dough is not subjected to any kind of mechanical action, the system drastically reduces the oxidation of the raw materials, which maintains the bright yellow color of semolina. In addition, compared to the traditional mixer, the belt mixer is easier and faster to clean, as stated by the device manufacturer.

3.5. From Kneading to Shaping

Shaping or forming aims at creating a well-defined shape (Table 1) and represents the heart of the pasta-making process. It can take place in two ways, by extrusion under pressure or by roll-sheeting. The former involves the kneading of the dough into a cylinder through a screw that compresses and pushes the mass towards the die, where pressure can reach values of 10 MPa or more. The size and the design of the screw can vary according to the manufacturing companies. Generally, screws are divided into three sections: the feeding section where the "lumps" of dough are pushed towards the transfer section and then to the extrusion section. During this flow, the dough undergoes a spiral movement favoring the kneading. At a macroscopic level, the mass acquires compactness, but the gluten network can undergo stretching and stresses of high intensity, especially in the final section of the extruder, before the dough passes through the die [44]. The second approach used to shape the dough involves rolling the dough through passages in cylinders that gradually and lightly reduce the thickness of the dough until a sheet of the desired thickness is obtained. During sheeting, dough is subjected to pressure for a very short time, i.e., only when it passes into the gap between the two cylinders; then the dough can immediately relax and recover from the deformation. Of the two processes, extrusion is the preferred approach at an industrial level not only for its higher productivity but also for its versatility; through extrusion, in fact, more than 200 different pasta shapes can be obtained. For this reason, the extrusion process is more studied than lamination. Indeed, most of the analyzed works focus exclusively on the study of some variables of the extrusion phase or on the comparison between extrusion and lamination.

The use of unconventional raw materials and/or incorrect hydration of dough affect this operation, as the variables involved during shaping are greatly influenced by the amount of water in the dough, which, as discussed in the previous paragraph, must be optimized based on the physicochemical characteristics of the raw material, including particle size, content of damaged starch and presence of fiber.

If the hydration step affects extrusion, the latter can irreversibly break down the protein network, resulting in its disruption during cooking especially when poor-quality raw materials are used. At the same time, improper extrusion conditions can cause starch swelling and gelatinization, due to the heat generated by shear stress. These setbacks can be limited by keeping the extrusion temperature below 50 °C and selecting semolina varieties having high starch gelatinization temperatures to delay starch swelling and solubilization and to decrease interference with protein reticulation [45,46].

3.6. The Effect of Extrusion Variables on Pasta Quality

Among the extrusion variables, the pressure (measured in the final part of the extrusion cylinder) and SME are useful for evaluating the overall process. They are correlated with and influenced by the same variables, including the level of hydration, the speed of the screw and the extrusion temperature. Since, as is known, the pressure varies during the advancement of the dough along the screw (reaching the maximum value near the die), studies usually consider the SME parameter. Specifically, the focus is on the relationship between hydration level and SME. An overly hydrated dough, being less compact, would require a lower SME and would not pose sufficient resistance, inside the extrusion cylinder, to promote protein aggregation and therefore a satisfying formation of gluten [43]. On a macroscopic level, a low SME, as seen above, reduces the density of spaghetti [36,43]. Water unbound to proteins and other hydrophilic (macro)molecules would be in a free state, making it easier to evaporate during the subsequent drying phase; this phenomenon would reduce density. This hypothesis could be confirmed by the study of the distribution and mobility of water inside spaghetti using NMR techniques [47,48].

The variables of the extrusion process (pressure, speed and SME) appear to be unrelated [43] or weakly correlated (r = 0.31–0.44) [36] to the diameter of spaghetti, suggesting that other factors are responsible for the determination of that characteristic. As previously discussed, in addition to hydration, the formulation also influences SME. In particular, the presence of bran or oil seeds reduces SME values; in fact, the presence of lipids helps lubricate the dough on the extrusion screw. As the dough poses less resistance to extrusion, it forms spaghetti with a smaller diameter [36,39]. As reported by de la Peña et al. [36], the diameter of spaghetti inversely affects the amount of material released into the cooking water.

The extrusion temperature also influences the quality of pasta in terms of cooking losses. Indeed, the increase in temperature in the extrusion cylinder from 35 °C to 70 °C leads to an increase in cooking losses up of to 250% [49]. If semolina proteins denature while the mass undergoes mixing and kneading, the denatured proteins are no longer able to interact in this phase with each other to create a protein network capable of retaining the starch granules during cooking. At high temperatures (about 70 °C) during extrusion, the increase in the level of hydration (from 44 to 48%) and in the rotation speed of the screw (from 15 to 30 rpm) has a positive effect on the final characteristics of pasta [49]. The high hydration, combined with the high speed of the screw, in fact, reduces extrusion time, thus limiting the damage that the high temperature could cause to proteins and their ability to aggregate. As is well known, temperatures between 40–50 °C are considered optimal for the pasta-making process of semolina, as they are not associated with significant denaturation of proteins and starch gelatinization but facilitate the extrusion of the dough by decreasing its viscosity. These considerations were also confirmed in the study by Debbouz and Doetkott [35]. Applying an experimental design and considering different levels of hydration (30–32–34%), water temperature (35–45–55 °C), mixing time (3–5–10 min), extrusion temperature (35–45–55 °C) and screw speed (20–25–30 rpm), the authors highlighted how all the variables have a significant effect on pasta quality. The hydration level and the temperature of the extrusion cylinder are the variables with the greatest influence. In particular, pasta cooking losses are reduced at hydration levels between 31.5 and 32% and extrusion temperatures between 45 and 50 °C.

Optimal extrusion conditions vary according to the formulation and how the design of the experimental approach optimizes the process. This holds true for various formulations. For example, for the production of wheat spaghetti enriched with soy flour, the best product (in terms of color and cooking behavior) is obtained when about 57 g of flour, 12 g of soy and 31 g of water are extruded at 35 °C and 40 rpm [50]. In the case of semolina and millet pasta (50:50), the optimal process conditions are as follows: extrusion temperature = 70 °C; hydration level = 30%; extrusion speed: 12 rpm; screw speed/feeding speed ratio = 10 [51].

As regards die extrusion, it is known how the coating material of the die inserts affects the appearance of the pasta: Teflon gives the product a smooth and bright yellow appearance, while bronze inserts produce a rough surface [52]. Furthermore, the use of a bronze die has the disadvantages of lowering extrusion pressure and die extrusion speed as well as a more rapid consumption of the part in contact with the dough [53]. Bronze-extruded spaghetti is more porous and therefore more fragile (breaking strength decreases by 20–30%) than Teflon-extruded products [54]. Furthermore, the rougher surface of the bronze-extruded spaghetti, together with its greater porosity, favors the deposition of eggs by *Sitophilus oryzae* (L.) (*Coleoptera Dryophthoridae*) and therefore is a more likely place for insects to incubate compared to Teflon-extruded pasta [54].

3.7. Type of Shaping: Extrusion vs. Sheeting

Some research compared the effect of the type of shaping on the structure and quality of the dough. Among these, the study by Zardetto and Dalla Rosa [55] involved fresh pasteurized pasta (76% semolina, 19% egg, 5% water) produced by extrusion or lamination. The results show that fresh pasta obtained by extrusion absorbs more water during cooking and releases a greater quantity of dry substance than pasta obtained by lamination. Extrusion does not form a continuous and homogeneous protein network as occurs for lamination. Furthermore, the mechanical stress exerted by the screw leads to the partial degradation of the starch and probably also to the formation of components (reducing sugars) capable of contributing to the Maillard reaction. In fact, higher furosine levels were found in extruded than in laminated fresh pasta. Pasta obtained by extrusion generally shows higher consistency values than laminated pasta, but cooking reduces the differences between the two types of product, making them more similar. From a molecular point of view, the cooking of extruded pasta promotes the formation of bonds between proteins, an indication that the extrusion process does not lead to the complete formation of a network but to the exposure of thiol groups that interact with each other during the cooking phase. In laminated pasta, on the other hand, the gluten network is well formed, as shown by the high resistance to disintegration (evaluated by sensory analysis) and low adhesiveness (instrumentally evaluated) of cooked pasta [56]. However, the differences observed at the structural level by Zardetto and Dalla Rosa [55] do not imply sensory differences and are probably not perceived by most consumers, as they are probably masked by egg proteins.

Lastly, the study by Carini et al. [57] compared different shaping processes (extrusion, rolling, and vacuum lamination) using a simpler dough system, consisting exclusively of semolina and water (70:30). The macroscopic characteristics of pasta (color, cooking losses, and firmness) seem to depend on the process, while the water status or how the water interacts with the biopolymers (the ability to retain frozen water and water mobility) was only slightly influenced by processing conditions [57]. Specifically, the extrusion process, due to the greater mechanical stress it requires, seems to facilitate the interactions between water and biopolymers, resulting in a more extensible product. On the other hand, the less stressful conditions for lamination result in a structure that is less compact and less extensible but better able to retain solids during cooking, confirming the results of Zardetto and Dalla Rosa [55]. The application of a vacuum to the lamination process seems to improve the quality indicators of fresh pasta, resulting in a product characterized by a more yellow color and with extensibility and consistency similar to fresh pasta obtained by extrusion. The application of a vacuum during lamination may have eliminated the air contained in the dough as a result of lamination by better compacting the biopolymers and facilitating the interactions between them and those with water.

The different effect of the shaping processes on the quality of pasta is even more evident when using common wheat, as it is less able to stand the physical stresses occurring inside the press. The greater compactness of the structure obtained by extrusion corresponds to longer cooking times and slower water absorption, which, however, does not translate into better cooking behavior for dry pasta. This feature is clearly linked to the different organization of proteins. The gluten network, in fact, appears more continuous in the case of laminated pasta, probably thanks to the lower stress and the action of the rollers that more effectively align the protein fibrils. The result is a firm pasta without stickiness. Finally, in the case of a semolina-based formulation enriched with buckwheat (25%), the preferred technology involves the extrusion of a sheet whose thickness is gradually reduced by rolling [56]. This process, in fact, seems to create a structure that is compact (as suggested by the slower hydration kinetics) and at the same time continuous (as suggested by the slower gelatinization of the starch granules), resulting in a product with lower cooking losses, greater firmness and less of a tendency to disintegrate during cooking.

3.8. Drying

Particular attention is paid to the final step of the pasta-making process: the drying step. As is well-known, the drying process gives dry pasta its final characteristics of physical and chemical stability and allows its shelf life to be extended. The overall cooking quality of the final product (high degree of firmness, low stickiness and low cooking loss) is the result of several simultaneous phenomena within pasta, whose extent depends on both raw material characteristics and the temperature–moisture conditions applied during drying.

The variables that regulate this phase (temperature, relative humidity, and time), in fact, can be modified by proposing various combinations (and as many drying cycles) in order to promote the coagulation of proteins and improve the cooking behavior of pasta (Table 1). In particular, the physicochemical modifications of the main macromolecules control pasta cooking behavior in an opposite way. When protein coagulation in the continuous network prevails, the starch material is trapped within the network and the cooked pasta will be firm with no stickiness on the surface and consequent bulkiness. On the contrary, when the protein network is not strong and elastic enough, the starch swells and gelatinizes before protein coagulation takes place.

Over the years, the scientific community's interest has changed as summarized in Figure 3.

The focus of studies shifted from the effects of high- and low-temperature drying cycles on the denaturation of proteins and pasta quality [58–60], also in relation to heat damage (1980–2000; see the review by De Noni and Pagani [30]), to the effect of drying on starch characteristics (2000–2005; Padalino et al. [61]), including aspects relating to digestibility (in the last 15 years; see the review by Petitot et al. [17]).

As regards the effect of drying temperature on pasta quality, high temperature drying cycles (>65 °C) are effective in improving the sensory characteristics of pasta [61], especially in the case of pasta made with semolina low in protein [59,62]. The same effect was not evident when pasta made from semolina with strong gluten was prepared [59]. Using multiple regression analysis, D'Egidio et al. [62] showed that stickiness played the most significant role relative to firmness and bulkiness in the case of pasta dried at a low temperature, whereas at a high temperature the three sensory attributes had a similar importance.

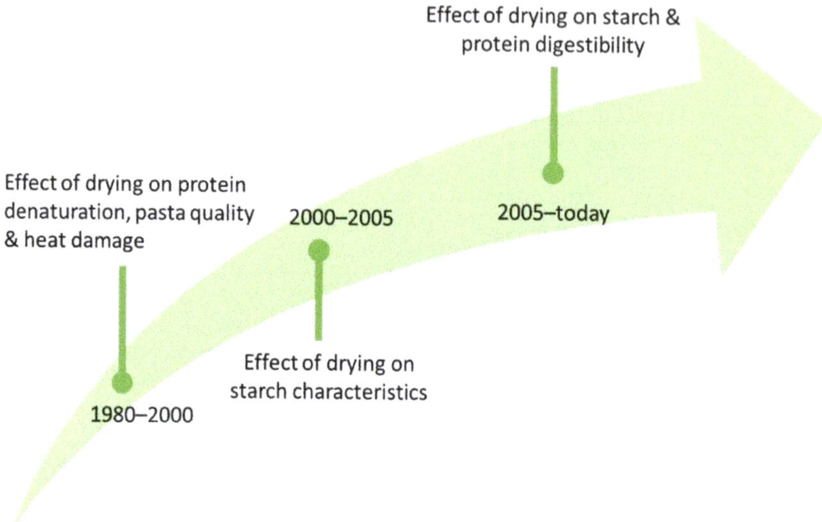

Figure 3. Trends in the research topics on pasta drying over the past 40 years.

Countless studies have addressed the issue of starch digestion in pasta, in view of its relevance to controlling glycemia, but only a few studies have addressed the issue of protein digestion in pasta. Some of these reports have addressed protein-digestibility issues related either to the use of different wheat varieties or to the impact of processing on it [17,63–66]. However, none of these studies appears to have fully addressed the complexity of the protein pattern in the raw material, as well as the relevance of protein–protein and protein–starch interactions in these complex matrices, either before or after processing. Moreover, conflicting results have been obtained due to the use of different methodologies as well as different pasta-making conditions. A great amount of variation can be seen in the drying conditions (i.e., time, temperature and relative humidity) of published pasta studies, making it difficult to compare findings obtained from various laboratories as pointed out by Murray et al. [32] The meta-analysis work carried out by Mercier et al. [20] on the relationship between the production process and the quality of enriched pasta confirmed what has already been studied for pasta from semolina of various qualities [62]: drying pasta at temperatures above 60 °C can partially compensate for the weakening of the structure of pasta (which is attributed to the enrichment and dilution of gluten) due to the reinforcing effect provided by protein coagulation. Wholewheat spaghetti dried at a low temperature (40 °C) had higher cooking loss but better overall appearance, mechanical strength and cooked firmness than wholewheat spaghetti dried at a high temperature (70 °C) [38]. Similar findings were found when comparing the quality of wholewheat pasta dried at 60 °C or 85 °C: a low temperature was effective in decreasing cooking loss and increasing firmness, even if differences in texture could not be detected using a trained panel [67,68].

Findings on the relation between the products of the Maillard reaction (i.e., advanced glycation end products, AGEs, such as the ε-pyrrole-lysine pyrraline or ε 2-formyl-5-hydroxymethyl-pyrrolaldehyde) and protein digestibility [64,69] as well as the onset of some diseases [70] have brought researchers' attention back to the investigation of heat damage. Pasta dried at a low temperature had low amounts of furosine, which is the most widely used marker for assessing the extent of the Maillard reaction [27,71]. Many pasta producers stress the importance of drying conditions, specifically the use of slow and/or low-temperature drying cycles. Unfortunately, this terminology is not sufficient to provide clear and univocal indications of pasta quality and/or the intensity of heat damage.

The survey carried out on more than 60 pasta samples available on the Italian market highlighted that the furosine level was greater than 300 mg/100 g protein for almost all pasta produced at industrial scale [71]. These values have been found, surprisingly, even in some "artisan pasta". Moreover, sensory analyses showed that low heat damage (furosine <250 mg/100 g protein) is not a guarantee of good cooking quality. Besides protein content, particle size distribution and, consequently, damaged starch content also greatly affect the furosine levels of pasta samples, even if the same drying cycle is applied [71].

Using wholegrain semolina instead of refined semolina led to increased furosine content, affecting sensory traits. Indeed, pasta with high furosine content (i.e., dried using high-temperature drying cycles) is perceived to be more bitter than pasta with low furosine content (i.e., dried using low-temperature drying cycles) [24]. On the contrary, in pasta made from whole common wheat, drying conditions did not have a significant impact on either taste or flavor (as assessed by descriptive analysis) [68,69].

3.9. New Trends in Drying Systems

Most innovations related to the drying stage have aimed at reducing drying times, without affecting pasta quality. In this context, recent work has been carried out on the use of microwaves (either alone or in combination with air drying). The process of drying pasta by microwaves has proven to be very efficient, not only as regards shortening the drying time but also because it is possible to have a final product without fissures, with higher firmness and a lower degree of gelatinization than pasta dried by hot air [72–74]. It increased the cooking resistance of pasta as well as its cooking time. Moreover, similar total organic matter values suggest that the cooking quality of samples dried differently was comparable [72–74].

More recently, the effect of vacuum drying (where moisture removal from food products occurs under low pressure) on pasta quality has been assessed in semolina pasta at lab scale [75,76]. Compared with conventional drying, vacuum drying is characterized by a lower drying temperature and a higher drying rate (i.e., water evaporation occurs more readily). The enhanced moisture transfer may lead to the prevention of surface barrier formation that causes internal stress within the product. Therefore, the use of vacuum-drying may reduce internal stress and prevent structural deterioration, resulting in better cooking quality (i.e., high water absorption and hardness, low cooking loss and adhesiveness) [75,76]. Moreover, since moisture is removed in the absence of oxygen, oxidative degradations, e.g., browning or fat oxidation, are minimized, resulting in a pasta with a bright yellow color [75,76].

At the industrial level, new drying lines capable of reducing the drying time to about 3 h for long pasta and less than 2 h for short pasta are available, with a significant reduction in the size of the plant. Although the superiority in quality of the product obtained from these systems is claimed by the company that manufactures the drying equipment, no data has been shared with the scientific community. Indeed, most of the studies on processing are mainly conducted by the manufacturers of pasta and/or pasta plants, and thus are subjected to company regulations related to privacy.

4. Knowledge Gaps and Perspectives

In this section, the main knowledge gaps related to pasta-making process are summarized.

Since each step in the pasta-making process impacts on the quality of the final product, it is extremely important to know how process variables and pasta properties relate in order to better predict and control product quality. The first steps of the pasta-making process—hydration of semolina and shaping of the dough by extrusion under pressure or roll-sheeting—have so far received less attention than the drying phase. The greater interest in the latter is justified by the modifications (which are well known and quantified) induced by temperatures above 60 °C on both proteins and starch properties and their great impact on pasta quality at both sensory (e.g., texture) and nutritional (e.g., heat damage) levels. A second reason for the apparent minor interest in the hydration and shaping phases is linked to the difficulties that their monitoring entails. In fact, the low humidity (between 30 and 32%) of the mixing system and, consequently, its low degree of smoothness inside the press, makes it difficult to study dough behavior during extrusion. Moreover, the mixture is uneven in temperature and viscosity [77]; these differences can be found not only at the entry of the cylinder towards the die but also at the cross section of the cylinder (in fact, the mass near the walls of the cylinder is colder and with higher consistency than the mass closer to the core of the screw).

Finally, to further complicate observations, process variables (first of all the extrusion pressure) are affected by dough properties (i.e., moisture, temperature, viscosity) and any change in one of the processing variables influences all the others in an interdependent way. In other words, when a parameter changes, the system responds in a very complex way. A further aspect regards the high degree of heterogeneity of extrusion systems due to their different specifications (geometry and pitch of the screw, single- or twin-screw extruder, etc.) that could have different repercussions on the workability of the mixture and on the characteristics of the finished product. Some studies applied prediction models of dough behavior by modifying extrusion variables [77]. However, these works are limited to the study of the process without relating it to the characteristics of the finished product. Finally, there are no studies evaluating the effect of mechanical and structural changes (for example, screw geometry, single- or twin-screw extruder, etc.) on pasta quality.

Moreover, among the studies focusing on the extrusion step, none associates process conditions with the nutritional quality of the finished product, in terms of digestibility and/or the formation of resistant starch. This aspect is left to the reformulation of the product using modified starches or raw materials rich in amylose. In this context, Camelo-Méndez et al. [78] summarizes the effect of different ingredients on the starch digestibility of pasta.

Further gaps come from the pasta quality evaluation side. Most of the studies aimed at understanding the relationship between processing conditions and pasta quality assessed the quality of the final products by evaluating changes in color, cooking loss and texture evaluated by instrumental analysis rather than sensory analysis. Besides requiring less time for the analysis, other factors account for the preference of instrumental tests: (1) a sensory evaluation testing facility should be set up to minimize the interactions occurring between participants; (2) consumer-based sensory evaluation measures liking of foods and requires large numbers of individuals; and (3) descriptive analysis requires trained tasters to evaluate the intensities of attributes found in foods [79].

Finally, most of the studies devoted to understanding the relationships between process conditions and pasta quality considered only semolina as the raw material to be used. Although it is easy to understand the reasons for this choice, worldwide (with the exception of Italy, France and Greece) hard wheat flour is the main raw material used for dry pasta. Indeed, it is widely available and less expensive than durum wheat. However, despite the great interest in describing the bread-making performance of common wheat, it is still unknown what features common wheat should have and what processing parameters should be adopted to obtain dried pasta of desirable quality.

5. Conclusions

Dry pasta can be considered an iconic Italian food and is nowadays appreciated around the world for its nutritional and sensory features, as well as for its versatility. Although an established technology, the pasta-making process needs to be optimized, taking into consideration changes in lifestyle and consumer awareness. A healthy diet, resilience and sustainability are the keywords of the era we are living in. Thus, recent interest in fiber-enriched formulations, as well as in underexploited grains necessitates the re-examination, re-thinking and re-adjustment of the conditions currently used for preparing pasta. Attention should be paid to extrusion to ensure the formation of a protein structure that can withstand cooking.

In this context, because the extrusion process is of great interest from an industrial point of view due to its high productivity and versatility, more resources must be allocated to the study and optimization of this phase of the process. More than this would be an opportunity to further the growing interest in alternative raw materials to satisfy increasing nutritional demands and foster environmental, social and economic sustainability, especially in light of future climatic changes that may limit wheat availability and/or deteriorate its quality.

Author Contributions: A.B., investigation, writing—original draft preparation; M.A.P., writing—review and editing; A.M., conceptualization, writing—original draft preparation, writing—review and editing, supervision. All authors have read and agreed to the published version of the manuscript.

Funding: This research received no external funding.

Acknowledgments: This study was partially supported by FAVA S.p.A. (Cento, Italy), in the frame of the Genesis (sviluppo di una nuova GENErazione di SIStemi modulari e interconnessi per produzione di pasta) project (Industria Sostenibile del MISE-DM 24/5/2017—"Accordi per l'Innovazione").

Conflicts of Interest: The authors have none to declare.

References

1. Giacco, R.; Vitale, M.; Riccardi, G. Pasta: Role in diet. In *The Encyclopedia of Food and Health*; Caballero, B., Finglas, P., Toldrá, F., Eds.; Elsevier Ltd.: Amsterdam, The Netherlands, 2016; pp. 242–245.
2. IPO. The World Pasta Industry Status Report—International Pasta Organization. 2014. Available online: http://www.internationalpasta.org (accessed on 5 May 2021).
3. Decreto del Presidente della Repubblica n.146. Regolamento per la revisione della normativa sulla produzione e commercializzazione di sfarinati e paste alimentari, a norma dell'articolo 50 della legge 22 febbraio 1994. *Gazz. Uff.* **2001**, *117*, 6–12.
4. Sissons, M. Role of durum wheat composition on the quality of pasta and bread. *Food* **2008**, *2*, 75–90.
5. Padalino, L.; Mastromatteo, M.; Lecce, L.; Spinelli, S.; Contò, F.; Del Nobile, M.A. Effect of durum wheat cultivars on physico-chemical and sensory properties of spaghetti. *J. Sci. Food Agric.* **2014**, *94*, 2196–2204. [CrossRef] [PubMed]
6. Hong, E.; Lee, S.Y.; Jeong, J.Y.; Park, J.M.; Kim, B.H.; Kwon, K.; Chun, H.S. Modern analytical methods for the detection of food fraud and adulteration by food category. *J. Sci. Food Agric.* **2017**, *97*, 3877–3896. [CrossRef]
7. Giannetti, V.; Mariani, M.B.; Marini, F.; Biancolillo, A. Effects of thermal treatments on durum wheat pasta flavour during production process: A modelling approach to provide added-value to pasta dried at low temperatures. *Talanta* **2021**, *225*, 121955. [CrossRef]
8. Dello Russo, M.; Spagnuolo, C.; Moccia, S.; Angelino, D.; Pellegrini, N.; Martini, D. Nutritional Quality of Pasta Sold on the Italian Market: The Food Labelling of Italian Products (FLIP) Study. *Nutrients* **2021**, *13*, 171. [CrossRef]
9. Angelino, D.; Martina, A.; Rosi, A.; Veronesi, L.; Antonini, M.; Mennella, I.; Vitaglione, P.; Grioni, S.; Brighenti, F.; Zavaroni, I.; et al. Glucose- and lipid-related biomarkers are affected in healthy obese or hyperglycemic adults consuming a whole-grain pasta enriched in prebiotics and probiotics: A 12-week randomized controlled trial. *J. Nutr.* **2019**, *149*, 1714–1723. [CrossRef]
10. Ciccoritti, R.; Taddei, F.; Nicoletti, I.; Gazza, L.; Corradini, D.; D'Egidio, M.G.; Martini, D. Use of bran fractions and debranned kernels for the development of pasta with high nutritional and healthy potential. *Food Chem.* **2017**, *225*, 77–86. [CrossRef] [PubMed]
11. Oliviero, T.; Fogliano, V. Food design strategies to increase vegetable intake: The case of vegetable enriched pasta. *Trends Food Sci. Technol.* **2016**, *51*, 58–64. [CrossRef]
12. Di Monaco, R.; Cavella, S.; Di Marzo, S.; Masi, P. The effect of expectations generated by brand name on the acceptability of dried semolina pasta. *Food Qual. Prefer.* **2004**, *15*, 429–437. [CrossRef]
13. Laureati, M.; Conte, A.; Padalino, L.; Del Nobile, M.A.; Pagliarini, E. Effect of fiber information on consumer's expectation and liking of wheat bran enriched pasta. *J. Sens. Stud.* **2016**, *31*, 348–359. [CrossRef]
14. Pagani, M.A.; Lucisano, M.; Mariotti, M. Traditional Italian products from wheat and other starchy flours. In *Handbook of Food Products Manufacturing*; Hui, Y.H., Ed.; John Wiley Ltd.: Hoboken, NJ, USA, 2007; pp. 327–388.
15. Chiavaroli, L.; Kendall, C.W.; Braunstein, C.R.; Mejia, S.B.; Leiter, L.A.; Jenkins, D.J.; Sievenpiper, J.L. Effect of pasta in the context of low-glycaemic index dietary patterns on body weight and markers of adiposity: A systematic review and meta-analysis of randomised controlled trials in adults. *BMJ Open* **2018**, *8*, e019438. [CrossRef]
16. Granfeldt, Y.; Björck, I. Glycemic response to starch in pasta: A study of mechanisms of limited enzyme availability. *J. Cereal Sci.* **1991**, *14*, 47–61. [CrossRef]
17. Petitot, M.; Abecassis, J.; Micard, V. Structuring of pasta components during processing: Impact on starch and protein digestibility and allergenicity. *Trends Food Sci. Technol.* **2009**, *20*, 521–532. [CrossRef]
18. Melini, V.; Melini, F.; Acquistucci, R. Phenolic compounds and bioaccessibility thereof in functional pasta. *Antioxidants* **2020**, *9*, 343. [CrossRef]
19. Wahanik, A.L.; Chang, Y.K.; Clerici, M.T.P.S. How to make pastas healthier? *Food Rev. Int.* **2018**, *34*, 52–69. [CrossRef]

20. Mercier, S.; Moresoli, C.; Mondor, M.; Villeneuve, S.; Marcos, B. A meta-analysis of enriched pasta: What are the effects of enrichment and process specifications on the quality attributes of pasta? *Compr. Rev. Food Sci. Food Saf.* **2016**, *15*, 685–704. [CrossRef]
21. Li, M.; Zhu, K.X.; Guo, X.N.; Brijs, K.; Zhou, H.M. Natural additives in wheat-based pasta and noodle products: Opportunities for enhanced nutritional and functional properties. *Compr. Rev. Food Sci. Food Saf.* **2014**, *13*, 347–357. [CrossRef]
22. Krishnan, M.; Prabhasankar, P. Health based pasta: Redefining the concept of the next generation convenience food. *Crit. Rev. Food Sci.* **2012**, *52*, 9–20. [CrossRef]
23. Fuad, T.; Prabhasankar, P. Role of ingredients in pasta product quality: A review on recent developments. *Crit. Rev. Food Sci.* **2010**, *50*, 787–798. [CrossRef]
24. Marti, A.; Cattaneo, S.; Benedetti, S.; Buratti, S.; Abbasi Parizad, P.; Masotti, F.; Iametti, S.; Pagani, M.A. Characterization of whole grain pasta: Integrating physical, chemical, molecular, and instrumental sensory approaches. *J. Food Sci.* **2017**, *82*, 2583–2590. [CrossRef]
25. Cecchini, C.; Menesatti, P.; Antonucci, F.; Costa, C. Trends in research on durum wheat and pasta, a bibliometric mapping approach. *Cereal Chem.* **2020**, *97*, 581–588. [CrossRef]
26. Bresciani, A.; Giordano, D.; Vanara, F.; Blandino, M.; Marti, A. High-amylose corn in gluten-free pasta: Strategies to deliver nutritional benefits ensuring the overall quality. *Food Chem.* **2021**, *353*, 129489. [CrossRef]
27. Marti, A.; Pagani, M.A. What can play the role of gluten in gluten free pasta? *Trends Food Sci. Technol.* **2013**, *31*, 63–71. [CrossRef]
28. Marti, A.; Caramanico, R.; Bottega, G.; Pagani, M.A. Cooking behavior of rice pasta: Effect of thermal treatments and extrusion conditions. *LWT—Food Sci. Technol.* **2013**, *54*, 229–235. [CrossRef]
29. Cabrera-Chávez, F.; de la Barca, A.M.C.; Islas-Rubio, A.R.; Marti, A.; Marengo, M.; Pagani, M.A.; Bonomi, F.; Iametti, S. Molecular rearrangements in extrusion processes for the production of amaranth-enriched, gluten-free rice pasta. *LWT—Food Sci. Technol.* **2012**, *47*, 421–426. [CrossRef]
30. De Noni, I.; Pagani, M.A. Cooking properties and heat damage of dried pasta as influenced by raw material characteristics and processing conditions. *Crit. Rev. Food Sci.* **2010**, *50*, 465–472. [CrossRef]
31. Sicignano, A.; Di Monaco, R.; Masi, P.; Cavella, S. From raw material to dish: Pasta quality step by step. *J. Sci. Food Agric.* **2015**, *95*, 2579–2587. [CrossRef]
32. Murray, J.C.; Kiszonas, A.M.; Morris, C.F. Pasta production: Complexity in defining processing conditions for reference trials and quality assessment methods. *Cereal Chem.* **2017**, *94*, 791–797. [CrossRef]
33. Marti, A.; D'Egidio, M.A.; Pagani, M.A. Pasta: Quality testing methods. In *Encyclopedia of Food Grains*; Wrigley, C.W., Cork, H., Seetharaman, K., Faubion, J., Eds.; Academic Press: Waltham, MA, USA, 2015; pp. 161–165.
34. Deng, L.; Manthey, F.A. Effect of single-pass and multipass milling systems on whole wheat durum flour and whole wheat pasta quality. *Cereal Chem.* **2017**, *94*, 963–969. [CrossRef]
35. Debbouz, A.; Doetkott, C. Effect of process variables on spaghetti quality. *Cereal Chem.* **1996**, *73*, 672–676.
36. De la Peña, E.; Manthey, F.A. Effect of formulation and dough hydration level on extrusion, physical and cooked qualities of nontraditional spaghetti. *J. Food Process Eng.* **2017**, *40*, e12301. [CrossRef]
37. De la Peña, E.; Manthey, F.A.; Patel, B.K.; Campanella, O.H. Rheological properties of pasta dough during pasta extrusion: Effect of moisture and dough formulation. *J. Cereal Sci.* **2014**, *60*, 346–351. [CrossRef]
38. Manthey, F.A.; Schorno, A.L. Physical and cooking quality of spaghetti made from whole wheat durum. *Cereal Chem.* **2002**, *79*, 504–510. [CrossRef]
39. Yalla, S.R.; Manthey, F.A. Effect of semolina and absorption level on extrusion of spaghetti containing non-traditional ingredients. *J. Sci. Food Agric.* **2006**, *86*, 841–848. [CrossRef]
40. La Gatta, B.; Rutigliano, M.; Padalino, L.; Conte, A.; Del Nobile, M.A.; Di Luccia, A. The role of hydration on the cooking quality of bran-enriched pasta. *LWT—Food Sci. Technol.* **2017**, *84*, 489–496. [CrossRef]
41. Dexter, J.E. Grain, paste products: Pasta and Asian noodles. In *Principles and Applications*; Scott Smith, J., Hui, Y.H., Eds.; Blackwell Publishing: Hoboken, NJ, USA, 2004; pp. 249–271.
42. Carini, E.; Vittadini, E.; Curti, E.; Antoniazzi, F.; Viazzani, P. Effect of different mixers on physicochemical properties and water status of extruded and laminated fresh pasta. *Food Chem.* **2010**, *122*, 462–469. [CrossRef]
43. Manthey, F.A.; Yalla, S.R.; Dick, T.J.; Badaruddin, M. Extrusion properties and cooking quality of spaghetti containing buckwheat bran flour. *Cereal Chem.* **2004**, *81*, 232–236. [CrossRef]
44. Sarghini, F.; Cavella, S.; Torrieri, E.; Masi, P. Experimental analysis of mass transport and mixing in a single screw extruder for semolina dough. *J. Food Eng.* **2005**, *68*, 497–503. [CrossRef]
45. Marti, A.; Seetharaman, K.; Pagani, M.A. Rheological approaches suitable for investigating starch and protein properties related to cooking quality of durum wheat pasta. *J. Food Qual.* **2013**, *36*, 133–138. [CrossRef]
46. Marti, A.; D'Egidio, M.G.; Dreisoerner, J.; Seetharaman, K.; Pagani, M.A. Temperature-induced changes in dough elasticity as a useful tool in defining the firmness of cooked pasta. *Eur. Food Res. Technol.* **2014**, *238*, 333–336. [CrossRef]
47. Bonomi, F.; D'Egidio, M.G.; Iametti, S.; Marengo, M.; Marti, A.; Pagani, M.A.; Ragg, E.M. Structure–quality relationship in commercial pasta: A molecular glimpse. *Food Chem.* **2012**, *135*, 348–355. [CrossRef] [PubMed]
48. Diantom, A.; Curti, E.; Carini, E.; Boukid, F.; Mattarozzi, M.; Vodovotz, Y.; Careri, M.; Vittadini, E. A multi-scale approach for pasta quality features assessment. *LWT—Food Sci. Technol.* **2019**, *101*, 285–292. [CrossRef]

49. Abecassis, J.; Abbou, R.; Chaurand, M.; Morel, M.H.; Vernoux, P. Influence of extrusion speed, temperature, and pressure in the extruder and on pasta quality. *Cereal Chem.* **1994**, *71*, 247–253.
50. Nasehi, B.; Mortazavi, S.A.; Razavi, S.M.A.; Mahallati, M.N.; Karim, R. Optimization of the extrusion conditions and formulation of spaghetti enriched with full-fat soy flour based on the cooking and color quality. *Int. J. Food Sci. Nutr.* **2009**, *60*, 205–214. [CrossRef] [PubMed]
51. Jalgaonkar, K.; Jha, S.K.; Mahawar, M.K.; Yadav, D.N. Pearl millet based pasta: Optimization of extrusion process through response surface methodology. *J. Food Sci. Technol.* **2019**, *56*, 1134–1144. [CrossRef] [PubMed]
52. Mercier, S.; Des Marchais, L.; Villeneuve, S.; Foisy, M. Effect of die material on engineering properties of dried pasta. *Procedia Food Sci.* **2011**, *1*, 557–562. [CrossRef]
53. Trasca, T.I.; Groza, I.; Rinovetz, A.; Rivis, A.; Radoi, B. The study of the behavior of polytetrafluorethylene dies for pasta extrusion comparative with bronze dies. *Rev. Mater. Plast.* **2007**, *44*, 307–309.
54. Lucisano, M.; Pagani, M.A.; Mariotti, M.; Locatelli, D.P. Influence of die material on pasta characteristics. *Food Res. Int.* **2008**, *41*, 646–652. [CrossRef]
55. Zardetto, S.; Dalla Rosa, M. Effect of extrusion process on properties of cooked, fresh egg pasta. *J. Food Eng.* **2009**, *92*, 70–77. [CrossRef]
56. Marti, A.; Fongaro, L.; Rossi, M.; Lucisano, M.; Pagani, M.A. Quality characteristics of dried pasta enriched with buckwheat flour. *Int. J. Food Sci. Technol.* **2011**, *46*, 2393–2400. [CrossRef]
57. Carini, E.; Vittadini, E.; Curti, E.; Antoniazzi, F. Effects of different shaping modes on physico-chemical properties and water status of fresh pasta. *J. Food Eng.* **2009**, *93*, 400–406. [CrossRef]
58. Zweifel, C.; Handschin, S.; Escher, F.; Conde-Petit, B. Influence of high-temperature drying on structural and textural properties of durum wheat pasta. *Cereal Chem.* **2003**, *80*, 159–167. [CrossRef]
59. Cubadda, R.E.; Carcea, M.; Marconi, E.; Trivisonno, M.C. Influence of gluten proteins and drying temperature on the cooking quality of durum wheat pasta. *Cereal Chem.* **2007**, *84*, 48–55. [CrossRef]
60. Bruneel, C.; Pareyt, B.; Brijs, K.; Delcour, J.A. The impact of the protein network on the pasting and cooking properties of dry pasta products. *Food Chem.* **2010**, *120*, 371–378. [CrossRef]
61. Padalino, L.; Caliandro, R.; Chita, G.; Conte, A.; Del Nobile, M.A. Study of drying process on starch structural properties and their effect on semolina pasta sensory quality. *Carbohydr. Polym.* **2016**, *153*, 229–235. [CrossRef] [PubMed]
62. D'Egidio, M.G.; Mariani, B.M.; Nardi, S.; Novaro, P. Viscoelastograph measures and total organic matter test: Suitability in evaluating textural characteristics of cooked pasta. *Cereal Chem.* **1993**, *70*, 67–72.
63. Simonato, B.; Curioni, A.; Pasini, G. Digestibility of pasta made with three wheat types: A preliminary study. *Food Chem.* **2015**, *174*, 219–225. [CrossRef]
64. Stuknytė, M.; Cattaneo, S.; Pagani, M.A.; Marti, A.; Micard, V.; Hogenboom, J.; De Noni, I. Spaghetti from durum wheat: Effect of drying conditions on heat damage, ultrastructure and in vitro digestibility. *Food Chem.* **2014**, *149*, 40–46. [CrossRef]
65. De Zorzi, M.; Curioni, A.; Simonato, B.; Giannattasio, M.; Pasini, G. Effect of pasta drying temperature on gastrointestinal digestibility and allergenicity of durum wheat proteins. *Food Chem.* **2007**, *104*, 353–363. [CrossRef]
66. Petitot, M.; Brossard, C.; Barron, C.; Larré, C.; Morel, M.H.; Micard, V. Modification of pasta structure induced by high drying temperatures. Effects on the in vitro digestibility of protein and starch fractions and the potential allergenicity of protein hydrolysates. *Food Chem.* **2009**, *116*, 401–412. [CrossRef]
67. West, R.; Seetharaman, K.; Duizer, L.M. Effect of drying profile and whole grain content on flavour and texture of pasta. *J. Cereal Sci.* **2013**, *58*, 82–88. [CrossRef]
68. West, R.; Duizer, L.; Seetharaman, K. The effect of drying and whole grain content on the pasting, physicochemical and qualitative properties of pasta. *Starch-Stärke* **2013**, *65*, 645–652. [CrossRef]
69. Seiquer, I.; Díaz-Alguacil, J.; Delgado-Andrade, C.; López-Frías, M.; Muñoz Hoyos, A.; Galdó, G.; Navarro, M.P. Diets rich in Maillard reaction products affect protein digestibility in adolescent males aged 11–14 y. *Am. J. Clin. Nutr.* **2006**, *83*, 1082–1088. [CrossRef] [PubMed]
70. Uribarri, J.; Woodruff, S.; Goodman, S.; Cai, W.; Chen, X.; Pyzik, R.; Yong, A.; Striker, G.E.; Vlassara, H. Advanced glycation end products in foods and a practical guide to their reduction in the diet. *J. Am. Diet. Assoc.* **2010**, *110*, 911–916. [CrossRef]
71. Marti, A.; Bottega, G.; Patacca, C.; Pagani, M.A. Survey on heat damage of dry pasta and its relations with characteristics of raw material and process conditions. *Tec. Mol.* **2013**, *64*, 286–294.
72. De Pilli, T.; Giuliani, R.; Derossi, A.; Severini, C. Study of cooking quality of spaghetti dried through microwaves and comparison with hot air dried pasta. *J. Food Eng.* **2009**, *95*, 453–459. [CrossRef]
73. Altan, A.; Maskan, M. Microwave assisted drying of short-cut (ditalini) macaroni: Drying characteristics and effect of drying processes on starch properties. *Food Res. Int.* **2005**, *38*, 787–796. [CrossRef]
74. Berteli, M.N.; Marsaioli, A., Jr. Evaluation of short cut pasta air dehydration assisted by microwaves as compared to the conventional drying process. *J. Food Eng.* **2005**, *68*, 175–183. [CrossRef]
75. Piwińska, M.; Wyrwisz, J.; Kurek, M.A.; Wierzbicka, A. Effect of drying methods on the physical properties of durum wheat pasta. *CYTA—J. Food* **2016**, *14*, 523–528. [CrossRef]
76. Piwińska, M.; Wyrwisz, J.; Kurek, M.; Wierzbicka, A. Hydration and physical properties of vacuum-dried durum wheat semolina pasta with high-fiber oat powder. *LWT—Food Sci. Technol.* **2015**, *63*, 647–653. [CrossRef]

77. Sarghini, F.; Romano, A.; Masi, P. Experimental analysis and numerical simulation of pasta dough extrusion process. *J. Food Eng.* **2016**, *176*, 56–70. [CrossRef]
78. Camelo-Méndez, G.A.; Ferruzzi, M.G.; González-Aguilar, G.A.; Bello-Pérez, L.A. Carbohydrate and phytochemical digestibility in pasta. *Food Eng. Rev.* **2016**, *8*, 76–89. [CrossRef]
79. Duizer, L.M.; Walker, S.B. The application of sensory science to the evaluation of grain-based foods. In *Encyclopedia of Food Grains*; Wrigley, C.W., Cork, H., Seetharaman, K., Faubion, J., Eds.; Academic Press: Waltham, MA, USA, 2015; pp. 144–153.

foods

MDPI

Article

Physicochemical and Sensorial Characterization of Artisanal Pasta from the Occitanie Region in France

Loubnah Belahcen [1], Denis Cassan [2], Elodie Canaguier [2], Marie-Hélène Robin [3,4], Yuna Chiffoleau [5], Marie-Françoise Samson [2,†] and Gwénaëlle Jard [1,*,†]

1 Food Science and Nutrition Department, INP EI-Purpan, Univ. Toulouse, CEDEX 3, 31076 Toulouse, France
2 IATE, Univ. Montpellier, INRAE, Institut Agro, CEDEX 2, 34060 Montpellier, France
3 AGIR, INRAE, 31326 Castanet Tolosan, France
4 Plant Science and Agronomy Department, INP EI-Purpan, Univ. Toulouse, CEDEX 3, 31076 Toulouse, France
5 INNOVATION, Univ. Montpellier, CIRAD, INRAE, Institut Agro, CEDEX 2, 34060 Montpellier, France
* Correspondence: gwenaelle.jard@purpan.fr; Tel.: +33-7-62-06-77-79
† These authors contributed equally to this work.

Abstract: Artisanal pasta made from wheat or underutilized cereal flours has grown in popularity with the expansion of the local and short food chains. Artisanal pasta makers do not use the same raw materials or production processes, leading to great variability in the final product. The purpose of the study is to determine the physicochemical and sensory characteristics of artisanal pasta made from durum wheat flour. Seven brands of fusilli pasta manufactured in the Occitanie region (France) were selected and analyzed in terms of their physicochemical composition (protein and ash content in dry samples), cooking properties (optimal cooking time, water absorption, and cooking loss), sensory characteristics (Pivot profile), and consumer appreciation. Differences in the physicochemical characteristics of the dry pasta samples partly explain the variations in pasta characteristics measured after cooking. The Pivot profile varied among pasta brands, but no major differences in hedonic properties were identified. To our knowledge, this is the first time that artisanal pasta made from flour has been characterized in terms of its physicochemical and sensory properties, which highlights the diversity of products on the market.

Keywords: artisanal pasta; cooking quality; Pivot profile; sensory analysis

Citation: Belahcen, L.; Cassan, D.; Canaguier, E.; Robin, M.-H.; Chiffoleau, Y.; Samson, M.-F.; Jard, G. Physicochemical and Sensorial Characterization of Artisanal Pasta from the Occitanie Region in France. *Foods* **2022**, *11*, 3208. https://doi.org/10.3390/foods11203208

Academic Editor: Mike Sissons

Received: 5 September 2022
Accepted: 6 October 2022
Published: 14 October 2022

1. Introduction

According to French law [1], pasta products must be made exclusively from durum wheat semolina, with precise criteria for particle size, color, and ash and protein contents [2]. Since the end of the 1990s, in response to consumer demand for products perceived as healthier and more respectful of the environment, local and short production chains have been developed [3,4]. This context has favored the growth of the so-called "artisanal" pasta in food markets, with a corresponding expansion of the literature on these products since 2010 [5]. Artisanal pasta is made by small-scale processors collaborating with farmers (in cooperatives or associations) from locally produced cereals, or by farmers–processors processing their own cereals, both selling directly and locally the products. In France, more than 250,000 tons of pasta are produced each year [6], including an unspecified portion of artisanal pasta. Artisanal pasta differs from the so-called "industrial" pasta in terms of the raw materials, the milling and pasta-making processes used, leading to different nutritional, culinary and sensory qualities [7].

The types and diversity of wheats used affect the quality of the pasta produced. In the Occitanie region, artisanal pasta is produced from different durum wheat varieties, from modern breeding programs (when these are well adapted to the area or to organic farming), participatory plant breeding projects (e.g., LA1823 [8]) or from ancient wheat varieties (e.g., Bidi17). Some farmers grow other underutilized wheat species, such as einkorn, emmer, spelt, or rivet wheat. Because of the low production volume, but mostly to control all the steps in the production chain, many farmers prefer to grind their grains using stone mills a more traditional approach considered to yield healthier products.

Artisanal pasta is produced from semi- or wholemeal wheat flour instead of semolina because it is easier for small-scale processors and farmers to process it locally. Milling and especially particle size are known to have a considerable effect on pasta cooking quality [9,10] and color [11,12]. The characteristics of the end-product depend also on the drying process used. Artisanal pasta is commonly dried at low temperature (<50 °C). High temperature drying cycles (>65 °C) improve the cooking quality of pasta made from semolina [13,14] but not of pasta made from wholemeal flour. Indeed, Manthey and Schorno (2002) [15] found that wholewheat spaghetti dried at a low temperature had a better appearance and firmness than the samples dried at a high temperature despite lower cooking loss. West et al. (2013) [16] found, for wholemeal macaroni, that a short high-temperature drying process did not improve cooking quality, particularly in terms of the cooking losses measured. Unsurprisingly, the absence of standard manufacturing practices for artisanal pasta leads to a high variability in product quality.

The sensory variability of artisanal pasta can be characterized by descriptive sensory analysis (DA). The ISO 7304-1 (2016) and ISO 7304-2 (2008) [17,18] standards define three criteria (firmness, stickiness, and starch release) with which to assess the texture of pasta. A sensory attribute lexicon for dried long pasta was also recently created [19], consisting of 35 terms. However, this standardized method is time-consuming and may be unsuitable to describe artisanal short pasta made from durum wheat flour.

The so-called "alternative" sensory analysis methods, such as the Pivot profile [20], are therefore interesting as a first approach to products that have never been previously characterized by sensory analysis [21].

The objective of the present study is, therefore, to identify the physicochemical and sensory characteristics of artisanal pasta processed and sold in local food systems. To our knowledge, the overall quality of artisanal pasta has never been previously evaluated, particularly in terms of sensory appreciation.

2. Material and Methods

2.1. Raw Materials

Seven samples of fusilli-shaped pasta made from durum wheat flour were collected from a geographically representative distribution of pasta producers in the Occitanie region (southern France, Figure 1). These producers were chosen among the 30 registered pasta producers in the region to be representative of local artisanal pasta production, i.e., small scale (between 3 and 40 t/year), local production systems, pasta made from semi- or wholemeal flour and dried at a low temperature (between 40 and 50 °C, for 11 to 19 h).

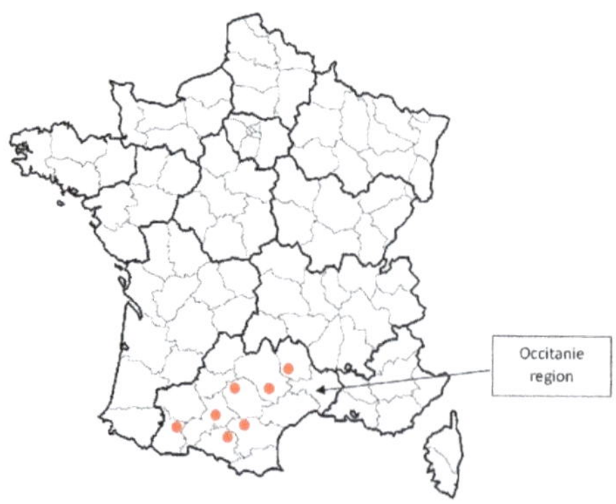

Figure 1. Geographical distribution of the pasta samples in Occitanie region.

Some of their characteristics are listed in Table 1.

Table 1. Characteristics of the studied pasta samples.

Sample Number	Unit of Production	Type of Agriculture	Type of Wheat Used	Type of Milling	Die Material
1	Small-scale cooperative processing unit	Organic	Rivet wheat *,o	Stone milling	Bronze
2	Individual small-scale farmer–processor	Organic	Bidi17 o, Senatore Capelli o, LA1823 m	Stone milling	Bronze
3	Association of farmers and small-scale miller and pasta maker	Conventional	Voilur m, Anvergur m	Stone milling	Teflon
4	Association of farmers and small-scale miller and pasta maker	Organic	Atoudur m	Stone milling	Bronze
5	Individual small-scale farmer–processor	Organic	LA1823 m, Anvergur m	Roller milling	Teflon
6	Association of farmers and small-scale miller and pasta maker	Organic	Own mix of durum wheat (~40 varieties) mixed at sowing time	Stone milling	Bronze
7	Association of farmers and small-scale pasta maker	Conventional	Miradoux m, Anvergur m	Roller milling	Teflon

o: old durum wheat; m: modern durum wheat as defined by Mefleh et al., 2019 [22]. *: Rivet wheat (*Triticum turgidum* L. ssp. turgidum) is considered a specific type of half-vitreous durum wheat different from other forms of durum wheat (*Triticum turgidum* L. ssp. durum).

2.2. Physicochemical Characterization of Dry Pasta

The following properties were measured on dry samples. The **dry matter** content of the pasta was measured in triplicate according to the approved ISO 712:2009 method [23].

Total mineral content was determined in triplicate according to the approved ISO 2171:2010 method [24].

The L*, a*, b **color** coordinates of the pasta were determined using a CR410 chroma meter (Konica Minolta, Roissy, France). The pasta samples were ground (Perten Lab mill 3303, Perkin Elmer, Haguenau, France) and the powder obtained was placed in a homogenous layer inside the black box of the chroma meter. The L* component quantifies brightness from dark (L* = 0) to bright (L* = 100), a* redness, from red (+a*) to green (−a*), and b* yellowness, from yellow (+b*) to blue (−b*). Color measurements were performed in triplicate. A yellowness index (YI) was also calculated using the Francis and Clydesdales formula [25]: $YI = 142.86\, b^*/L^*$. The **dimensions of the pasta** (length, width, and thickness of the pasta spirals) were measured with a caliper for 5 pasta samples. All samples were observed with an AZ100M multizoom microscope (Nikon Europe, Amsterdam, The Netherlands) with white LED epi-illumination at a low magnification (×2).

2.3. Physciochemical Characterization of Cooked Pasta

2.3.1. Cooking Behavior of the Pasta

Optimal cooking times (OCTs) were determined in triplicate according to the approved AACC Method 66-50.01 [26]. Briefly, the pasta was cooked in boiling, demineralized and salted water (7 $g \cdot L^{-1}$) and the OCT was defined as the time taken for the white line in the core to disappear when the pasta was crushed between two plexiglass plates, indicating that the starch had gelatinized.

Water absorption was measured as the weight gain of the pasta after cooking, expressed as a percentage of the dry weight. Briefly, 100 g of pasta was cooked at OCT in 5 L of water with 7 $g \cdot L^{-1}$ salt. The pasta was drained, rinsed twice with tap water, and the residual water was absorbed with a paper towel before weighing. This procedure was performed twice and the average value was retained.

Cooking losses, i.e., dry matter losses during cooking, were calculated for each pasta sample as follows. Briefly, 8 g of dry pasta was cooked at OCT in 300 mL of water (hardness 15 ± 1 °F) in a beaker. The cooked pasta was then freeze-dried for 72 h using a Beta 2-8 LSCbasic device (Christ, Osterode am Harz, Germany) and weighed. Cooking losses were calculated as the difference in dry matter weight between the uncooked and freeze-dried cooked pasta, expressed as a percentage of the dry matter weight before cooking.

2.3.2. Texture Analyses

A TA-XTplus texture analyzer (Stable Micro Systems, Scarsdale, Godalming, United Kingdom) equipped with a TA-93WST wire mesh extrusion fixture was used to evaluate the rheological properties of the pasta after cooking. The test involved extruding a 100 g sample of cooked (OCT) and drained pasta through a wire mesh screen. The plunger height was calibrated beforehand to 110 mm above the wire mesh at the base of the extrusion cylinder. The test was conducted in compression, at 5 mm·s^{-1} and the target distance was 105 mm. The bulk of the pasta was first compacted before being extruded through the wire mesh. The average **extrusion force** was measured on the force versus time curve between 16 and 21 s, corresponding to the last 25 mm of the stroke. The test was performed twice for two samples of each brand of pasta (2 cooked samples × 2 tests per sample = 4 replicates per product).

2.3.3. Protein Content and Protein Profile of Freeze-Dried Cooked Pasta

Freeze-dried cooked pasta was ground with an A10 basic mill (IKA, Staufen, Germany). The moisture content of the resulting powder was determined using AACC method 44-15.02 [26], and the **total protein** content was determined using the Kjeldahl method as described in AACC method 46-12.01 [26] with 5.7 s as the conversion factor. To determine the protein profiles, proteins were extracted following Morel et al. (2000) with modifications [27]. Freeze-dried and ground cooked pasta samples (160 mg) were suspended in 20 mL of sodium phosphate buffer (0.1 M, pH 6.9) containing 1% (w/v) sodium dodecyl sulfate (SDS). The suspension was stirred for 80 min at 60 °C. After centrifugation (39,000× g; 30 min; 20 °C), the supernatant containing SDS soluble proteins was collected and stored (−20 °C) until analysis. The pellet was re-suspended in 5 mL SDS–sodium phosphate buffer containing 20 mM dithioerythritol (DTE) and sonicated for 3 min at 7.5 watts. The new supernatant was stored until analysis. The proteins recovered after the different extraction steps were separated by size-exclusion high-performance liquid chromatography (SE-HPLC) using a TSKgel G4000 SWXL column (7.8 mm i.d. × 30 cm, TOSO BIOSCIENCE GmbH, Griesheim, Germany), following Dachkevitch and Autran (1989) [28] on an Alliance system (Waters, Saint Quentin en Yvelines, France). The proteins were eluted at ambient temperature with 0.1 M sodium phosphate buffer (pH 6.9) containing 0.1% (w/v) SDS at a flow rate of 0.7 mL·min^{-1} and the absorbance was measured at 214 nm. The fraction of **SDS-soluble proteins** was obtained from the area under the first peak in the chromatograms obtained, and the fraction of **DTE-soluble proteins** (i.e., after DTE reduction and sonication) from the area under the second. Both areas were converted into protein contents and the non-extracted protein fraction was calculated by subtracting the sum of the SDS-soluble and DTE-soluble protein contents from the Kjeldahl total protein content. When the fraction of non-extracted protein was negative, due to high recovery, this value was forced to 0 and the sum of the SDS soluble and DTE extracted protein fractions was corrected to reach 100%.

2.3.4. In Vitro Pasta Digestibility Tests

The digestibility of cooked pasta samples was evaluated by measuring the rate of proteolysis in vitro using the Protein Digestibility Assay kit (Neogen, Auchincruive, UK), with a few modifications to the standard procedure. For each brand of pasta, proteolysis was carried out on two 250 mg samples, according to manufacturer specifications and trypsin/chymotrypsin digestion was conducted for 4 h. Digestion was stopped by immersing the tubes in boiling water. After cooling, the tubes were centrifuged at 4696× g for 15 min at 15 °C. After centrifugation, the supernatants were set aside and the pellets frozen. Protein digestibility was then estimated by determining the amount of nitrogen remaining in the pellets from two extractions of the same sample using the Kjeldahl method. The extent of proteolysis after 1 h of peptic digestion followed by 4 h of tryptin/chymotryptin digestion was expressed as the percentage of the initial protein content of the sample that remained after digestion.

2.4. Sensory Analysis

2.4.1. Descriptive Sensory Analysis

Pivot profiling was carried out as described by Thuillier et al. (2015) [20]. Fifty-seven panelists aged 20 to 50 years (58% female) were recruited. Among the panelists, 28.5% consumed regularly artisanal pasta, 34.6% consumed artisanal pasta once in a while, and 36.7% never consumed artisanal pasta. At the beginning of each session, participants confirmed they were willing and consented to participate and that they did not have any food allergies.

Ten-gram samples of cooked pasta (OCT) were served to each panelist on white plates, at room temperature. Sensory analysis was carried out in separate boxes under white light. The samples were coded with three-digit random numbers and were presented in sequential monadic order by pair (one sample and one pivot). A Latin square design was used to balance the sample order.

Before the start of the sensory analysis session, pasta brand 3 (Pasta 3) was chosen as the pivot because it was the most "central product" in terms of color, texture, and taste. Participants evaluated the samples according to three criteria (visual appearance, in-mouth texture, and flavor) with free comments (attributes) as "more" or "less" than the pivot. Negations and hedonic comments were avoided.

All attributes were listed and categorized and then grouped by synonyms after discussions with panelists and a dictionary of synonyms was generated using the software TASTEL (version 2015.2, ABT informatique, Rouvroy-sur-Marne, France) The main descriptors were selected by frequency of citation. The number of negative comments were subtracted from the number of positive comments for each descriptor and the resulting scores were adjusted to obtain only positive scores. A contingency table was obtained, and the pivot was integrated by attributing scores of 0 before adjustment, as described by Fonseca et al. (2016) [29].

2.4.2. Ranking Test

A ranking test was performed according to NF ISO 8587 to classify the pasta samples in terms of hedonic properties [30]. Sixty-five pasta consumers aged 19 to 60 years (55% female) were recruited. At the beginning of each session, participants confirmed they were willing and consented to participate and that they did not have any food allergies. Ten-gram samples of cooked pasta (OCT) were served to each panelist on white plates, at room temperature. Sensory analysis was carried out in separate boxes under white light. Samples coded with three-digit random numbers were presented simultaneously to the participants and ranked from most (rank 1) to least preferred (rank 7).

2.5. Statistical Analyses

All statistical analyses were performed with XLSTAT (Addinsoft, Paris, France). The results were considered statistically significant at $p < 0.05$.

The results for the **physicochemical characteristics of dry and cooked pasta** (moisture, ash, color, firmness, OCT, and water absorption) were compared using Kruskal–Wallis tests. For these analyses, multiple pairwise comparisons were performed with the Conover/Iman test. Correlations between parameters were measured using Pearson correlation matrices and principal component analysis (PCA).

A sensory map was generated from the **Pivot profile** contingency table using correspondence analysis (CA). Correlations between descriptors and products were investigated using a global chi-squared test and, if this was significant, chi-squared tests were performed cell-by-cell as previously described by Fonseca et al. (2016) [29].

The **ranking test** data were analyzed using Friedman's test.

3. Results

The physicochemical characteristics of the different brands of pasta (dry and cooked samples) are listed in Table 2.

3.1. Dry Pasta Characterization

The moisture content of the pasta was below 12.5% in all cases, in accordance with French regulations. Mineral contents were high and ranged from 1.00 to 1.88 % DM, with some samples containing more than the maximum of 1.3 % recommended in France for durum wheat semolina and pasta [1].

The shape of the fusilli varied between brands. Some were shorter than others (approximately 25 mm for pasta 1, 2, 4, and 6 versus approximately 30 mm for pasta 3, 5, and 7). Samples 2, 3, 5, and 7 were more regular in shape and narrower than samples 1, 4, and 6. The surfaces of pasta samples 1, 2, 4, and 6 were irregular and rough, a typical characteristic of pasta prepared with bronze extrusion dies. White specks were observed on pasta samples 2, 3, 5, and 7 (Figure 2), probably due to insufficient hydration [31]. Samples 2, 3, 6, and 7 also had black/brown specks (black point disease and bran particles).

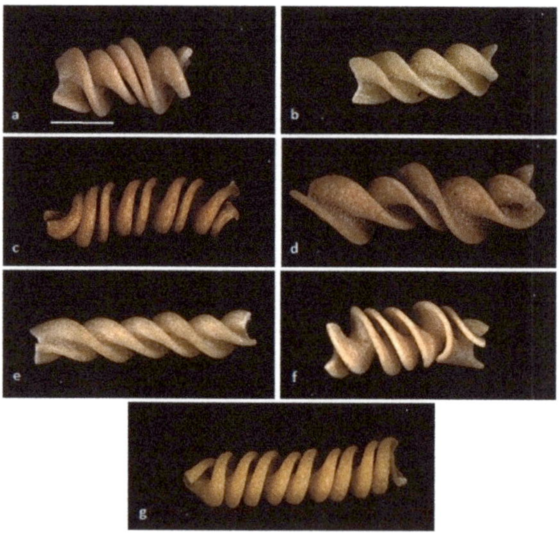

Figure 2. Diversity of shape and appearance of different brands of fusilli pasta: photographs of (**a**) pasta 1, (**b**) pasta 2, (**c**) pasta 3, (**d**) pasta 4, (**e**) pasta 5, (**f**) pasta 6, and (**g**) pasta 7 (white bar = 1 cm).

In terms of L*, a*, b* characteristics (Table 2), the brightness (L*) of the pasta ranged from 66.88 to 75.78. Samples 2 and 7 were brighter, samples 3 and 4 darker, and samples 1, 5, and 6 intermediate. The redness (a*) component ranged from 2.03 to 4.76 and none of the pasta samples were similar in this respect. Yellowness (b*) ranged from 19.66 to 24.22, with samples 3 and 7 being the most yellow and samples 1 and 2 being the least yellow. The highest YI was measured in pasta 3 and the lowest in pasta 5.

Table 2. Main characteristics of the dry and cooked pasta.

Samples	Dry Pasta						Cooked Pasta									
	Moisture (%)	Ash (g/100 g DM)	Color				Firmness (N)	OCT (min)	Water Absorption (%)	Cooking Losses (%)	Protein Content (g/100 g DM)	SDS-Soluble Protein (%)	DTE-Soluble Protein (%)	Unextractable Protein (%)	Protein Content After Proteolysis (%)
			L*	a*	b*	Yellowness Index									
Pasta 1	8.79 ^{a}	1.00 ^{a}	70.09 ^{bc}	3.95 ^{d}	19.93 ^{a}	40.62 ^{bc}	89.80 ^{a}	11.50 ^{f}	158.3 ^{abc}	11.20	11.25	24.59	60.41	14.99	79.34
Pasta 2	8.30 ^{ab}	1.56 ^{abc}	74.46 ^{de}	2.84 ^{b}	19.66 ^{a}	37.73 ^{a}	99.75 ^{a}	8.85 ^{cd}	166.1 ^{bc}	11.47	11.32	22.62	66.80	10.58	80.29
Pasta 3	8.12 ^{bc}	1.63 ^{bc}	67.46 ^{ab}	4.76 ^{g}	23.26 ^{c}	49.25 ^{e}	61.06 ^{c}	7.16 ^{a}	196.3 ^{d}	14.25	11.92	28.02	71.96	0.03	80.82
Pasta 4	7.38 ^{d}	1.66 ^{bc}	66.88 ^{a}	4.24 ^{e}	22.15 ^{bc}	47.31 ^{cd}	91.05 ^{a}	7.53 ^{bc}	123.9 ^{a}	14.02	13.39	39.82	57.40	2.78	64.37
Pasta 5	8.11 ^{bc}	1.30 ^{ab}	67.36 ^{a}	4.44 ^{f}	22.62 ^{bc}	47.96 ^{de}	94.57 ^{a}	6.00 ^{a}	152.7 ^{ab}	11.50	10.76	31.71	58.30	9.99	79.82
Pasta 6	7.99 ^{cd}	1.88 ^{c}	74.01 ^{cd}	2.95 ^{c}	20.62 ^{ab}	39.80 ^{ab}	63.88 ^{bc}	11.00 ^{ef}	196.7 ^{d}	15.01	9.84	27.62	60.98	11.40	81.84
Pasta 7	7.18 ^{d}	1.29 ^{ab}	75.78 ^{e}	2.03 ^{a}	24.22 ^{c}	43.77 ^{bc}	73.43 ^{b}	9.83 ^{de}	189.4 ^{cd}	12.56	13.07	21.43	61.63	16.93	75.35
Mean	7.98	1.47	70.86	3.60	21.78	43.96	81.93	8.84	169.0	12.86	11.65	27.97	62.50	9.53	77.40
SD	0.55	0.29	3.82	1.00	1.74	3.56	15.58	2.05	26.9	1.56	1.26	6.29	5.15	6.13	6.10

$^{a–f}$ Mean values labeled with the same letter(s) in the same column are not significantly different ($p > 0.05$) according to Conover/Iman tests.

3.2. Cooking Behavior of the Pasta

The optimal cooking times ranged from 6 to 11.5 min. Pastas 5 and 3 had the shortest OCTs, pastas 6 and 1 the longest, and the other three brands had OCTs close to the overall mean (Table 2).

No significant differences in texture were observed among pasta samples 1, 2, 4, and 5 and among samples 3, 6, and 7. The first group was firmer (mean firmness, 93.79 N) than the second (mean firmness, 66.12 N). Water absorption ranged from 123.9% to 196.7%, with, as for the firmness, two groups: samples 3, 6, and 7 with a mean water absorption of 194.1% and samples 1, 2, 4, and 5 with a mean water absorption of 150.2%. Pasta 4 had the lowest water absorption and was among the firmest.

Pasta cooking losses ranged from 11.20 to 15.01 % with an average value of 12.86 $\pm$ 1.56%.

3.3. Protein Profile and Protein In Vitro Digestion

The protein content of the pasta ranged from 9.84 to 13.39 g/100 g DM, with a mean of 11.65 $\pm$ 1.26 g/100 g DM. Pasta 6 had a much lower protein level than that expected for durum wheat pasta. The percentages of SDS-soluble, DTE-soluble (after sonication), and unextractable proteins are reported in Table 2. All the samples of cooked pasta had high concentrations of aggregated proteins, as shown by the high percentages of DTE-soluble and unextractable proteins. Low SDS-soluble protein fractions may indicate the formation of additional disulfide bonds during processing and cooking. Pasta 4 had the highest concentration of SDS-soluble protein (39.82 %) and pasta 7 the lowest (21.43%), both indicating a high degree of protein aggregation during processing and cooking. All samples except for pasta 3 and 4 (0.03 and 2.78%, respectively) had high concentrations of unextractable proteins. In the protein digestibility assays, pasta 4 differed significantly from the other samples with a much lower level of proteins remaining after 5 h proteolysis (64.37%).

3.4. Correlation between Parameters

Several of the physical and chemical parameters in the Pearson matrix were significantly associated ($p < 0.05$). Cooking losses were significatively correlated with ash contents ($p = 0.02$). Firmness and water absorption were negatively correlated. The protein content in the cooked pasta was negatively correlated with YI and was positively correlated with protein content after 5 h proteolysis. Firmness was not associated with protein content. Cooking losses were negatively, but not significantly, correlated with firmness (correlation factor, -0.754; $p = 0.05$).

The first two principal components (Figure 3) accounted for 64.26% of the variability of the data. Pasta 4 was characterized by a high level of soluble proteins, high YI, low water absorption, and low remaining protein content after 5 h proteolysis. Pasta samples 1, 2, and 5 were distinguished by greater firmness, lower cooking losses, and lower mineral content. Samples 3 and 6 were softer, with higher water absorption, ash content, and cooking loses. Pasta 7 had intermediate values of a number of features (ash content, YI, firmness, and OCT).

3.5. Sensory (Pivot Profile) Analysis

The results of the Pivot profile analysis were summarized in a contingency table (Table 3) and a CA map (Figure 4). The words used by participants to describe the samples (1485) were analyzed and grouped by meaning. Fifteen descriptors accounted for 79% of the terms used and were classified as follows: seven terms describing the appearance of the pasta (bright, yellow, dark, structured, speckled, unstructured, and compact), five terms for the texture (hard, pasty, soft, grainy, and melting), and three terms for the flavor (dull, flavor intensity, and salty).

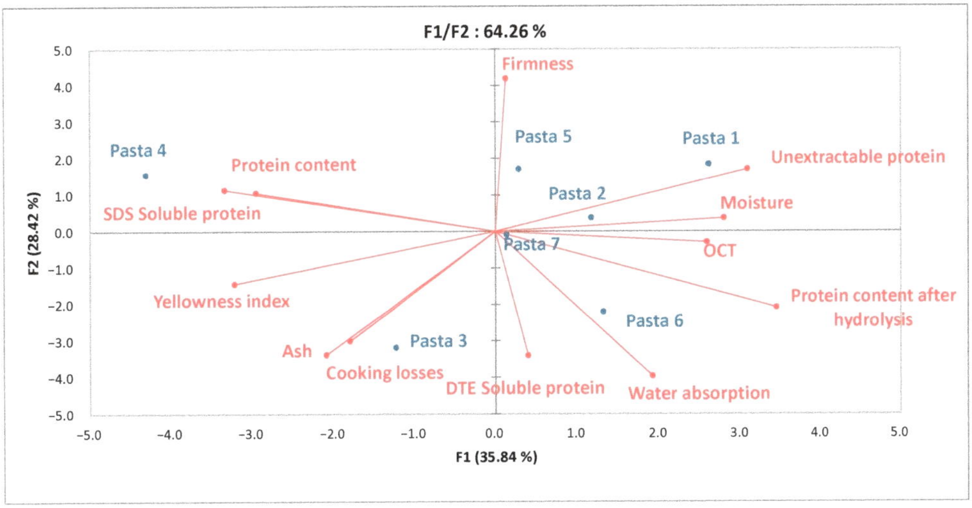

Figure 3. PCA plot of the pasta samples in terms of their physicochemical characteristics.

Table 3. Contingency table obtained from the Pivot analysis.

Category	Descriptors	Pasta 1	Pasta 2	Pasta 4	Pasta 5	Pasta 6	Pasta 7
Visual Appearance	Bright	6 (−)	52 (+)	11	0 (−)	2 (−)	46 (+)
	Yellow	43	61	61 (+)	46	52	37
	Dark	65 (+)	19 (−)	47	60	61	42
	Structured	38	43	63 (+)	40	38	48
	Speckled	61	50	51	57	61	58
	Unstructured	54	55	37	53	51	48
	Compact	51	47	42	48	50	48
Texture in mouth	Hard	43 (+)	5 (−)	53 (+)	19	27	12 (−)
	Pasty	55	46	13 (−)	60 (+)	51	35
	Soft	52	60	42	55	55	60
	Grainy	61	52	55	53	57	48
	Melting	42	57	51	54	53	56
Flavor	Dull	21 (−)	53 (+)	22 (−)	25	35	45 (+)
	Flavor intensity	70	35 (−)	74 (+)	76 (+)	64	49
	Salty	48	52	49	56	53	51

(−/orange color) and (+/green color) indicate significant results ($p < 0.05$, chi-squared test).

The chi-squared test performed on the contingency table was significant ($p < 0.001$); therefore, further chi-squared tests were performed on a cell-by-cell basis to investigate associations between descriptors and specific samples [29]. Results are shown alongside the data in Table 3. In terms of appearance, samples 2 and 7 were considered brighter and samples 1, 5, and 6 less bright. Pasta 4 was deemed more yellow and structured. There were no significant differences between samples in terms of the descriptors speckled, unstructured, and compact, which were therefore not discriminating for these products. In terms of texture, pastas 1 and 4 were classified as harder and pastas 2 and 7 less hard. Pasta 4 was judged to be less pasty and pasta 5 more pasty than the others. The descriptors grainy, soft, and melting were not differentiating for these samples, despite having been used frequently by participants. In terms of flavor, samples 2 and 7 were considered duller, whereas pastas 4 and 5 were described as being more intense in flavor. Finally, "salty" was not a differentiating descriptor in this analysis.

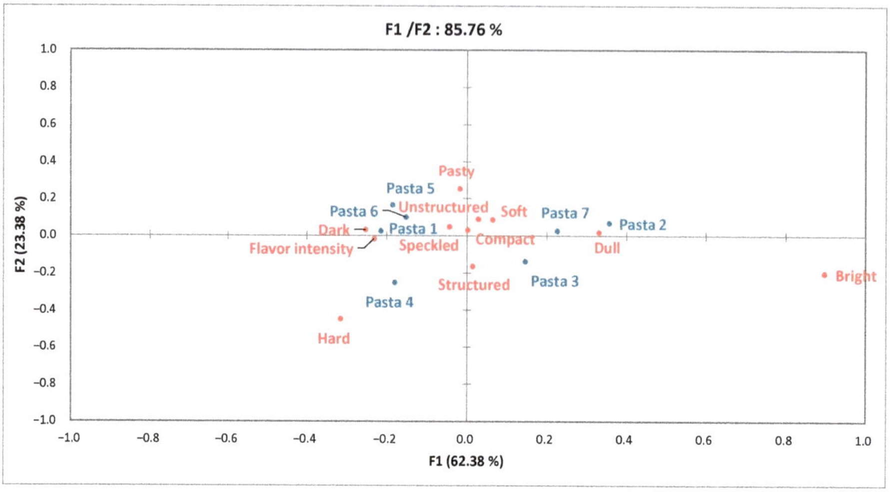

Figure 4. Correspondence analysis of the contingency table obtained from the Pivot analysis.

The first two dimensions of the CA accounted for 85.76% of the variability of the data. Pasta 3 (the pivot) was integrated into the contingency table for the map representation (Figure 4). Pasta 4 stood out from the other through its hard, non-pasty texture and more intense flavor. Samples 1, 5, and 6 formed a cluster and of more colorful and intensely flavored pasta. Samples 2 and 7 were less dark and were duller in flavor.

3.6. Sensory Appreciation

Pasta 6 was the least appreciated pasta and pasta 1 the most appreciated (Friedman test, Figure 5). There were no significant differences in rank between pasta samples 3, 4, 5, and 7 and pasta 1, or between this group and pasta 2.

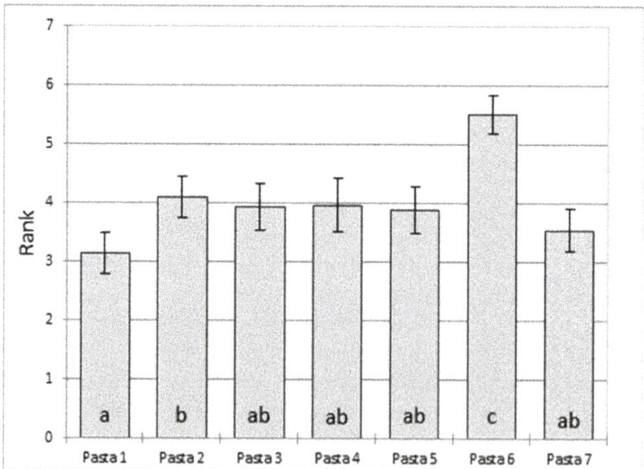

Figure 5. Results of ranking tests based on hedonic characteristics. a–c Mean values labeled with the same letter are not significantly different ($p > 0.05$).

4. Discussion

The aim of this study was to characterize artisanal pasta made from durum wheat flour produced in the Occitanie region in France. Samples of dry and cooked pasta were analyzed. The literature on industrially produced pasta quality is vast, especially for durum wheat spaghetti made from semolina [5,32,33]. To our knowledge, however, no study has previously been performed on the physicochemical properties and sensory quality of artisanal pasta made from durum wheat flour.

The cooking and the organoleptic qualities of pasta are known to depend on the physicochemical characteristics of the durum wheat flour or semolina used (ash, protein, and color) and on the specifics of the manufacturing process (milling, hydration, mixing, extrusion, and drying) [33]. The pasta-making process should lead to the formation of a protein network (gluten) that entraps starch granules and prevents their leaching during cooking to produce pasta with a compact structure [34]. The characteristics of cooked pasta (firmness, dry matter loss, and water absorption) can be explained in part by its physicochemical properties before cooking (especially its ash and protein content), whose variability explains the diversity of products encountered. The pasta samples studied here were produced from semi-whole or wholemeal flour, as reflected by their rather high ash contents (mean, 1.47%; range, from 1.00 to 1.88%). Industrially produced pasta is generally made from durum wheat semolina with low ash contents (0.6–0.9%) [35]. In accordance with the results from Pearson correlation matrix, the high ash content led to high cooking losses, high water absorption, and low firmness, probably because of bran particles weakening the protein network [5]. The protein content measured in our study was in average of 11.65 % DM, in the range of the industrial pasta protein content encountered in the literature [19,35]. A high protein content is required to produce pasta with good cooking quality [33]. The quantity but also the quality of proteins affects textural properties [22]. Indeed, pasta with a low protein content and less gluten will not have a protein network capable of preventing the leaching of starch granules during cooking and will therefore be stickier. A high protein content is a key parameter in the drying process even more impactful at low temperatures. D'Egidio et al. (1990) found indeed that protein content and gluten quality both played a crucial role if a low drying temperature was used, whereas with high-temperature drying, gluten quality was less important than protein content [36]. No direct correlation between pasta firmness and both quantity and quality of protein was observed in our study. Moreover, data obtained on the degree of gluten polymerization and its susceptibility to proteolysis after cooking suggest that pasta samples differ in the structure of their gluten networks as previously explained by Petitot et al. [27] and Bruneel et al. [37] with less accessible gluten networks corresponding to firmer pasta. It could be explained by different processing parameters (mixing, extrusion, and drying) that are well known to affect the gluten network in pasta and thus its firmness [5].

Color is an important purchasing criterion for pasta consumers and a discriminating factor between brands [38]. The studied samples varied in color from bright yellow to dull brown. No direct correlation was found between the brightness, redness, and yellowness (L^*, a^*, b^*) of the dry samples and the physicochemical variables considered. The color of pasta depends on several biochemical and technological factors, such as the quantity of yellow pigments and soluble brown pigments, the activity of enzymes such as polyphenol oxidase and peroxidase, the protein content, the ash content, and the particle size of the flour [39–41]. For example, pasta made from flour tends to be brighter and less yellow [12], but high ash and protein contents have a negative effect on brightness [42]. The observed variety of colors can therefore be explained by different combinations of these parameters.

Descriptive analyses have widely been used to study the relationship between pasta production parameters and sensory quality. The effects of new cereals (e.g., emmer) have been studied by Kucek et al. (2017), for example [43]. The impact of process parameters such as drying has been investigated by Padalino et al. (2016) and West et al. (2013) [14,16]. Many studies have also been performed on the use of alternative ingredients and their effect on nutritional and sensory characteristics [44].

The differences in the appearance of the pasta samples and their variable cooking qualities were highlighted by Pivot profile. This alternative sensory analysis method has previously been validated for the comparison of a set of products with respect to each other.

The reliability of our results is supported by the well-balanced nature of the panel (58.5% female, 41.5% male) with roughly a third each of regular, occasional, and not consumers of artisanal pasta (28.5, 34.6, and 36.7%, respectively). The panel was also sufficiently large (n = 57). Ares (2015) have shown indeed that at least 50 untrained panelists are required in this context to obtain reproducible results [45].

The Pivot profile method, based on free consumer descriptions, is simple for participants, fast, and robust [20], and has been validated for the sensory description of a set of products with respect to each other, in comparison with other alternative sensory analysis methods [29,46–48]. For example, Esmerino et al. (2017) [46] compared Greek yogurt samples using the Pivot profile, the projective mapping (PM), and the check-all-that-apply (CATA) questions and found that the Pivot profile was closer to similarity-based methods, such as projective mapping, than to verbal-based approaches, such as check-all-that-apply, suggesting that it is well-suited for general product descriptions. This alternative method was an appropriate choice for our study because (i) it allowed consumers to generate descriptors for previously undescribed products, and (ii) it revealed the main overall differences between the studied products. It does not provide detailed sensorial description of each pasta, on the contrary to a descriptive analysis using, for example, the lexicon from Irie et al. [19]. This lexicon consists of 35 attributes: 5 for visual appearance, 11 for aroma flavor, and 19 for texture. In comparison, the ratio of attributes in the Pivot profile suggests our panelists were more comfortable describing visual appearance (7/15 attributes) and texture in mouth (5/15 attributes). The proportion of flavor attributes (3/15 attributes) is similar to the one in Irie et al.'s lexicon (11/35 attributes). Our panelists chose less precise qualifiers (dull, flavor intensity, and salty) than found in Irie et al.'s lexicon (wheat aroma, wheat flavor, sweet aroma, roasted aroma, deteriorated grain, cinnamon, bran, pungent, corn, astringent, and chlorine), leading to a less precise sensory characterization. Significant differences between pasta types were nevertheless identified for more than half of these attributes (8/15 attributes). A previous study of spaghetti involving a trained panel found, as did we, that the main discriminating attributes were related to texture (firmness, elasticity, and stickiness) [35]. No correlation was observed between firmness and sensory attributes. This is probably because texture was defined by several attributes (pasty, hard, and compact) in the Pivot profile.

Since all the pasta samples have been mapped at the sensory level, it would be interesting to use the attributes generated by the Pivot method to analyze each type of pasta using the same descriptive approach as Khalil et al. (2022) for a variety of labneh, a typical Lebanon fermented milk [49]. This could provide elements to better understand the hedonic evaluation.

5. Conclusions

In the context of an evolving market and the growing popularity of locally sourced products, this study provides information on the physicochemical and sensory characteristics of artisanal pasta processed from durum wheat flour and sold locally by producers, a previously unstudied topic. The studied pasta samples were all sourced from a specific geographical area, but our results highlighting the variable physicochemical and sensory nature of artisanal pasta are generalizable to other territories. The variability observed was slightly correlated to physico-chemical characteristics but finally not perceived by the consumers. The cooking quality of artisanal pasta on some aspects (e.g., cooking losses) should be improved. Helping pasta producers optimize the cooking quality of these kinds of pasta will require identifying more specific relationships between input variables and product properties at the artisanal level.

Author Contributions: Conceptualization, G.J. and M.-F.S.; methodology, L.B., M.-F.S. and G.J.; validation, M.-F.S., G.J. and L.B.; formal analysis, M.-F.S., G.J., L.B., D.C. and E.C.; writing—original draft preparation, Y.C., M.-F.S., G.J. and L.B.; writing—reviewing and editing, M.-F.S., G.J., L.B. and M.-H.R.; funding acquisition, Y.C. All authors have read and agreed to the published version of the manuscript.

Funding: This research was funded by the Occitanie region (France) through the Activa Blé project (Grant "Research and society").

Institutional Review Board Statement: Ethical review and approval were waived for this study due to the non-biomedical nature of the study involving humans and the adherence to the Guidelines for Ethical and Professional Practices for the Sensory Analysis of Foods from the Institute for Food Science and Technology.

Informed Consent Statement: Informed consent was obtained from all subjects involved in the study.

Data Availability Statement: The data presented in this study are available on request from the corresponding author. The data are not publicly available due to adherence to the General Data Protection Regulation of the European Union.

Acknowledgments: The authors thank Lucille Gey for her contribution to data collection, Eva Sabouko and Théo Estève for tasting samples, and Aurélie Putois for taking photographs. The authors also thank Dominique Desclaux for her help in the reviewing the manuscript.

Conflicts of Interest: The authors declare no conflict of interest. The funders had no role in the design of the study; in the collection, analyses, or interpretation of data; in the writing of the manuscript, or in the decision to publish the results.

References

1. Agriculture Ministry. *Ministère de l'Agriculture Décret N°55-1175 Du 31 Août 1955*; Agriculture Ministry: Paris, France, 1997.
2. Codex Alimentarius. *Norme Pour la Semoule et la Farine de Blé Dur*; Codex Alimentarius 1991, CXS 178; FAO: Rome, Italy, 2019.
3. Feldmann, C.; Hamm, U. Consumers' perceptions and preferences for local food: A review. *Food Qual. Prefer.* **2015**, *40*, 152–164. [CrossRef]
4. EIP-AGRI. *Focus Group on Innovative Short Food Supply Chain Management: Final Report*; EIP-AGRI: Brussels, Belgium, 2015.
5. Bresciani, A.; Pagani, M.A.; Marti, A. Pasta-Making Process: A Narrative Review on the Relation between Process Variables and Pasta Quality. *Foods* **2022**, *11*, 256. [CrossRef]
6. Union of Organisations of Manufacturers of Pasta Products in the European Union. Statics-World Pasta Production. Available online: https://www.pasta-unafpa.org/newt/unafpa/default.aspx?IDCONTENT=102 (accessed on 5 May 2022).
7. Chiffoleau, Y.; Echchatbi, A.; Rod, J.; Gey, L.; Akermann, G.; Desclaux, D.; Jard, G.; Kessari, M.; Moinet, K.; Peres, J.; et al. Quand l'innovation sociale réoriente l'innovation technologique dans les systèmes agroalimentaires: Le cas des chaînes locales autour des blés. *Innovations* **2021**, *64*, 41–63. [CrossRef]
8. Chiffoleau, Y.; Desclaux, D. Participatory plant breeding: The best way to breed for sustainable agriculture? *Int. J. Agric. Sustain.* **2006**, *4*, 119–130. [CrossRef]
9. Sacchetti, G.; Cocco, G.; Cocco, D.; Neri, L.; Mastrocola, D. Effect of semolina particle size on the cooking kinetics and quality of spaghetti. *Proc. Food Sci.* **2011**, *1*, 1740–1745. [CrossRef]
10. Sarkar, A.; Fu, B.X. Impact of Quality Improvement and Milling Innovations on Durum Wheat and End Products. *Foods* **2022**, *11*, 1796. [CrossRef] [PubMed]
11. Deng, L.; Elias, E.M.; Manthey, F.A. Influence of Durum Genotype on Whole Wheat and Traditional Spaghetti Qualities. *Cereal Chem.* **2017**, *94*, 857–865. [CrossRef]
12. Hidalgo, A.; Fongaro, L.; Brandolini, A. Wheat flour granulometry determines colour perception. *Food Res. Int.* **2014**, *64*, 363–370. [CrossRef]
13. Cubadda, R.E.; Carcea, M.; Marconi, E.; Trivisonno, M.C. Influence of Gluten Proteins and Drying Temperature on the Cooking Quality of Durum Wheat Pasta. *Cereal Chem.* **2007**, *84*, 48–55. [CrossRef]
14. Padalino, L.; Caliandro, R.; Chita, G.; Conte, A.; del Nobile, M.A. Study of drying process on starch structural properties and their effect on semolina pasta sensory quality. *Carbohydr. Polym.* **2016**, *153*, 229–235. [CrossRef]
15. Manthey, F.A.; Schorno, A.L. Physical and Cooking Quality of Spaghetti Made from Whole Wheat Durum. *Cereal Chem.* **2002**, *79*, 504–510. [CrossRef]
16. West, R.; Duizer, L.; Seetharaman, K. The effect of drying and whole grain content on the pasting, physicochemical and qualitative properties of pasta. *Starch* **2013**, *65*, 645–652. [CrossRef]
17. *ISO 7304-1:2016*; AFNOR ISO—Semoule de Blé Dur et Pâtes Alimentaires—Appréciation de La Qualité de Cuisson Par Analyse Sensorielle—Partie 1. ISO: Geneva, Switzerland, 2016.
18. *ISO 7304-2:2008*; AFNOR ISO—Pâtes Alimentaires Produites à Partir de Semoule de Blé Dur—Appréciation de La Qualité de Cuisson Par Analyse Sensorielle—Partie 2. ISO: Geneva, Switzerland, 2008.
19. Irie, K.; Maeda, T.; Kazami, Y.; Yoshida, M.; Hayakawa, F. Establishment of a sensory lexicon for dried long pasta. *J. Sens. Stud.* **2018**, *33*, e12438. [CrossRef]
20. Thuillier, B.; Valentin, D.; Marchal, R.; Dacremont, C. Pivot© profile: A new descriptive method based on free description. *Food Qual. Prefer.* **2015**, *42*, 66–77. [CrossRef]

21. Valentin, D.; Chollet, S.; Lelievre-Desmas, M.; Abdi, H. Quick and dirty but still pretty good: A review of new descriptive methods in food science. *Int. J. Food Sci. Technol.* **2012**, *47*, 1563–1578. [CrossRef]
22. Mefleh, M.; Conte, P.; Fadda, C.; Giunta, F.; Piga, A.; Hassoun, G.; Motzo, R. From ancient to old and modern durum wheat varieties: Interaction among cultivar traits, management, and technological quality. *J. Sci. Food Agric.* **2018**, *99*, 2059–2067. [CrossRef]
23. *AFNOR NF EN ISO 712:2010*; Céréales et Produits Céréaliers, Détermination de La Teneur En Eau. ISO: Geneva, Switzerland, 2010.
24. *AFNOR NF EN ISO 2171:2010*; Céréales, Légumineuses et Produits Dérivés, Dosage Du Taux de Cendres Par Incinération. ISO: Geneva, Switzerland, 2010.
25. Hirschler, R. Whiteness, yellowness, and browning in food colorimetry: A critical review. In *Color in Food*; Taylor & Francis: Abingdon, UK, 2012; pp. 93–103. ISBN 978-1-4398-7693-0.
26. AACC International. *Approved Methods of Analysis (11th Edition), Method 44-15.02 Moisture—Air-Oven Methods, 46-12.01 Crude Protein—Kjeldahl Method, Boric Acid Modification, 66-50.01 Pasta and Noodle Cooking Quality—Firmness*; AACC International: St. Paul, MN, USA, 2009.
27. Petitot, M.; Brossard, C.; Barron, C.; Larré, C.; Morel, M.-H.; Micard, V. Modification of pasta structure induced by high drying temperatures. Effects on the in vitro digestibility of protein and starch fractions and the potential allergenicity of protein hydrolysates. *Food Chem.* **2009**, *116*, 401–412. [CrossRef]
28. Dachkevitch, T.; Autran, J.C. Prediction of Baking Quality of Bread Wheats in Breeding Programs by Size-Exclusion High-Performance Liquid Chromatography. *Cereal Chem.* **1989**, *66*, 448–456.
29. Fonseca, F.G.; Esmerino, E.A.; Filho, E.R.T.; Ferraz, J.P.; da Cruz, A.G.; Bolini, H. Novel and successful free comments method for sensory characterization of chocolate ice cream: A comparative study between Pivot profile and comment analysis. *J. Dairy Sci.* **2016**, *99*, 3408–3420. [CrossRef]
30. *AFNOR NF ISO 8587:2007*; Analyse Sensorielle-Méthodologie-Classement Par Rang. ISO: Geneva, Switzerland, 2007.
31. Samson, M.-F.; Leygue, J.-P.; Lullien-Pellerin, V.; Abécassis, J.; Braun, P. Les Qualités Du Grain et Des Produits Pour Satisfaire Les Attentes Du Consommateur. In *Blé dur: Synthèse des Connaissances pour une Filière Durable*; Abécassis, J., Massé, J., Allaoua, A., Eds.; QUAE: Paris, France, 2021; pp. 239–276. ISBN 978-2-8179-0414-6.
32. De Noni, I.; Pagani, M.A. Cooking Properties and Heat Damage of Dried Pasta as Influenced by Raw Material Characteristics and Processing Conditions. *Crit. Rev. Food Sci. Nutr.* **2010**, *50*, 465–472. [CrossRef]
33. Sicignano, A.; di Monaco, R.; Masi, P.; Cavella, S. From raw material to dish: Pasta quality step by step. *J. Sci. Food Agric.* **2015**, *95*, 2579–2587. [CrossRef]
34. Leygue, J.-P.; Samson, M.-F.; Barron, C.; Lullien-Pellerin, V.; Abécassis, J.; Cuq, B.; Crépon, K.; Braun, P. La Transformation: Le Grain Du Silo à l'assiette Du Consommateur. In *Blé Dur: Synthèse des Connaissances pour une Filière Durable*; Abécassis, J., Massé, J., Allaoua, A., Eds.; QUAE: Paris, France, 2021; pp. 239–276. ISBN 978-2-8179-0414-6.
35. Padalino, L.; Mastromatteo, M.; Lecce, L.; Spinelli, S.; Contò, F.; Del Nobile, M.A. Effect of durum wheat cultivars on physico-chemical and sensory properties of spaghetti. *J. Sci. Food Agric.* **2013**, *94*, 2196–2204. [CrossRef]
36. D'Egidio, M.G.; Mariani, B.M.; Nardi, S.; Novaro, P.; Cubadda, R. Chemical and Technological Variables and Their Relationships: A Predictive Equation for Pasta Cooking Quality. *Cereal Chem.* **1990**, *67*, 275–281.
37. Bruneel, C.; Pareyt, B.; Brijs, K.; Delcour, J.A. The impact of the protein network on the pasting and cooking properties of dry pasta products. *Food Chem.* **2010**, *120*, 371–378. [CrossRef]
38. Di Monaco, R.; Cavella, S.; di Marzo, S.; Masi, P. The effect of expectations generated by brand name on the acceptability of dried semolina pasta. *Food Qual. Prefer.* **2004**, *15*, 429–437. [CrossRef]
39. Cabas-Lühmann, P.A.; Manthey, F.A. Environment during grain filling affects pasta color. *Cereal Chem.* **2020**, *97*, 967–980. [CrossRef]
40. Joubert, M.; Morel, M.-H.; Lullien-Pellerin, V. Pasta color and viscoelasticity: Revisiting the role of particle size, ash, and protein content. *Cereal Chem.* **2018**, *95*, 386–398. [CrossRef]
41. Wang, K.; Fu, B.X. Inter-Relationships between Test Weight, Thousand Kernel Weight, Kernel Size Distribution and Their Effects on Durum Wheat Milling, Semolina Composition and Pasta Processing Quality. *Foods* **2020**, *9*, 1308. [CrossRef]
42. Cabas-Lühmann, P.A.; Manthey, F.A.; Elias, E.M. Variations of colour, polyphenol oxidase and peroxidase activities during the production of low temperature dried pasta in various durum wheat genotypes. *Int. J. Food Sci. Technol.* **2021**, *56*, 4700–4709. [CrossRef]
43. Kucek, L.K.; Dyck, E.; Russell, J.; Clark, L.; Hamelman, J.; Burns-Leader, S.; Senders, S.; Jones, J.; Benscher, D.; Davis, M.; et al. Evaluation of wheat and emmer varieties for artisanal baking, pasta making, and sensory quality. *J. Cereal Sci.* **2017**, *74*, 19–27. [CrossRef]
44. Mercier, S.; Moresoli, C.; Mondor, M.; Villeneuve, S.; Marcos, B. A Meta-Analysis of Enriched Pasta: What Are the Effects of Enrichment and Process Specifications on the Quality Attributes of Pasta? *Compr. Rev. Food Sci. Food Saf.* **2016**, *15*, 685–704. [CrossRef]
45. Ares, G. Methodological challenges in sensory characterization. *Curr. Opin. Food Sci.* **2015**, *3*, 1–5. [CrossRef]
46. Esmerino, E.A.; Filho, E.R.T.; Carr, B.T.; Ferraz, J.P.; Silva, H.L.; Pinto, L.P.; Freitas, M.Q.; da Cruz, A.G.; Bolini, H.M. Consumer-based product characterization using Pivot Profile, Projective Mapping and Check-all-that-apply (CATA): A comparative case with Greek yogurt samples. *Food Res. Int.* **2017**, *99*, 375–384. [CrossRef]

47. Deneulin, P.; Reverdy, C.; Rébénaque, P.; Danthe, E.; Mulhauser, B. Evaluation of the Pivot Profile©, a new method to characterize a large variety of a single product: Case study on honeys from around the world. *Food Res. Int.* **2018**, *106*, 29–37. [CrossRef]
48. Brand, J.; Valentin, D.; Kidd, M.; Vivier, M.; Næs, T.; Nieuwoudt, H. Comparison of Pivot profile© to frequency of attribute citation: Analysis of complex products with trained assessors. *Food Qual. Prefer.* **2020**, *84*, 103921. [CrossRef]
49. Abi Khalil, R.; Yvon, S.; Couderc, C.; Belahcen, L.; Jard, G.; Sicard, D.; Bigey, F.; El Rammouz, R.; Nakhoul, P.A.; Eutamène, H.; et al. Microbial Communities and Main Features of Labneh Ambaris: A Traditional Lebanese Fermented Goat's Milk Product. *J. Dairy Sci.* **2022**, *in press*.

foods

MDPI

Article

Drying Behavior of Bulgur and Its Effect on Phytochemical Content

Sfayhi Terras Dorra *, Dridi Farah, Hadjyahia Nesrine, Allouch Wafa and Zarroug Youkabed

Field Crop Laboratory (LR16INRAT02), National Institute of Agronomic Research of Tunisia, University of Carthage, Rue Hedi Karray, Ariana 2049, Tunisia; dridi.farah8@gmail.com (D.F.); nesrinehajyahia@gmail.com (H.N.); allouchwaf@gmail.com (A.W.); zarrougyoukabed@yahoo.fr (Z.Y.)

* Correspondence: sfaihi.dorra@iresa.agrinet.tn; Tel.: +216-71-230-024

Abstract: The objective of this study was to determine the influence of two types of dryers (hot air oven and vacuum dryer) and the yellow berry percentage (1.75%, 36.25%, 43.25%) on the drying process and phytochemical content of bulgur. Results showed that the Midilli model successfully described the moisture diffusion during drying at 60 °C in all bulgur samples, where an increase in yellow berry percentage generated an increase in moisture content. Effective diffusion coefficient (D_{eff}) increased significantly ($p \leq 0.05$) from 7.05×10^{-11} to 7.82×10^{-11} ($m^2.s^{-1}$) and from 7.73×10^{-11} to 7.82×10^{-11} ($m^2.s^{-1}$) for the hot air oven and vacuum dryer, respectively. However, it decreased significantly with a decrease of yellow berry percentage. It was concluded that the vacuum dryer provided faster and more effective drying than the hot air oven. Total polyphenol (TPC), total flavonoid (TFC), and yellow pigment contents (YPC) of bulgur were investigated. TPC ranged between 0.54 and 0.64 (mg GAE/g dm); TFC varied from 0.48 to 0.61 (mg QE/g dm). The YPC was found to be between 0.066 and 0.079 (mg ß-carotene/100g dm). Yellow berry percentage positively and significantly affected the TPC, TFC, and YPC contents due to the hard separation of the outer layers from the starchy grain during the debranning step.

Keywords: bulgur; wholegrain; yellow berry; drying; phytochemical

Citation: Dorra, S.T.; Farah, D.; Nesrine, H.; Wafa, A.; Youkabed, Z. Drying Behavior of Bulgur and Its Effect on Phytochemical Content. *Foods* **2022**, *11*, 1062. https://doi.org/10.3390/foods11071062

Academic Editor: Mike Sissons

Received: 11 February 2022
Accepted: 16 March 2022
Published: 6 April 2022

1. Introduction

Bulgur is a famous dish in Central Asia, Turkey, the Middle East, and North Africa [1]. It is considered a highly nutritious food [2], since it contains high dietary fiber content [3,4] and a high amount of vitamins and minerals such as phosphorus, zinc, potassium, and selenium. It also has a low glycemic index [5]. Bulgur is composed of 9–13% water, 10–16% protein, 1.2–1.5% fat, 76–78% carbohydrate, 1.2–1.4% ash, and 1.1–1.3% fiber [6]. Generally, bulgur is made from hard wheat (*Triticum durum*) [7], which results in its yellow color and higher protein content compared to the other wheat types [8,9]. However, other grains can be used to produce bulgur, such as bitter and sweet lupin [10], barley [11], soybean [12], and chickpea [13]. The quality of durum wheat affects the bulgur quality. In fact, a positive correlation has been determined between bulgur yield and the thousand kernel weight [14], but no studies have shown a relationship between yellow berry percentage and bulgur quality. As known, yellow berry (yb) is a physiological disorder, mainly found in durum wheat. It is defined as the poor development of endosperm [15], where soil with nitrogen insufficiency is the main cause [16]. This disorder is characterized by starchy spots that can cover small areas up to the entire grain [17,18]. The grain becomes less vitreous, starchy, softer, opaque, and light-colored [17,19,20]. Lopez-Ahumada et al. [18] reported that grains with yellow berry have higher starch content than normal grains, which affects crystallinity percent. Dexter et al. [20] reported that yellow berry grains have high moisture and low protein content compared to normal vitreous grains. A negative correlation has been found between protein content and yellow berry percentage [21,22].

Bulgur production involves several steps: cooking, drying, tempering, and debranning [23–25]. Due to the importance of drying, several researchers have tried to model moisture diffusion in parboiled wheat [26–28] and wheat [29,30]. After cleaning, the grains are cooked in boiling water until the starch is completely gelatinized. Bayram [31] proposed 40 min as the optimum cooking time, where the starch is gelatinized without any deformation of the wheat kernel. Additionally, Sfayhi-Terras et al. [23] determined that 43 min is the ideal cooking time to generate high-quality bulgur. During cooking, time and temperature are considered the most critical parameters that have an impact on the dimensions, volume, and crease of the wheat kernel [24,32]. The boiled wheat is then dried to decrease moisture content from 45% to 10% (d.b). After drying, the dried, parboiled wheat is debranned, which involves removing the grain outer layers by abrasion and friction [33].

Since the cooking and drying operations may significantly alter the color, yield, chemical composition, nutritive quality, and physical properties of bulgur [11,13,34–36], many works have studied the effect of each processing step on bulgur quality. Hayta [37] investigated the effects of different drying methods (solar, sun, microwave, tray drying) on yield and protein extractability. Among the drying methods, the yield of the sun-dried sample was the lowest. However, these methods did not affect the extraction of protein. Kadakal et al. [38] studied the effect of cooking (in a beaker at 90 and 100 °C, and autoclave at 121 °C) and drying (in a hot air oven at 60, 70, and 80 °C, and in open-air sun-drying) on the water-soluble vitamins of bulgur. It was shown that hot air oven drying at 60 °C does not affect the water-soluble vitamin contents, unlike drying in sun-drying and hot air oven drying at 80 °C.

It is well-known that during the drying process, temperature directly affects the nutritional quality of products. Yilmaz and Koca [39] reported that autoclave cooking and hot air drying at 60 °C presented the highest retention of total phenolic content and total yellow pigment than both autoclave cooking/hot air drying and microwave cooking and drying. Although extensive work has been carried out on drying, limited literature is available on the variation of bioactive phytochemicals in bulgur such as total phenolic, flavonoid, and yellow pigment contents during drying, and no work was found that studied the effect of yellow berry percent on the drying behavior and quality of bulgur.

The objective of the present work is to investigate the effect of yellow berry percent and dryer type on drying behavior, to find a suitable drying model, to determine the effective diffusivity coefficient, and to study the change of the bioactive components of bulgur during the drying operation.

2. Materials and Methods

2.1. Material

For this study, bulgur was prepared from Tunisian durum wheat (*Triticum durum*), Maali variety, for one cultivation with high quality. Three different samples from this variety were used. The difference was in the yellow berry percentage (yb) and the thousand-kernel weight (TKW). The yellow berry percentages were 1.75%, 36.25%, and 43.25%, and the TKW were 53.8 g, 53.9 g, 48.6 g, respectively. The moisture, protein, and ash content of these samples were 11.0 ± 0.5% (d.b), 13.0 ± 0.4% (d.b), and 1.7 ± 0.4% (d.b), respectively. The thousand-kernel weight (TKW) was determined using the Numigral Chopin (Chopin, Villeneuve-la-Garenne, France). Yellow berry percentage was determined by inspecting 50 kernels sliced using a Pohl farinothome (Chopin, Asnières-sur-Seine, France). Moisture content before debranning was determined according to the AACC-approved method 44-15A [40]. Grain protein was evaluated using a Near-Infrared Spectroscopy System (Perten-Inframatic-8600, Hamburg, Germany) [41]. Ash content was evaluated according to ICC Standard 104/1 [42].

2.2. Bulgur Processing

The grain was cleaned with distilled water for 1 min to remove any adhesive particles stuck to the surface of the kernels. Then, it was cooked in boiling water at 100 °C for 42–53 min until the entire grain starch was gelatinized. The cooking time was determined using the center cutting method [31]. Precooked grain (100 g) was dried at 60 °C for

180 min. During the drying operation, 5 g was collected at 15 min intervals. Two types of natural convective air dryers were used for dehydration of precooked grain: a hot air oven (Venticell 404-ECO line, München, Germany) and a vacuum dryer (Monferrina EC50, Castell'Alfero, Italy), where the Hr of the air was fixed at 80%. After cooking, each sample was debranned with an abrasive laboratory mill (Strong-Scott, England) at a constant speed of 830 rpm for 1.6min [23,43]. The debranned grains were separated from the debranned part with a sieve of 1.04 mm set inside the apparatus. For this study, bulgur was considered the recovered sample.

2.3. Moisture Content

The moisture loss from the parboiled wheat during drying was determined every 15 min for 180 min. The experiments were conducted in duplicate and average values were taken. The moisture content of samples was calculated using Equation (1):

$$M_t = \frac{(W_0 + W_t) - W_f}{W_t} * 100 \quad (1)$$

where W_0 is the initial weight (g), W_t is the weight of the sample (g) at any drying time (t), and W_f is the final weight (g). M_t is the moisture content of the wheat samples at the different drying times.

2.4. Phytochemical Analysis

Before extraction, samples were ground by a grinder (CT 293 Cyclotec, Foss, Hamburg, Germany), then separated using a sieve of 0.8 mm. According to the procedure by Mau et al. [44], the phenolic compounds were extracted with 25 mL of 80% methanol using a 2.5 g sample. The extraction solvent and the sample were mixed in an orbital shaker for 30 min at ambient temperature and then stored in the dark for 24 h at 4 °C. The mixture was filtered through Ashless Wattman paper (No. 4). The filtrate obtained was concentrated under vacuum by rota-vapor (60 °C). Thus, the extracts obtained were collected, weighed, stored at 4 °C, and protected from light. For further analysis, 1 mg of the extract was dissolved in 1 mL of methanol.

2.4.1. Total Polyphenol Content (TPC)

Total polyphenol content was determined according to the procedure from Dewanto et al. [45], using a modification of the Folin–Ciocalteu method. The absorbance was measured at 760 nm using a spectrophotometer (Onda V-10 Plus, Capri, Italy), and the results were expressed as milligram gallic acid equivalents per gram of sample dry matter (mg GAE/g dm).

2.4.2. Total Flavonoid Content (TFC)

Total flavonoid content was determined by using the modified method from Dewanto et al. [45]. The absorbance was measured at 510 nm using a spectrophotometer (Onda V-10 Plus, Capri, Italy). The results were expressed as milligram quercetin equivalents per g of sample dry matter (mg QE/g dm).

2.4.3. Yellow Pigment Content (YPC)

Yellow pigment content was determined according to the norm ISO 11052 [46]. Ten grams of samples were extracted with 50 mL water-saturated butanol (ratio 6:2). The mixture was homogenized and kept for 16 h at room temperature. Then, it was filtered in conical bottles. The absorbance was measured at 440 nm using a spectrophotometer (Onda V-10 Plus, Capri, Italy). The results were expressed as milligram beta carotene equivalents per g of sample dry matter (mg ß-carotene/100 g dm).

2.5. Modeling of the Drying Process

Among the mathematical models, Lewis, Henderson and Pabis, Logarithmic and Midilli models were employed to describe the drying kinetics of the parboiled wheat (Table 1).

Table 1. Mathematical models used.

Model Name	Model Equation	Reference
Lewis	$MR = \exp(-kt)$	[47]
Henderson and Pabis	$MR = a * \exp(-kt)$	[48]
Logarithmic	$MR = a * \exp(-kt) + b$	[49]
Midilli	$MR = a * \exp(-kt^n) + bt$	[50]

By noting the moisture content every 15 min in the different dryers, moisture ratios and drying rates of samples were calculated by Equations (2) and (3), respectively. The drying experiments were carried out for 180 min. The simplified equation of Rayaguru and Routray [51] was used to determine the moisture ratio (MR):

$$MR = \frac{M_t}{M_0} \tag{2}$$

where M_t is the moisture content at any time (%) and M_o is the initial moisture content (%) of the samples.

The drying rate (DR) of parboiled wheat samples was calculated using Equation (3) [52]:

$$DR = \frac{M_{t+dt} - M_t}{dt} \tag{3}$$

where MR_{t+dt} and MR_t are moisture ratios at the time (t + dt) and t (dimensionless), t is the drying time (min).

2.6. Effective Diffusion Coefficient

The simplified solution of Fick's diffusion was used [53]:

$$MR = \frac{6}{\pi^2} \exp(-\frac{D_{eff} * \pi^2 * t}{R_e^2}) \tag{4}$$

where n is the positive integer, D_{eff} is the effective moisture diffusion coefficient (m^2s^{-1}), t is drying time (s), and R_e is the average radius of wheat (2.21×10^{-1} m). Equation (4) can be written in logarithmic form:

$$\ln(MR) = \ln\frac{6}{\pi^2} - \frac{D_{eff} * \pi^2 * t}{R_e^2} \tag{5}$$

The effective diffusion coefficient is calculated from the slope of Equation (5), which is obtained from the graph describing the change in ln (MR) values with drying time.

$$\text{Slope} = \frac{D_{eff} * \pi^2}{R_e^2} \tag{6}$$

2.7. Statistical Analysis

Sigma plot 14.5 (Systat Software, Inc., San Jose, CA, USA) was used to present all drying data. The variance analysis (ANOVA) was executed using the significance level of ($p < 0.05$) using SPSS software (version 23.0) (IBM Software, New York, NY, USA). The results were followed with letters in case of the existence of a significant difference.

3. Results

3.1. Drying Kinetics and Modeling

By noting the weight loss during the drying process, moisture ratio (MR) change over time in the different dryers was determined using Equation (2) and then presented in Figure 1. Examining Figure 1, the moisture ratio (MR) decreased with time, in both dryers, and then reached a plateau. A significant difference ($p \leq 0.05$) was found between the two dryers, where MR was significantly lower for the vacuum dryer compared to the hot air oven. According to ANOVA results, the yellow berry percentage had a significant ($p \leq 0.05$) effect in terms of variation of moisture ratio. The highest moisture ratio was obtained for bulgur at 43.25 yb%.

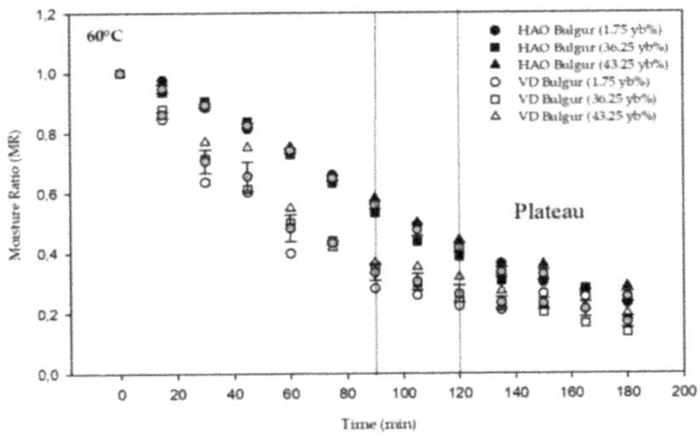

Figure 1. Variation of moisture ratio (MR) with time in different dryers at 60 °C. HAO: hot air oven; VD: vacuum dryer.

A comparison between the slopes of the drying curves for the declining phase (Figure 1) in both dryers (P_1 hot air oven, P_2 vacuum dryer) was realized. It can be seen from Table 2 that a significant difference ($p \leq 0.05$) in slope values was found between oven-dried and vacuum-dried bulgur, where P_2 was found to be the smallest. Moreover, to reach the plateau (Figure 1), the vacuum dryer required a shorter time (90 min) than the oven dryer (120 min).

However, the three vacuum-dried slopes, as well as the hot air oven-dried slopes, were found to be significantly ($p \leq 0.05$) different (Table 3). This result indicates that the variation of yellow berry percent has a significant effect on the drying behavior of bulgur.

Table 2. The effect of dryer type on the drying curves at 60 °C.

	Slope	Bulgur 1.75 yb%	Bulgur 36.25 yb%	Bulgur 43.25 yb%
Hot air oven	P_1	$-5.51 \times 10^{-3} \pm 1.98 \times 10^{-5\,a}$	$-5.35 \times 10^{-3} \pm 0.68 \times 10^{-5\,a}$	$-4.71 \times 10^{-3} \pm 5.94 \times 10^{-5\,a}$
Vacuum dryer	P_2	$-7.48 \times 10^{-3} \pm 2.85 \times 10^{-5\,b}$	$-6.97 \times 10^{-3} \pm 1.46 \times 10^{-5\,b}$	$-6.56 \times 10^{-3} \pm 7.90 \times 10^{-5\,b}$

Mean values with a row followed by different letters are significantly different ($p < 0.05$).

Table 3. The effect of yellow berry percent on the drying curves at 60 °C.

	Hot Air Oven (P_1)	Vacuum Dryer (P_2)
Bulgur 1.75 yb%	$-5.51 \times 10^{-3} \pm 1.98 \times 10^{-5\,c}$	$-7.48 \times 10^{-3} \pm 2.85 \times 10^{-5\,c}$
Bulgur 36.25 yb%	$-5.35 \times 10^{-3} \pm 0.68 \times 10^{-5\,b}$	$-6.97 \times 10^{-3} \pm 1.46 \times 10^{-5\,b}$
Bulgur 43.25 yb%	$-4.71 \times 10^{-3} \pm 5.94 \times 10^{-5\,a}$	$-6.56 \times 10^{-3} \pm 7.90 \times 10^{-5\,a}$

Mean values with a row followed by different letters are significantly different ($p < 0.05$).

Using Equation (3), the drying rate (DR) variation with time was determined and is represented in Figure 2. As can be seen from Figure 2, the drying rate (DR) in both dryers decreases over time. Only one phase was noted—the falling rate period. The drying rate of vacuum-dried samples was slightly lower than the drying rate of oven-dried bulgur (Figure 2).

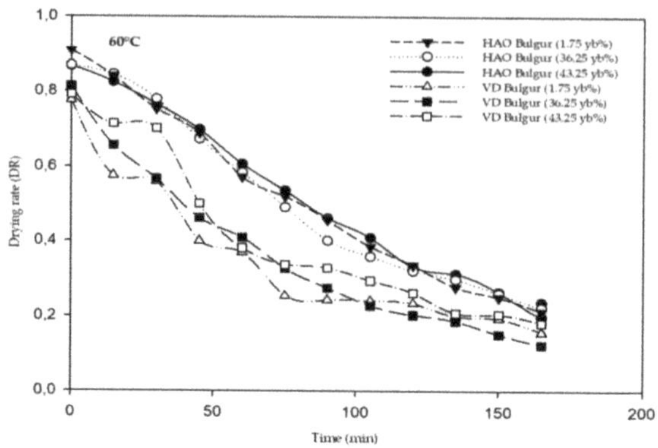

Figure 2. Drying rate curves in the different dryers at 60 °C. HAO: hot air oven; VD: vacuum dryer.

The moisture ratio (MR) was fitted to the four models listed in Table 1 and presented in Figure 3. As standard error (StdErr) and residual sum of squares (RSS) values approach zero the closer the prediction is to the experimental data. The drying models were compared based on their R^2 to assess their respective goodness of fit. Accordingly, all the tested models had high coefficient of determination (R^2) values in the range 0.95–0.99 and 0.94–0.99 for the hot air oven and vacuum dryer, respectively. Among the used models, the Midilli model had the highest R^2 values and the lowest StdErr and RSS values for the hot air oven and vacuum dryer as shown in Table 4.

Table 4. Parameters of the four drying models.

Samples	Model	Hot Air OvenDrying				Vacuum Drying			
		Parameters	R^2	Std Err	RSS	Parameters	R^2	Std Err	RSS
Bulgur 1.75 yb%	Lewis	K 6.82×10^{-3}	0.9570	0.0004	0.0381	K 1.17×10^{-2}	0.9486	0.0599	0.0431
	Henderson and Pabis	K 7.64×10^{-3} a 1.08	0. 9760	0.0440	0.0213	K 1.13×10^{-2} a 9.95×10^{-1}	0.9512	0.0610	0.0409
	Logarithmic	K 2.23×10^{-3} a 2.43 b -1.41×10^{-1}	0.9937	0. 0236	0. 0056	K 1.80×10^{-2} a 8.60×10^{-1} b 1.59×10^{-1}	0.9742	0.0465	0.0216
	Midilli	K 6.93×10^{-4} a 1.00 b 3.30×10^{-4} n 1.51	0.9943	0.0324	0.0095	K 9.51×10^{-3} a 1.00 b 9.72×10^{-4} n 1.12	0.9885	0.0462	0.0192
Bulgur 36.25 yb%	Lewis	K 7.12×10^{-3}	0.9392	0.0661	0.0306	K 1.12×10^{-2}	0.9978	0.0132	0.0021
	Henderson and Pabis	K7.92×10^{-3} a 1.08	0.9570	0.0600	0.0216	K 1.13×10^{-2} a 1.01	0.9981	0.0128	0.0018
	Logarithmic	K 4.08×10^{-3} a 1.61 b-5.64×10^{-1}	0.9830	0.0414	0.0086	K 1.17×10^{-2} a 1.00 b 1.56×10^{-2}	0.9982	0.0131	0.0017
	Midilli	K 4.48×10^{-4} a 9.80×10^{-1} b 1.10×10^{-3} n 1.79	0.9961	0.0197	0.0035	K 8.24×10^{-3} a 1.00 b 2.64×10^{-4} n 1.08	0.9994	0.0117	0.0012
Bulgur 43.25 yb%	Lewis	K 6.64×10^{-3}	0.9604	0.0548	0.0420	K 9.68×10^{-3}	0.9778	0.0004	0.0198
	Henderson and Pabis	K 7.32×10^{-3} a 1.07	0.9719	0. 0478	0. 0298	K 9.85×10^{-3} a 1.01	0.9782	0.0420	0.0194
	Logarithmic	K 2.48×10^{-3} a 2.32 b -1.28	0. 9794	0.0427	0.0218	K 1.06×10^{-2} a 9.84×10^{-1} b 3.60×10^{-2}	0.9786	0.0436	0.0190
	Midilli	K 2.55×10^{-4} a 9.79×10^{-1} b 7.75×10^{-4} n 1.60	0.9992	0.0087	0.0007	K 3.07×10^{-3} a 9.84×10^{-1} b 8.04×10^{-4} n 1.31	0.9918	0.0403	0.0146

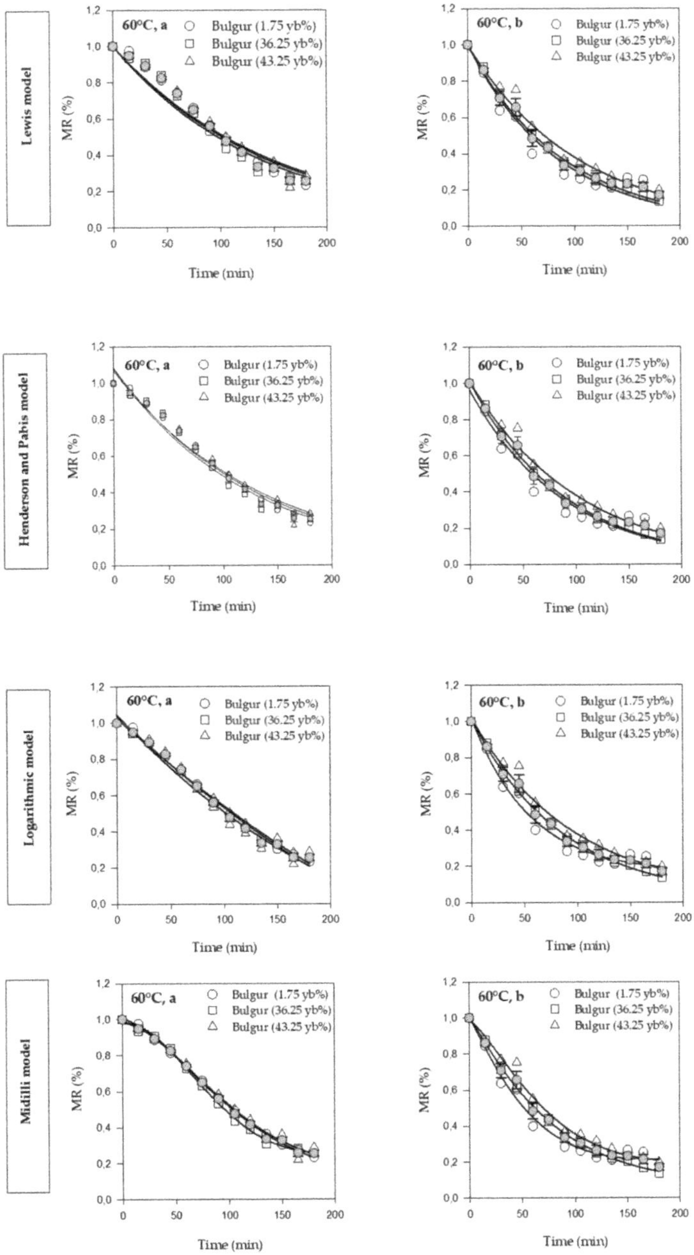

Figure 3. Simulated moisture ratio during drying of parboiled wheat for different dryers ((**a**): hot air oven; (**b**): vacuum dryer) at 60 °C.

The effect of dryer type on the drying rate constant k of the Midilli model value can be seen in Table 4. When comparing the k values of the hot air oven with the vacuum dryer, the k values increased from 6.93×10^{-4} to 9.51×10^{-3}, from 4.48×10^{-4} to 8.24×10^{-3}, and from 2.55×10^{-4} to 3.07×10^{-3} for bulgur 1.75 yb%, 36.25 yb%, and 43.25 yb%, respectively. Using Equation (6), the effective diffusion coefficient was determined.

Drying at 60 °C, the effective diffusion coefficient D_{eff} values varied from $6.86 \times 10^{-11} \pm 4.52 \times 10^{-21}$ to 7.05×10^{-11} 3.17×10^{-22} ($m^2.s^{-1}$), and from $7.73 \times 10^{-11} \pm 4.74 \times 10^{-22}$ to $7.82 \times 10^{-11} \pm 7.05 \times 10^{-22}$ ($m^2.s^{-1}$) for the hot air oven and the vacuum dryer, respectively (Table 5). A significant difference ($p \leq 0.05$) in the values of the effective diffusion coefficient values was found where the vacuum dryer presented the highest D_{eff} values (Table 5).

Table 5. The effect of dryer type on the effective diffusion coefficient of bulgur.

		Bulgur 1.75 yb%	Bulgur 36.25 yb%	Bulgur 43.25 yb%
D_{eff} (m^2/s)	Hot air oven	$7.05 \times 10^{-11} \pm 3.17 \times 10^{-22}$ a	$6.86 \times 10^{-11} \pm 2.05 \times 10^{-21}$ a	$6.86 \times 10^{-11} \pm 4.52 \times 10^{-21}$ a
	Vacuum dryer	$7.82 \times 10^{-11} \pm 7.05 \times 10^{-22}$ b	$7.73 \times 10^{-11} \pm 4.74 \times 10^{-22}$ b	$7.73 \times 10^{-11} \pm 1.81 \times 10^{-20}$ b

Mean values with a row followed by different letters are significantly different ($p < 0.05$).

Meanwhile, drying at the temperature of 60 °C, the ANOVA showed that yellow berry percentage also significantly affects the effective diffusion coefficient (Table 6). An increase in yellow berry percentage generates a decrease in D_{eff} value.

Table 6. The effect of yellow berry percentage on the effective diffusion coefficient of bulgur.

	Samples	Hot Air Oven	Vacuum Dryer
D_{eff} (m^2/s)	Bulgur 1.75 yb%	$7.05 \times 10^{-11} \pm 3.17 \times 10^{-22}$ b	$7.82 \times 10^{-11} \pm 7.05 \times 10^{-22}$ b
	Bulgur 36.25 yb%	$6.86 \times 10^{-11} \pm 2.05 \times 10^{-21}$ a	$7.73 \times 10^{-11} \pm 4.74 \times 10^{-22}$ a
	Bulgur 43.25 yb%	$6.86 \times 10^{-11} \pm 4.52 \times 10^{-21}$ a	$7.73 \times 10^{-11} \pm 1.81 \times 10^{-20}$ a

Mean values with a row followed by different letters are significantly different ($p < 0.05$).

3.2. Phytochemicals Content of Bulgur

The variation of total polyphenol content (TPC) and total flavonoid content (TFC) in bulgur during drying is presented in Figure 4a,b, respectively. It can be seen that TPC and TFC decreased over time during drying at 60 °C. After 3 h of drying at 60 °C, the TPC varied from $0.57 \pm 3.20 \times 10^{-5}$ to $0.62 \pm 5.5 \times 10^{-4}$ (mg GAE/g dm), and from $0.54 \pm 3.46 \times 10^{-4}$ to $0.64 \pm 1.9 \times 10^{-5}$ (mg GAE/g dm) for the hot air oven and vacuum dryer, respectively. The TFC of bulgur ranged from $0.48 \pm 4.5 \times 10^{-4}$ to $0.59 \pm 9 \times 10^{-5}$ (mg QE/g dm) and from $0.49 \pm 6.9 \times 10^{-5}$ to $0.61 \pm 1.11 \times 10^{-4}$ (mg QE/g dm) for the hot air oven and vacuum dryer, respectively. During drying, no significant difference was determined between the two drying methods.

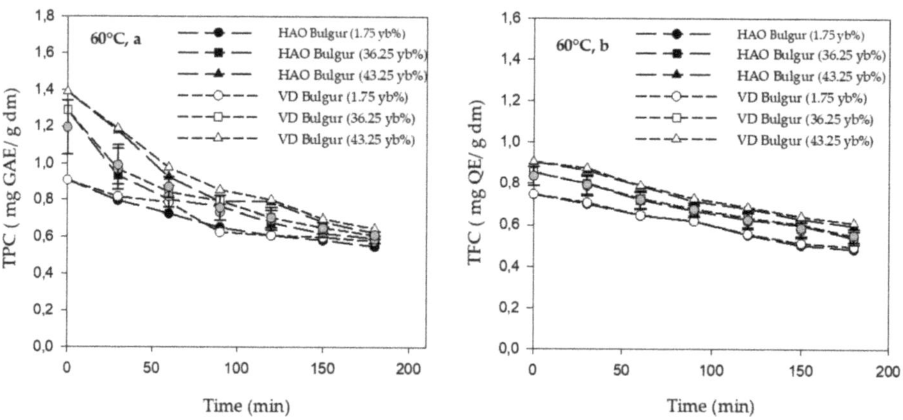

Figure 4. Total polyphenol content (**a**) total flavonoid content (**b**) in bulgur samples at 60 °C. HAO: hot air oven; VD: vacuum dryer.

Examining Figure 5, the yellow pigment content decreased over time. Comparing YPC in the different dryers, no significant difference was found. The YPC ranged from 0.066 ± 0.419 to $0.075 \pm 1.5 \times 10^{-4}$ (mg ß-carotene/100g dm), and from $0.073 \pm 1.9 \times 10^{-5}$ to $0.079 \pm 3.09 \times 10^{-4}$ (mg ß-carotene/100g dm) for the hot air oven and vacuum dryer, respectively.

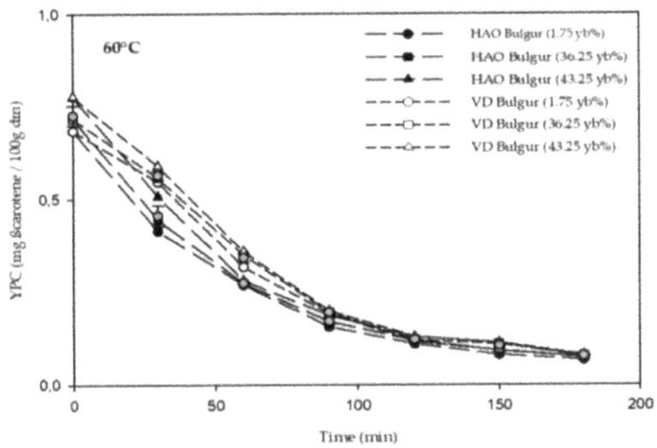

Figure 5. Total yellow pigment content in bulgur samples at 60 °C. HAO: hot air oven; VD: vacuum dryer.

According to ANOVA results, yellow berry percentage had a positive significant effect on the TPC, TFC, and YPC values ($p \leq 0.05$) where bulgur 43.25 yb% samples, in the vacuum dryer and hot air oven, had the highest TPC, TFC, and YPC contents, whereas the bulgur 1.75 yb% samples presented the lowest contents.

4. Discussion

Drying is an important step in bulgur processing since it directly affects the quality [37]. Traditionally, bulgur is spread onto a flat surface and left to dry under the sun for 8–10 h to decrease moisture content from 45 to 10% (dry basis). Several drying methods have been studied, such as the microwave drying method [37]. However, when this technique is assisted by spouted bed drying, the bulgur has a more porous microstructure and lower water absorption capacity, inducing a decrease in drying time [54]. Savas and Basman [35] used infrared treatment at various power levels and periods as an alternative bulgur-drying technique. The results showed that infrared dried samples were similar to sun-dried samples in terms of quality, but that drying time is shorter, thus indicating that infrared drying is a promising technique for the future.

The present paper examines the drying behavior of bulgur using hot air and vacuum dryers. The drying curves obtained from the variation of moisture ratios with time were found to be similar to the drying curves observed by Yildirim [28], who established general equations describing the moisture ratio of parboiled wheat during drying at different temperatures for the different dryers used. Concerning the variation of drying rate with time, only one phase was noted: the falling rate period. The absence of a constant drying period was also reported for parboiled wheat drying in Mohapatra and Rao [27] and Yildirim [28]. Thus, the entire drying process only takes place during the falling rate period, which indicates that moisture diffusion was the governing factor [55] for deciding the drying behavior of bulgur. Comparing the two dryers, the results have confirmed that the vacuum dryer provides faster drying compared to the hot air oven dryer where the Midilli model successfully predicted the drying behavior of bulgur. The vacuum dryer showed the highest D_{eff} value, which can be explained by the easy evaporation of moisture and a higher mass transfer, confirming a faster drying behavior. This result is in agreement with Yildirim [28], who showed that the vacuum dryer was found to be the fastest compared to the convective air and forced-air dryers, and drying time was shortened with the increase of temperature. This confirms that vacuum dryers tend to work faster than other drying methods, reducing the processing time [56].

On the other side, the results have shown that yellow berry disorder had a significant effect on the drying behavior of bulgur, where a higher percent induces high MR content, low value from the Midilli model drying rate constant k, and low value of D_{eff}. This result could be explained by the fact that yellow berry disorder induces a high starch content in wheat where the starchy granules are reported to have a larger diameter and high crystallinity percent than vitreous grains [18,57]. A positive correlation was found between the gelatinization enthalpy (ΔH) and crystallinity percent [58]. The ΔH exhibits the loss of the molecular double helical [59], which induces the stability of the structure and enhances the resistance of the granules to gelatinization [60]. Thus, excess water absorption is required to destabilize the structure generating the gelatinization of the starches [61]. Hence, high yellow berry percentage generates higher moisture content, which induces slower drying. As a result, the drying rate constant and the effective diffusion coefficient decrease with the increase of yellow berry percentage.

Bulgur is considered a practical food [2] since it contains several bioactive compounds. Many works have reported the presence of ferulic acid, gallic acid, 3.4hydroxybenzoic acid, epicatechin, caffeic acid, p-hydroxybenzoic acid, p-coumaric acid, syringic acid, and low amounts of chlorogenic acid in bulgur [2,39,62,63]. The effect of drying on phytochemical content is studied in this work. The obtained values of our samples are in agreement with studies by Caba et al. [2] and Ertas [64]. In fact, Caba et al. [2] investigated the composition of bioactive components of industrial bulgur samples in which the TPC varied between 0.553 and 0.621 (mg GAE/g dm), whereas Ertas [64] studied twelve industrial bulgur samples, four homemade bulgur samples produced in Turkey, and one laboratory-made sample. TPCs of the industrial, homemade, and laboratory-made bulgur samples were found to be between 0.449 and 0.968, 0.632 and 1.173, and 0.986 (mg GAE/g dm), respectively. Concerning the flavonoid contents, the obtained values were lower than those reported in Yüksel et al. [65] where the flavonoid content of bulgur flour was found to be 105.88 (mg catechin/100 g sample). This decrease in TFC in the bulgur samples might be due to the difference in wheat species used, the different bulgur production techniques, and the use of quercetin instead of catechin. In fact, according to Morel et al. [66], the catechin had a bigger effect than quercetin.

It is important to note that the values of TPC and TFC of bulgur are lower compared to wholegrain wheat since these compounds are mainly localized in the bran of durum wheat, and bulgur is defined as a debranned precooked wheat grain [67–72].

Carotenoid content in wheat bran was higher than endosperm [68], since the yellow pigments are more concentrated in the outer layers than the inner layers [73]. Lutein is the major and predominant carotenoid and is responsible for the bulgur's distinct yellow color [74,75]. Other carotenoids, such as zeaxanthin, b-cryptoxanthin, and ß-carotene were also found [76]. A significant correlation was found between yellow pigment content (YPC) in bulgur and cultivar cooking methods as well as their interactions [36]. The moisture content of wheat and abrasion time was also found to significantly affect the total carotenoid content [77]. The carotenoid pigment and lipoxygenase activity are responsible for b* of the grain [23]. Therefore, due to the Maillard reaction, in the presence of heat applied at cooking and drying treatments, the pigments are degraded, which generates discoloration of the bulgur [78]. The obtained results are slightly higher than what was reported in the study of A.K. Elvice and Hazim Ozkaya [36], where the average YPC in coarse and fine bulgur samples was 3.14 (μg/g). This difference is probably due to the different wheat species and different bulgur production processes.

Thermal treatments, such as drying, have been reported to negatively affect the phytochemical content (polyphenol, flavonoid, and carotenoid) in bulgur, which causes its decrease [39,79]. It is important to mention that despite this decrease, no significant difference was found between the two drying methods. This might be due to the use of the same low temperature of 60 °C, which was reported to have the highest retention of total phenolic and yellow pigment content in the bulgur [25,39].

The effect of yellow berry percent on the phytochemical content can be explained by the difference in the hardness structure of the endosperm. In fact, the debranning process removes only the outer layers of the grains, which allows the recovery of intact kernels. Due to the high starch content (high yellow berry percentage), the texture becomes soft [17,19] and therefore could affect the peeling of the outer layers of grain where they are not totally removed from the grain, compared to those debranned from grains with low yellow berry percentage [17]. Therefore, the presence of the outer layers induces high TPC, TFC, and YPC contents in bulgur during debranning.

5. Conclusions

In this study, two different drying methods (hot air oven and vacuum drying) were used in bulgur production using three durum wheat samples at different yellow berry percentages (1.75%, 36.25%, and 43.25%). The drying behavior of the bulgur was successfully described by the Midilli model. Comparing both dryers, a significant difference ($p \leq 0.05$) was found in terms of the variation of moisture ratio and drying rate over time. The vacuum dryer presented the highest D_{eff} and k values, confirming faster and more effective drying than the hot air oven. Yellow berry percentage had a significant effect ($p \leq 0.05$) on the bulgur's drying behavior. Results showed the presence of a strong correlation between high starch content and moisture content, where an increase in yellow berry percentage generates an increase in MR and a decrease in D_{eff} and k values.

Drying at 60 °C decreases bulgur phytochemical content, where no significant difference was observed between the two types of dryers. However, yellow berry disorder had a positive effect on preserving the phytochemical content in bulgur because theyremainin^{2}the outer layers after the debranning process and induce a higher bulgur quality.

Author Contributions: Writing, review and editing; S.T.D. Writing and formal analysis; D.F. Software; H.N. Formal analysis; A.W. Review; Z.Y. All authors have read and agreed to the published version of the manuscript.

Funding: This research was supported by the Ministry of Higher Education and Scientific Research of Tunisia.

Data Availability Statement: The datasets generated for this study are available on request to the corresponding author.

Conflicts of Interest: The authors declare no conflict of interest. The funders had no role in the design of the study; in the collection, analyses, or interpretation of data; in the writing of the manuscript, or in the decision to publish.

References

1. Miskelly, D. Optimisation of End-Product Quality for the Consumer. In *Cereal Grains*, 2nd ed.; Woodhead Publishing: Cambridge, UK, 2017; pp. 653–688. [CrossRef]
2. Caba, Z.T.; Boyacioglu, M.H.; Boyacioglu, D. Bioactive healthy components of bulgur. *Inter. J. Food Sci. Nutr.* **2012**, *63*, 250–256. [CrossRef] [PubMed]
3. Bayram, M.; Öner, M.D. Bulgur milling using roller, double disc and vertical disc mills. *J. Food Eng.* **2007**, *79*, 181–187. [CrossRef]
4. Dreher, M.L. Dietary fiber: Overview. In *Handbook of Dietary Fiber*; Sungsoo, S., Ed.; Marcel Dekker: New York, NY, USA, 2001; pp. 21–36.
5. Jenkins, D.J.A.; Wolever, T.M.S.; Jenkins, A.L.; Giordano, C.; Giudici, S.; Thompson, L.U.; Kalmusky, J.; Josse, R.G.; Wong, G.S. Low glycemic response to traditionally processed wheat and rye products: Bulgur and pumper nickel bread. *Am. J. Clin. Nutr.* **1986**, *43*, 516–520. [CrossRef] [PubMed]
6. Yousif, S.I.; Bayram, M.; Kesen, S. Characterization of volatile compounds of Bulgur (AntepType) produced from durum wheat. *J. Food Qual.* **2018**, *2018*, 8564086. [CrossRef]
7. Balci, F.; Bayram, M. Bulgur cooking process: Recovery of energy and waste water. *J. Food Eng.* **2020**, *269*, 109734. [CrossRef]
8. Bayram, M. Bulgur around the world. *Cereal Foods World* **2000**, *45*, 80–82.
9. Bayram, M.; Kaya, A.; Öner, M.D. Changes in properties of soaking water during production of soy-bulgur. *J. Food Eng.* **2004**, *61*, 221–230. [CrossRef]
10. Yorgancilar, M.; Bilgiçli, N. Chemical and nutritional changes in bitter and sweet lupin seeds (*Lupinus albus* L.) during bulgur production. *J. Food Sci. Technol.* **2014**, *51*, 1384–1389. [CrossRef]

11. Köksel, H.; Edney, M.J.; Özkaya, B. Barley bulgur: Effect of processing and cooking on chemical composition. *J. Cereal Sci.* **1999**, *29*, 185–190. [CrossRef]
12. Bayram, M.; Öner, M.D.; Eren, S. Influence of soaking on the dimensions and colours of soybean for bulgur production. *J. Food Eng.* **2004**, *61*, 331–339. [CrossRef]
13. Ertas, N.; Türker, S. Bulgur processes increase nutrition value: Possible role in in-vitro protein digestability, phytic acid, trypsin inhibitor activity and mineral bioavailability. *J. Food Sci. Technol.* **2014**, *51*, 1401–1405. [CrossRef] [PubMed]
14. Özboy, Ö.; Köksel, H. An application of linear regression technique for predicting bulgur yield and quality of wheat cultivars. *Nahr. Food* **2002**, *46*, 21–24. [CrossRef]
15. Ramirez-Wong, B.; Rodriguez-Félix, F.; Torres-Chavez, I.P.; Medina-Rodriguez, C.L.; Matus-Barba, E.A.; Ledesma-Osuna, A.I. Effects of Nitrogen and irrigation on gluten protein composition and their relationship to "Yellow Berry" disorder in wheat (*Triticum aestivum*). *Pak. J. Bot.* **2014**, *46*, 1797–1804.
16. Anderson, W.K. Grain yields responses of barley and durum wheat to split nitrogen applications under rained conditions in a Mediterranean environment. *Field Crop Res.* **1985**, *12*, 191–202. [CrossRef]
17. Herbert, F.R. Yellow berry in hard winter wheat. *J. Agric. Res.* **1919**, *18*, 155–168.
18. López-Ahumada, G.A.; Ramírez-Wong, B.; Torres-Chávez, P.; Bello-Pérez, L.; Figueroa-Cardenas, J.; Garzón-Tiznado, J.; Gomez-Aldapa, C. Physicochemical characteristics of starch from bread wheat (*Triticum aestivum*) with "yellow berry". *Starch-Stärke* **2010**, *62*, 517–523. [CrossRef]
19. Ammiraju, J.S.S.; Dholakia, B.B.; Jawdekar, G.; Santra, D.K.; Gupta, V.S.; Roder, M.S.; Singh, H.; Lagu, M.D.; Dhaliwal, H.S.; Rao, V.S.; et al. Inheritance and identification of DNA markers associated with yellow berry tolerance in wheat (*Triticum aestivum* L.). *Euphytica* **2002**, *123*, 229–233. [CrossRef]
20. Dexter, J.E.; Marchylo, B.A.; Macgregor, A.W.; Tkachuk, R. The structure and protein composition of vitreous, piebald and starchy durum-wheat kernels. *J. Cereal Sci.* **1989**, *10*, 19–32. [CrossRef]
21. Behera, U.K.; Sharma, A.R.; Pandey, H.N. Sustaining of wheat-soybean cropping system through integrated nutrient management practices on the vertisols on central India. *Plant Soil* **2007**, *297*, 185–199. [CrossRef]
22. Rodriguez-Félix, F.; Ramirez-Wong, B.; Torres-Chavez, P.I.; Álvarez-Avilés, A.; Moreno-Salazar, S.; Eugenia Renteria-Martinez, M.; Arturo Bello-Pérez, L. Yellow Berry, Protein and Agronomic Characteristics in bread wheat under different conditions of nitrogen and irrigation in north west Mexico. *Pak. J. Bot.* **2014**, *46*, 221–226.
23. Sfayhi Terras, D.; Djebbali, K.; Jedidi, E. Influence of process parameters on bulgur quality. *Qual. Assur. Saf. Crops Foods* **2019**, *11*, 431–439. [CrossRef]
24. Sfayhi Terras, D.; Hadjyahia, N.; Zarroug, Y. Effect of soaking time and temperature on water absorption capacity and dimensional changes of bulgur. *Cereal Chem.* **2021**, *98*, 851–857. [CrossRef]
25. Stone, A.K.; Wang, S.; Tulbek, M.; Koksel, F.; Nickerson, M.T. Processing and quality aspects of bulgur from *Triticum durum*. *Cereal Chem.* **2020**, *97*, 1099–1110. [CrossRef]
26. Chakraverty, A.; Kaushal, R.T. Determination of optimum drying conditions and development of drying equations for thin layer drying of parboiled wheat. *Am. Mark. Assoc.* **1982**, *13*, 33–36.
27. Mohapatra, D.; Rao, P.S. A thin layer drying model of parboiled wheat. *J. Food Eng.* **2005**, *66*, 513–518. [CrossRef]
28. Yildirim, A. Kinetics and thermodynamic properties of parboiled burgos wheat (*Triticum durum*) in Turkey during drying. *Appl. Ecol. Environ. Res.* **2018**, *16*, 495–510. [CrossRef]
29. Becker, H.A. A study of diffusion in solids of arbitrary shape, with application to the drying of wheat kernel. *J. Appl. Polym. Sci.* **1959**, *1*, 212–226. [CrossRef]
30. Watson, E.L.; Bhargava, V.K. Thin layer studies on wheat. *Can. J. Agri. Eng.* **1974**, *16*, 18–22.
31. Bayram, M. Determination of the cooking degree for bulgur production using amylase/iodine, centre cutting and light scattering methods. *Food Control* **2006**, *17*, 331–335. [CrossRef]
32. Bayram, M.; Öner, M.D.; Eren, S. Effect of cooking time and temperature on the dimensions and crease of the wheat kernel during bulgur production. *J. Food Eng.* **2004**, *64*, 43–51. [CrossRef]
33. Rios, G.; Pinson-Gadais, L.; Abecassis, J.; Zakhia-Rozis, N.; LillienPellerin, V. Assessment of dehulling efficiency to reduce deoxynivalenol and Fusarium level in durum wheat grains. *J. Cereal Sci.* **2009**, *49*, 387–392. [CrossRef]
34. Bilgiçli, N. Effects of cooking and drying processes on physical, chemical and sensory properties of legume based bulgur. *J. Food Process. Preserv.* **2008**, *33*, 590–604. [CrossRef]
35. Savas, K.; Basman, A. Infrared drying: A promising technique for bulgur production. *J. Cereal Sci.* **2016**, *68*, 31–37. [CrossRef]
36. Evlice, A.K.; Özkaya, H. Effects of wheat cultivar, cooking method, and bulgur type on nutritional quality characteristics of bulgur. *J. Cereal Sci.* **2020**, *96*, 103–124. [CrossRef]
37. Hayta, M. Bulgur quality as affected by drying methods. *J. Food Sci.* **2001**, *67*, 2241–2243. [CrossRef]
38. Kadakal, Ç.; Ekinci, R.; Yapar, A. The effect of cooking and drying on the water-soluble vitamins content of Bulgur. *Food Sci. Technol. Inter.* **2007**, *13*, 349–354. [CrossRef]
39. Yilmaz, V.A.; Koca, A.F. Effect of different production techniques on bioactive compounds and antioxidant capacity of einkorn (*Triticum monococcum* L.) and durum (*Triticum turgidum* subsp. *durum*) bulgur. *J. Sci. Food Agri.* **2016**, *97*, 269–277. [CrossRef] [PubMed]

40. American Association of Cereal Chemists. *Approved Methods*, 10th ed.; American Association of Cereal Chemists (AACC): Eagan, MN, USA, 2000.
41. Sgrulletta, D.; DeStefanis, E. Simultaneous evaluation of quality parameters of durum wheat (*Triticum durum*) by near infrared spectroscopy. *Ital. J. Food Sci.* **1997**, *9*, 295–301.
42. International Association for Cereal Science and Technology. Standard Methods of the ICC. In *Determination of Ash in Cereals and Cereal Products*; Schaifer: Detmold, Germany, 1990.
43. Smirani, N.; Sfayhi, D. Bulgur modeling drying kinetic. In Proceedings of the 15 the International Cereal and Bread Congress, Istanbul, Turkey, 18–21 April 2016.
44. Mau, J.L.; Chao, G.R.; Wu, K.T. Antioxidant properties of methanolic extracts from several ear mushrooms. *J. Agri. Food Chem.* **2001**, *49*, 5461–5467. [CrossRef]
45. Dewanto, V.; Wu, X.; Adom, K.K.; Liu, R.H. Thermal processing enhances the nutritional value of tomatoes by increasing total antioxidant activity. *J. Agric. Food Chem.* **2002**, *50*, 3010–3014. [CrossRef]
46. Determination of Yellow Pigment Content. In *Durum Wheat Flour and Semolina*; International Standardization Organization (ISO): Berlin, Germany, 1994; p. 11052.
47. Bruce, D.M. Exposed-layer barley drying, three models fitted to new data up to 150 °C. *J. Agri. Eng. Res.* **1985**, *32*, 337–348. [CrossRef]
48. Henderson, S.M.; Pabis, S. Grain drying theory II: Temperature effects on drying coefficients. *J. Agri. Eng. Res.* **1961**, *6*, 169–174.
49. Toğrul, H. Suitable drying model for infrared drying of carrot. *J. Food Eng.* **2006**, *77*, 610–619. [CrossRef]
50. Midilli, A.; Kucuk, H.; Yapar, Z.A. New model for single-layer drying. *Dry Technol.* **2002**, *20*, 1503–1513. [CrossRef]
51. Rayaguru, K.; Routray, W. Mathematical modeling of thin layer drying kinetics of stone apple slices. *Inter. Food Res. J.* **2012**, *19*, 1503–1510.
52. Kavak Akpinar, E. The Development of a Cyclone Type Dryer for Agricultural Products. Ph.D. Thesis, Firat University, Elazığ, Turkish, 2002.
53. Crank, J. *The Mathematics of Diffusion*; Oxford University Press: Oxford, NY, USA, 1975.
54. Kahyaoglu, L.N.; Sahin, S.; Sumnu, G. Physical properties of parboiled wheat and bulgur produced using spouted bed and microwave assisted spouted bed drying. *J. Food Eng.* **2010**, *98*, 159–169. [CrossRef]
55. Singh, H.; Sodhi, N. Dehydration kinetics of onions. *J. Food Sci. Technol.* **2000**, *37*, 520–522.
56. Parikh, D.M. Vacuum Drying: Basics and application. *Chem. Eng.* **2015**, *122*, 48–54.
57. Lyon, T.L.; Keyser, A. Winter Wheat. In *Bulletin of the Agricultural Experiment Station of Nebraska*; Michigan Agricultural Experiment Station: East Lansing, MI, USA, 1905; Volume 17, pp. 1–28.
58. Singh, S.; Singh, N.; Isono, N.; Noda, T. Relationship of granule size distribution and amylopectin structure with pasting, thermal, and retrogradation properties in wheat starch. *J. Agric. Food Chem.* **2009**, *58*, 1180–1188. [CrossRef]
59. Cooke, D.; Michael, J.G. Loss of crystalline and molecular order during starch gelatinisation: Origin of the enthalpic transition. *Carbohydr. Res.* **1992**, *227*, 103–112. [CrossRef]
60. Kumar, R.; Khatkar, B.S. Thermal, pasting and morphological properties of starch granules of wheat (*Triticum aestivum* L.) varieties. *J. Food Sci. Technol.* **2017**, *54*, 2403–2410. [CrossRef] [PubMed]
61. Wootton, M.; Bamunuarachchi, A. Application of Differential Scanning Calorimetry to Starch Gelatinization. II. Effect of Heating Rate and Moisture Level. *Starch-Starke* **1979**, *31*, 262–264. [CrossRef]
62. Yu, L.; Cheng, Z. Antioxidant properties of wheat phenolic acids. In *Wheat Antioxidants*; Yu, L., Ed.; Wiley-Interscience: Hoboken, NJ, USA, 2008; pp. 54–72.
63. Sosulski, F.; Krysztof, K.; Lawrence, H. Free, esterified, and insoluble-bound phenolic acids. 3. Composition of phenolic acids in cereal and potato flours. *J. Agric Food Chem.* **1982**, *30*, 337–340. [CrossRef]
64. Ertaş, N. A comparison of industrial and homemade bulgur in turkey in terms of physical, chemical and nutritional properties. *Chem. Ind. Chem. Eng. Q.* **2017**, *23*, 341–348. [CrossRef]
65. Yüksel, A.N.; Oner, M.D.; Bayram, M. Usage of undersize bulgur flour in production of short-cut pasta-like couscous. *J. Cereal Sci.* **2017**, *77*, 102–109. [CrossRef]
66. Morel, I.; Lescoat, G.; Cogrel, P.; Sergent, O.; Pasdeloup, N.; Brissot, P.; Cillard, P.; Cillard, J. Antioxidant and Iron-Chelating activities of the flavonoids catechin, quercetin and diosmetin on Iron-Loaded rat hepatocyte. *Biochem. Pharmacol.* **1993**, *45*, 13–19. [CrossRef]
67. Ragaee, S.; Guzar, I.; Abdel-Aal, E.-S.M.; Seetharaman, K. Bioactive components and antioxidant capacity of Ontario hard and soft wheat varieties. *Can. J. Plant Sci.* **2011**, *92*, 19–30. [CrossRef]
68. Adom, K.; Liu, R. Antioxidant activity of grains. *J. Agric. Food Chem.* **2002**, *50*, 6182–6187. [CrossRef]
69. Beta, T.; Nam, S.; Dexter, J.E.; Sapirstein, H.D. Phenolic content and antioxidant activity of pearled wheat and roller milled fractions. *Cereal Chem.* **2005**, *82*, 390–393. [CrossRef]
70. Li, W.; Shan, F.; Sun, S.; Corke, H.; Beta, T. Free radicals cavenging properties and phenolic content of Chinese black-grained wheat. *J. Agric. Food Chem.* **2005**, *53*, 8533–8536. [CrossRef]
71. Moore, J.; Hao, Z.; Zhou, K.; Luther, M.; Costa, J.; Yu, L. Carotenoid, tocopherol, phenolic acid, and antioxidant properties of Maryland-grown soft wheat. *J. Agric. Food Chem.* **2005**, *53*, 6649–6657. [CrossRef] [PubMed]

72. Shewry, P.; Piironen, V.; Lampi, A.M.; Nystrom, L.; Li, L.; Rakszegi, M.; Fras, A.; Boros, D.; Gebruers, K.; Courtin, C.M. Phytochemical and fiber components in oat varieties in the health grain diversity screen. *J. Agric. Food Chem.* **2008**, *56*, 9777–9784. [CrossRef] [PubMed]
73. Hentschel, V.; Kranl, K.; Hollmann, J.; Lindhauer, M.G.; Böhm, V.; Bitsch, R. Spectrophotometric determination of yellow pigment content and evaluation of carotenoids by high-performance liquid chromatography in durum wheat grain. *J. Agric. Food Chem.* **2002**, *50*, 6663–6668. [CrossRef]
74. Balcı, F.; Bayram, M. Modification of mechanical polishing operation using preheating systems to improve the bulgur color. *J. Cereal Sci.* **2017**, *75*, 108–115. [CrossRef]
75. Burkhardt, S.; Böhm, V. Development of a new method for the complete extraction of carotenoids from cereals with special reference to durum wheat (*Triticum durum* Desf.). *J. Agri. Food Chem.* **2007**, *55*, 8295–8301. [CrossRef]
76. Zhou, K.; Yin, J.J.; Yu, L. Phenolic acid, tocopherol and carotenoid compositions and antioxidant functions of hard red winter wheat bran. *J. Agri. Food Chem.* **2005**, *53*, 3916–3922. [CrossRef]
77. Chen, Y.; Dunford, N.T.; Goad, C. Phytochemical composition of extracts from wheat grain fractions obtained by tangential abrasive dehulling. *Food Sci. Technol.* **2013**, *54*, 353–359. [CrossRef]
78. Özboy, Ö.; Köksel, H. The effects of bulgur production on various chemical properties of wheats. *J. Food* **1998**, *23*, 449–457.
79. Yilmaz, V.A. Effects of different cooking and drying methods on phenolicacids, carotenoids, and antioxidant activity of emmer (*Triticum turgidums* sp. *dicoccum) bulgur. Cereal Chemistry* **2019**, *96*, 1093–1102. [CrossRef]

foods

MDPI

Review

Development of Novel Pasta Products with Evidence Based Impacts on Health—A Review

Mike Sissons

Department of Primary Industries, Tamworth Agricultural Institute, 4 Marsden Park Road, Tamworth, NSW 2340, Australia; mike.sissons@dpi.nsw.gov.au

Abstract: Pasta made from durum wheat is a widely consumed worldwide and is a healthy and convenient food. In the last two decades, there has been much research effort into improving the nutritional value of pasta by inclusion of nonconventional ingredients due to the demand by health-conscious consumers for functional foods. These ingredients can affect the technological properties of the pasta, but their health impacts are not always measured rather inferred. This review provides an overview of pasta made from durum wheat where the semolina is substituted in part with a range of ingredients (barley fractions, dietary fibre sources, fish ingredients, herbs, inulin, resistant starches, legumes, vegetables and protein extracts). Impacts on pasta technological properties and in vitro measures of phytonutrient enhancement or changes to starch digestion are included. Emphasis is on the literature that provides clinical or animal trial data on the health benefits of the functional pasta.

Keywords: durum wheat; pasta; health benefits; functional pasta; functional food

1. Introduction

Carbohydrates in foods are an important source of energy for humans, with cereals, tubers and pulses being the main dietary sources. Pasta is a popular food worldwide known for its ease of preparation, good storage stability (dried form), low cost, simple preparation with a low glycaemic index (GI). Pasta consists mostly of carbohydrates (70–76%), protein (~10–14%), lipids (~1.8%), dietary fibre (~2.9%) and small amounts of minerals and vitamins [1]. Pasta is made from either semolina (derived from durum wheat (*Tritcum turgidum* var. *durum* Desf.) or common wheat flour (*aestivum*) usually when supply of durum is limited or the price too high) mixed with water and mechanical energy input (mixing, extrusion, lamination) to produce a crumbly dough (~28–32% *w*/*w* water) either on an industrial or artisan scale producing a fresh pasta, which can then be dried. However, pasta has low amounts of dietary fibre, vitamins, essential amino acids and minerals [1] and during milling to make semolina there is some loss of these components. Pasta can be considered a good vehicle for including bioactive ingredients (proteins, phytochemicals, minerals, vitamins, etc.) as recognised by the World Health Organization and U.S. Food and Drug Administration because in some situations, up to 10–15% of non-traditional ingredients can be added without major loss of pasta quality depending on the ingredient and pasta processing technology employed [2,3]. However, the benefit of the added ingredient purported to provide, can be limited with such a low incorporation level. While designing foods with biologically active compounds, the resultant food often has technological deficiencies, undesirable appearance and sensorial properties making them less attractive to consumers or simply uneconomic to manufacture.

Increasingly many consumers in more wealthy economies are more interested in food that provides a benefit to prevent or reduce nutritional related diseases than was the case a decade or two ago. The main so called "lifestyle or civilisation" diseases have been associated with a combination of excessive caloric intake, poor nutrient balance and lack of sufficient exercise include obesity, overweight, elevated blood pressure, elevated blood

Citation: Sissons, M. Development of Novel Pasta Products with Evidence Based Impacts on Health—A Review. *Foods* **2022**, *11*, 123. https://doi.org/10.3390/foods11010123

Academic Editor: Alessandra Marti

Received: 29 November 2021
Accepted: 30 December 2021
Published: 4 January 2022

cholesterol, cardiovascular disease (CVD), cancers, alimentary system disorders and type II diabetes mellitus (2-DM). These diseases afflict a large percentage of the population of westernized countries, with the trend continuing to worsen in developing nations and these diseases represent the main non-communicable cause of death [4]. The term functional food was developed first in Japan defined by the Food and Nutrition Board of the National Academy of Sciences as "any modified food or food ingredient that may provide a health benefit beyond that of the nutrients it contains" [5]. More common now is the consumer demand for food that not only provides additional health benefits but tastes good and has the texture and flavour close to the traditional food. Adding a functional ingredient will increase the cost of the food which, if a health benefit can be demonstrated, many consumers are likely to pay in discerning markets.

Increasing demand by a growing number of health-conscious consumers for healthy foods has garnered interest from food manufacturers and a plethora of studies exist in the literature today [4,6]. Within the last decade or so, there has been a trend towards manufacturers trying to improve the nutritional value or create a presumed health benefit by supplementing semolina with various ingredients either as a partial replacement of the semolina or a complete replacement [3]. This approach is a powerful strategy for improving diet and wellbeing. Typical strategies employed to create a functional pasta are summarised in Table 1.

Creating genetically modified wheat through transgenic or non-transgenic (TILLING, CRISPR-Cas9) or conventional breeding to modify starch, protein and lipid composition is an ambitious strategy gaining interest by breeders but is very much in its infancy. For example, a high amylose durum wheat was developed using TILLING showing that the GI of durum pasta, already low–medium, can be further reduced with the right starch biosynthetic enzyme mutation [7]. Beta-carotene has already been expressed in a genetically engineered rice cultivar, named Golden Rice to benefit people with vitamin A deficiency in developing countries [8]. Breeding durum wheat for enhanced functional or medicinal value is a new concept that has yet to take-off in commercial plant breeding programs and traditionally plant breeding of crop species has mostly focussed on yield, quality, biotic and abiotic tolerance but not food nutrition or health benefits. Research arising from the HEALTHGRAIN project between 2005–2010 described the genetic variation of key grain nutrient components and tools were developed for breeders for selecting cultivars with high levels of healthy compounds [9] but commercial application has been limited until recently. Some programs have begun work in this area and the reader is referred to the review by Yu and Tian [8]. Pasta made from pseudocereals (amaranth, buckwheat, quinoa) or blends with wheat flour are rich in micronutrients, phytonutrients, gluten free, with a more balanced amino acid profile than rice, wheat and maize, could be another route to have "ready-made" functional cereal products [10]. Ancient grains that can be readily made into flours can also provide some benefits although the evidence for their functional value is unclear [11]. Gluten free pasta can address the needs of the celiac person who must avoid gluten to prevent symptoms of this disease while gluten intolerant individuals, which are on the increase in society, will choose these products over gluten containing pasta. Simply adding various nutrients by substituting some of the semolina or adding them in the water used to make the dough at various levels (1–20%) of ingredient(s) is the most common strategy used and a focus of this review (Tables 1 and 2). To improve the total protein of pasta (beyond the typical 10–15% range) or improve the amount of essential amino acids lacking in pasta (lysine, threonine, methionine), common sources used for pasta include legumes, cereal germ, dairy powders (egg white, casein), bovine serum powders, fish proteins and microbial fermentation products [6]. During the milling of grains many vitamins are lost which can be overcome by adding vitamin-rich tissues such as spinach, tomatoes, mushrooms, calf liver, sunflower seeds, chicken or fish meat [6]. Cereal bran is a good source of fibre, and vegetable oils, seafood and fish oil are excellent sources of polyunsaturated fatty acids. See Table 2 for some studies relating to enhancing limiting nutrients in pasta.

Other avenues such as valorising cereal and non-cereal by-products is becoming more popular with a push to reduce food waste and create a circular food economy. Finally processing to modify the additive before its addition to the food, e.g., germination, fermentation, enzymic treatment, etc. is another approach to create a functional pasta. The aim of these strategies is to improve the nutritional and/or physiological functions from consuming such food on for example, gut health, immune system activity, mental status and/or to reduce the risk of specific pathologies for example cancer, cardiovascular disease, diabetes, and osteoporosis.

In designing a functional pasta consideration must be given to the form the food is consumed. Pasta is traditionally eaten cooked after boiling in water for several minutes. In processing the extrusion or lamination process introduces forces, the drying process uses high temperature and variable humidity, so all these can affect the functionality of the included active ingredient(s). Even during digestion in the human system, there can be a loss in efficacy of the active ingredient (bioavailability) [12]. An added challenge is to ensure that the added ingredient should have a minimal impact on the pasta quality, palatability and consumer appeal.

The present review aims to provide an overview of recently published scientific articles from the year 2000 to 2021 focussing on ingredients added to a durum wheat semolina –water formulation to make pasta with a functional benefit. However, where appropriate relevant pre-2000 references are included. Emphasis will be on more recent novel ingredients and evidence for demonstrated health benefits from animal or human trials rather than reports relying on the presumed benefit from only in vitro studies alone or assuming that the added ingredient's known medicinal value transfers to the pasta consumed, which is not necessarily the case. While it might be a simple matter of adding an ingredient into the semolina–water mix at various percentages, the active ingredients functionality cannot be assured.

Review of methods Recent reviews on the impacts on the technological quality of the pasta with added functional ingredients have been published in the last decade and the reader is referred to these to supplement this review [2–4,6,13–19]. Many of the studies mentioned in these reviews and other publications have been summarised in Table 2. Most of these have not shown evidence of the impact of the added ingredient on human health. A focus of this review was to present studies where some clinical evidence for the health benefit of the functional pasta (made only from durum wheat) is presented either in human or animal studies. For the other studies, a summary is provided in the form of a table (Table 2) where the author has decided to present relevant studies from the perspective of the raw ingredient(s) added to the pasta formulae, the likely active ingredient providing the purported health benefit, impact on the pasta quality and functional value with some prediction over the possible in vivo benefit if clinical studies were performed. Areas not covered in this review include the effect of functional pasta ingredients from ancient grains and pseudocereals [10]; gluten free pasta [20,21]; cereal and non-cereal by-products [17] and pasta made with agro-industrial by-products [13] to keep the scope of this review manageable. Database searches were supplemented by manual searches of the reference lists of included reports and previous reviews. Language was restricted to English only. The search strategy used was last updated on 5 October 2021 (Table 3). Studies using common wheat to prepare pasta were not included.

Table 1. Strategies to create pasta with added nutritional functionality.

Approach	Intention	Reference
Genetically modify the composition of the grain	Enhance a specific component of the grain	[7,8]
• conventional breeding		
• GMOs		
• Gene technologies—TILLING, CRISPR-Cas9		
Substitute semolina with various levels (1–20+%) of ingredient(s) with potential health value:	Functional foods: Increase protein content and quality; increase fibre, AO, phytochemicals, etc.	[2–4,6,13–19]
• Proteins, Fructo oligosaccharides, dietary fibre,		
• Prebiotics, ω-3 fatty acids, minerals, vitamins		
• Phytochemicals and probiotics		
• Legumes (chickpeas, red lentils, yellow peas, faba beans, soy)		
• Vegetables (pumpkin, zucchini, spinach, tomato)		
• Herbs (oregano leaves, parsley leaves)		
• Roots and tubers (cassava, sweet potato, beet, carrot)		
• Others (gums, resistant starch, modified starch, β-glucan, psyllium seed husk)		
Composite flours	Reduce cost of pasta by replacing some/all durum semolina with common wheat flour and other flours	[15]
• Common wheat and durum wheat		
• Other cereal flours and durum wheat		
Gluten free pasta	Gluten free diet, celiac diet	[20,21]
Processing to modify the additive before its addition to the food	Enhance ingredient nutritional value, remove anti-nutritional factors	
• Germination		[22]
• Fermentation		[6,23,24]
• Enzymic treatment		[6]
• Micronisation		[25,26]

Table 1. *Cont.*

Approach	Intention	Reference
Low value products/waste streams	Valorisation of cereal and noncereal by-products	
• Bran fractions		[25,27,28]
• Aleurone fractions		[29]
• Grape marc, fruit pomace		[30]
• Fish meal		[31,32]
• Whey		[15,33]
• Algae		[34]
Ancient grain/pseudocereals	Valorisation of underutilised grains	[10,15,35]
• einkorn		
• emmer wheat		
• kamut		
• spelt		
• buckwheat		
• quinoa		
• amaranth		

Table 2. Examples of pasta made with a combination of semolina and non-traditional ingredients.

Ingredient Added	Active Ingredient	Substitution Range	Impact on Pasta	Predicted Health Benefits from Data Presented	Reference
			Barley fractions or β-glucan		
Barley Balance®	β-glucan	0, 7.5, 15, 20	Provides AO, lowers IVSD, minimal impact on pasta making quality up to 7.5%	Lower GI, cholesterol reduction, SCFA production	[36]
Glucagel, Barley Balance® (BB)	β-glucan	0, 2, 4, 6, 8, 10	BB only reduced IVSD. Some impact on pasta making quality after 4%	Lower GI, cholesterol reduction, SCFA production	[37]
Barley β-glucan fibre fraction	β-glucan	0, 2.5, 5, 7.5, 10	lowers IVSD but reduced pasta making quality above 2.5%	Lower GI, cholesterol reduction, SCFA production	[38]

Table 2. *Cont.*

Ingredient Added	Active Ingredient	Substitution Range	Impact on Pasta	Predicted Health Benefits from Data Presented	Reference
Barley fractions	β-glucan	0, 5, 20, 40	Increased TDF, darker, acceptable sensory, lower total calories	unknown	[39]
Barley pearling fractions	β-glucan	50	Higher TDF, pasta darker with good cooking qualities	unknown	[40]
β-glucan enriched barley flour	β-glucan	40	Increased β-glucan to 5% in pasta; pasta quality comparable to control; higher AO and TPA; reduced IVSD	Could lower GI and enhance plasma oxidative defence	[41]
Barley hull air classification fractions	β-glucan	50% coarse Fr; 45% coarse Fr + 5% gluten; 95% coarse Fr + 5% gluten	Increased TDF and β-glucan; Higher AO and flavan-3-ols, TPA	Could enhance plasma oxidative defence	[42]
Soluble fibres: BB, psyllium, GR-inulin, HPX-inulin enriched pasta and doughs	Soluble fibres	15% individual fibre and dual combinations, 7.5% each	Pasta containing BB individually and in combination with psyllium showed an overall sensory acceptability comparable to control and in vivo glycaemic index reduction of 33–37%	Reduced pasta GI	[43]
Oat (1,3)(1,4) β-D-glucans	β-glucan	0, 5, 10, 15, 20	Oat β-glucan increased pasta water absorption, fat, TDF, and increased cooking loss >5%, minimal impact on appearance but sensory acceptable up to 15%. 10–15% Oat β-glucan and 5% additive of vital wheat gluten and xanthan gum yielded functional pasta containing 3.3–5.5 g β-glucans/100 g	Oat claim for lowering GI, lowering cholesterol	[44]
Other dietary fibres components					
Guar gum, CMC	Soluble fibres	CMC: 0, 0.25, 0.5, 0.75, 1.0, 1.5 GG: 0, 2.5, 5, 10, 15, 20	lowers IVSD with 20% GG but impacts pasta making quality; lowers IVSD with 1.5% CMC no impact on pasta making quality	Lower GI unknown level needed	[45]
Bran, pollard	Insoluble fibres	Bran: 0, 10, 20, 30 Pollard: 0, 10, 20, 30, 40, 50, 60	Up to 10% pollard can be tolerated minimal impact pasta quality with elevated AO, TDF. Bran had negative impacts pasta at all doses but with enhanced TDF, AO with no effect on IVSD	Higher TDF bowel health and transit	[27]
Commercial sources of pea fibre, Inulin, GG, locust bean gum, Xanthan gum, Bamboo fibre, HISol (B-glucan)	Non-starch polysaccharides, inulin	0, 2.5, 5, 7.5, 10	Increased cooking loss and reduced starch and protein and effects on texture varied with non-starch polysaccharides used and quantity with 5% the limit. Fresh pasta only used	Higher TDF bowel health and transit; lower GI likely, enhanced Ca absorption	[14]

Table 2. *Cont.*

Ingredient Added	Active Ingredient	Substitution Range	Impact on Pasta	Predicted Health Benefits from Data Presented	Reference
Debranning fractions (DF) and micronized debranned kernels (MK)	AO, phenolics	DF 30%; MK 100%	Higher content of phenolic compounds with minimal effects on pasta sensory properties	Higher TDF bowel health and transit	[46]
Phenolic extract	Phenolics	Phenolic extract liquid replaces water used in pasta making	Dough was weakened, pasta was more brown and sensory scores impaired (more bitter and salty)	Poor strategy to enhance phenolics in pasta	[25]
Long-chain inulin (HPX), short-chain inulin (GR), Glucagel, psyllium and oat material added individually and in combinations	Inulin, β-glucan, dietary fibre	15	Addition of DF individually and in combination increased pasta optimum cooking time, cooking loss, water absorption and a deterioration in texture and colour values compared to non-DF enriched control. Oat bran flour with another DF gave the best pasta while psyllium fibre was the worst	Higher TDF	[47]
Wheat, rice, barley, oat brans	Insoluble fibres	0, 5, 10, 15, 20, 25	Decreasing sensory acceptability and colour and increase in cooking loss with increasing dose, least impact with oat bran	unknown	[28]
Dephytinized rice, rye, wheat, oat	Insoluble fibres	20	A 1.7–2.9% increase in pasta TDF. Increased TPA andF AO and Ca, P, K, Mg, Zn with significantly reduced phytic acid content	Higher TDF bowel health and transit	[48]
Whole wheat durum pasta	Wholegrain components	Whole wheat vs. regular pasta	Whole wheat dough is weaker, pasta is reddish-brown with higher cooking loss, lower firmness in cooked product and reduced mechanical strength of dried compared to regular pasta	Potentially multiple benefits, likely lower GI	[49]
Micronized wheat bran with CMC, XG, locus bean gum	Insoluble fibres and gums	11.5	Egg tagliatelle pasta with added XG > 0.8% improved textural properties and CMC >0.6% to enhance yellowness was found to produce a healthier pasta product with higher content of fibre, minerals and vitamins and suitable quality	Higher TDF and potential health benefits from this	[26]
High fibre oat pasta	Soluble and insoluble fibre	10, 20	Oat fibre increased pasta TDF ~8% but increased water absorption and cooking loss, decreased brightness and firmness and impacts reduced using fine (volume mean diameter, µm 50.5) vs. medium (141) and coarse (249) oat powder	Oat claim for lowering GI, lowering cholesterol	[50]

Table 2. *Cont.*

Ingredient Added	Active Ingredient	Substitution Range	Impact on Pasta	Predicted Health Benefits from Data Presented	Reference
			Fish products and algae		
Spirulina microalgae enriched pasta	water-soluble pigments and phycocyanin and phenolic compounds	3	The technological properties of pasta were affected, but overall acceptability index (85.13%) not influenced by microspheres. Microencapsulated spirulina protects the microalgae's antioxidant potential	Benefits from AO	[34]
Pastas with added concentrates of flesh and skin from aquaculture seabass	Source of polyunsaturated fatty acids and minerals	concentrate fish flesh powder 10, concentrate fish skin powder, 20	Increased Ω-3 fatty acids in pastas with fish concentrates, decrease in the Ω6/Ω3 ratio that greatly exceeds current nutritional guidelines. All pastas showed a low valuation in negative attributes such as oil, or rancidity flavours. Main differences detected were colour, fishy flavour, odour, and texture (chewiness)	Possible improved cardiovascular health markers	[51]
Salmon fish (*Oncorhynchus tschawytscha*) powder (SFP) supplemented pasta	Antioxidants and other carotenoids	5, 10, 15, 20	SFP addition to pasta increased the release of phenolic compounds and AO activity from pasta during digestion to achieve higher levels than control pasta and also reduced the in vitro starch digestibility	Lowers GI	[31]
Pasta formulation was substituted with shrimp meat	Omega-3 polyunsaturated fatty acids	10, 20, 30	shrimp meat (*P. monodon*) can be added up to 20% without drastically affecting the sensory attributes of pasta with enhanced nutritional quality (protein, fat and ash content)	unknown	[32]
			Herbs		
Dried amaranth leaves and amaranth seed flour pasta	Peptides derived from protein, source AO, phenolic acids, flavonoids, carotenoids	amaranth seed flour, 21.25–50.97% and dried amaranth leaves, 0–5.61%.	Pasta with amaranth seed flour and dried amaranth leaves exhibited significantly higher content of protein, crude fibre, minerals with higher AO but panellists preferred pastas with low percentage levels of amaranth seed flour	Benefits from AO anti-hypertensive, anti-oxidant, antithrombotic, anti-proliferative, and anti-inflammatory activities	[35]

Table 2. *Cont.*

Ingredient Added	Active Ingredient	Substitution Range	Impact on Pasta	Predicted Health Benefits from Data Presented	Reference
Wild edible plants, *Pereskia aculeata* Miller or American gooseberry dried leaf flour enriched pasta	Source of protein and lysine, soluble fibre, minerals, vitamins	0, 10, 20	Improved pasta dietary fibre, calcium, iron compared to the control pasta. Enriched pasta presented a greenish fibrous appearance. Sensory evaluations indicated that pasta enriched with 10% did not affect consumer acceptance	constipation, obesity (high satiety due the dietary fibre content)	[52]
			Inulin addition		
Inulin enriched pasta	Inulin	0, 2.5, 5, 7.5, 10, 20	The higher molecular weight inulin had minimal impact on pasta quality and sensory properties until 20% while lower MW inulin had more negative impacts on pasta firmness, cooking loss, and sensory acceptability. IVSD was reduced in pasta with inulin higher MW inulin up to 5% but was increased with 20% inulin. Inulin enhanced the gluten structure in pasta with higher starch crystallinity	Lower GI	[53]
Fresh pasta with inulin (FRUTAFIT HD)	Inulin	0, 2.5, 5, 7.5, 10	Inulin was shown to influence the swelling index and firmness, but not the adhesiveness and elasticity of pasta products and lowered IVSD	Lower GI	[54]
			Legume addition		
Chickpea flour	phytic acid, sterols, tannins, carotenoids, as isoflavones	5–20	Increased protein content; sensory properties (colour, flavour and overall acceptability) improved up to 10%; >30% led to lasagne processing handling and cooking characteristics deterioration and soft mushy pasta	Higher quality protein with good balance of amino acids	[55]
Desi chickpea 'besan' flour	phytic acid, sterols, tannins, carotenoids, as isoflavones	0, 10, 15, 20, 25, 30	Up to 15% chickpea can be tolerated in spaghetti with acceptable pasta making quality	Higher lysine and protein content	[56]
Legume pasta (mung, soya, red spit lentil, chickpea)	Soluble and insoluble fibres	10	No negative impact on technological quality or IVSD	Higher TDF and potential health benefits from this	[57]

Table 2. *Cont.*

Ingredient Added	Active Ingredient	Substitution Range	Impact on Pasta	Predicted Health Benefits from Data Presented	Reference
Black chickpea flour and fermented black chickpea dough pasta	phytic acid, sterols, tannins, carotenoids, as isoflavones	5.6 (Black chickpea flour), 15 (black chickpea dough)	Fermentation enabled release of 20% of bound phenolic compounds in the dough, higher resistant starch and total free amino acids while antinutritional factors significantly decreased. Fortified pasta had higher in vitro protein digestibility (up to 38%) and higher AO levels. Fermentation reduced antinutritional elements in the black chickpea flour. Sensory acceptance while different to control described a peculiar but appreciated profile of the fortified samples, especially for the pasta including fermented black chickpea dough.	unknown	[58]
Lentil flour and CMC	proteins, dietary fibres, oligosaccharides, starch, polyphenols, fatty acids, and antioxidants and prebiotics	40 (lentil) 2 (CMC)	Lentil fortified spaghetti increased essential amino acids but caused a decrease in pasta quality (e.g., higher cooking loss, lower breaking energy) that was improved by adding CMC	unknown	[59]
Mexican common bean flour	proteins, vitamins, complex carbohydrates and minerals	0, 15, 30	The cooking time and water absorption decreased and cooking loss increased to unacceptable levels, firmness decreased and pasta was darker as a function of the bean flour percentage. Protein increased. Increases of furosine and marginal increases in phenolic contents in pasta	Benefits from TPA	[23]
Faba bean pasta	Essential amino acids	0, 30, 70, 100	Faba enriched pasta weakened the protein network that could be responsible for the increase in the in vitro protein digestion but led to high cooking loss and reduced resilience in cooked product. Very high temperature drying strengthened the protein structure of pasta, resulting in increased integrity and better resilience of pasta without altering their in vitro protein digestibility. Appreciation of legume pasta containing 80% or 100% was similar to that of commercial whole wheat pasta	unknown	[60]

Table 2. *Cont.*

Ingredient Added	Active Ingredient	Substitution Range	Impact on Pasta	Predicted Health Benefits from Data Presented	Reference
Pasta with added chickpea flour	Fibre, proteins	20, 40	Protein, ash, lipid, and dietary fibre and RS content increased by adding chickpea flour to the pasta. The starch hydrolysis index decreased as chickpea flour in the pasta increased, with a lower predicted glycaemic index than durum wheat-control pasta.	Lower GI	[61]
Yellow pea pasta	alkaloids, flavonoids, glycosides, isoflavones, phenols, phytosterols, phytic acid, protease inhibitors, saponins, tannins	0, 10, 20, 30	20% yellow pea flour had favourable sensory attributes, protein content, good texture, yellowness values, reduction in the glucose release and increased protein digestibility. Dough was weaker while product appearance similar to control	Lower GI	[62]
Pasta with split pea and faba bean	Fibre, protein, vitamins and minerals	35	Increased cooking loss, lower pasta breaking energy, altered sensory properties (higher hardness and fracturability). High drying temperature improved slightly but pasta redness increased to undesirable levels with very high T drying	Higher TDF and potential health benefits from this	[63]
Pasta with added germinated pigeon pea (*Cajanus cajan*)	low fat, fibre, proteins and starch, balanced of minerals	0, 5, 8, 10	Germination of pigeon pea reduced antinutritional components and increased vitamin B2, E and C. Good acceptability, higher protein, total available sugars, dietary fibre, micronutrients, and vitamins than pasta made from 100% semolina but impacts on pasta making quality (shorter cooking time, higher water absorption and cooking loss)	Vitamins, fibre, better protein balance	[22]
Corn gluten meal enriched pasta	High protein source	0, 5, 10	Corn gluten meal increased pasta protein content, had a similar cooked weight and cooking loss but was less firm with inferior colour compared with the control. The overall flavour quality score of the spaghetti decreased	Unknown	[64]

Table 2. *Cont.*

Ingredient Added	Active Ingredient	Substitution Range	Impact on Pasta	Predicted Health Benefits from Data Presented	Reference
			Lupin addition		
Lupin flour to replace semolina	High protein and fibre source	0, 10, 20, 30, 40, 50	Minimal impacts on pasta cooking loss and dry pasta colour and no difference in sensory acceptability up to 20% but α-galactosides and antinutritional factors like phytic acid, saponins, lectins and protease inhibitors reduce protein digestibility	unknown	[65]
Lupin protein isolate	Proteins, AO, TDF	0, 5, 17, 30	Lupin protein isolate increased protein up to 129%, reduced pasta cooking time, water absorption and cooked firmness while stickiness and cooking loss were increased. Lupin protein isolate made the dried pasta more red and yellow and decreased brightness. The percentage of starch digested under in vitro conditions was reduced using 17% lupin protein isolate	Reduced GI	[66]
α-galactosides free lupin flour	High protein and fibre source	0, 50, 80, 100	α-galactosides free lupin flour can improve pasta nutritional value without flatulent causing oligosaccharides	unknown	[22]
			Protein addition		
Lupin protein isolate	High protein and fibre source	0, 5, 17, 30	Lupin protein isolate increased protein, reduced cooking time, water absorption and firmness but stickiness and cooking loss increased making dried pasta duller	unknown	[66]
Durum bran protein concentrate	High in phytosterols, protein and EAA	0, 1 5, 10, 20	Pasta quality acceptable up to 10% and enriched in EAA	Benefits from better protein quality	[67–69]
Whey enriched pasta	High in protein and EAA	0, 20	Whey addition increased protein content, and pasta water uptake with minimal impact on sensory quality	unknown	[33]
Beef lung powder enriched pasta	High in protein and EAA, Fe	0, 10, 15, 20	Pasta had higher cooking loss, cooked pasta was firmer and much darker than control with reduced IVSD, higher Fe and protein content	Lowers GI	[70]

Table 2. *Cont.*

Ingredient Added	Active Ingredient	Substitution Range	Impact on Pasta	Predicted Health Benefits from Data Presented	Reference
Mustard protein isolate enriched pasta	High in protein and EAA	0, 2.5, 5, 10	Increased pasta protein while cooking loss, cooked weight and stickiness decreased and firmness increased while pasta is duller and more red	Unknown	[71]
Phaseolus vulgaris protein hydrolysate	angiotensin I-converting enzyme inhibitory activity (ACE) and AO	0, 5, 10	Pasta with bean had higher protein content with good sensory acceptability up to 10% with ACE and AO activity	BP regulation	[72]
Resistant starch					
Hi Maize™ RSII and Novelose 330™ (RSIII) enriched pasta	Resistant starch	RSII: 0, 10, 20, 50 RSIII: 0, 10, 20	Minimal impact on pasta quality using these ingredients up to 20% while increasing RS content of pasta, stable after cooking. Both RS reduced IVSD	Lower GI gut health benefits from RS	[73]
Hi Maize260™, Hi Maize1043™, RSII and Fibersym70™ (RSIV) enriched pasta	Resistant starch	0, 10, 20	RS addition had minimal impact on pasta quality and acceptability while reducing the IVSD	Lower GI gut health benefits from RS	[73,74]
Unripe banana fibre	Starch from unripe banana flour	0, 5, 10, 15, 20	Increased pasta RS, decreased gluten, was darker, higher cooking loss and firmness lower while sensory analysis found banana starch improved acceptability up to 15% but this analysis was limited	Unknown	[75]
Pastas with elderberry juice Concentrate (EJC) and Hi-maize starch or apple pectin	phenolic acids, anthocyanins, flavanols, carotenoids, vitamins and minerals, soluble DF	10 g Hi-maize starch, pectin or combination, and diluted elderberry juice concentrate (50 mL per 50 g flour)	Adding EJC to fettuccine pastas reduced the firmness, wettability and volume expansion of the fresh pastas, but Increased protein, total DF content, total antioxidant activity and total extracted TPA content	AO and TDF mostly from insoluble fibre	[76]
Vegetables					
Stems of *Opuntia ficus-indica* (cladodes), dried and ground and extracted (*Opuntia cladode* extract)	Rich in soluble fibre (arabinose, galactose, rhamnose, xylose and galacturonic acid)	0, 10, 20, 30 mL substituting the added water used to prepare pasta	Comparable quality and sensory acceptability using up to 10–20% *Opuntia cladode* extract. IVSD decreased with increasing level of *Opuntia cladode* extract and cholesterol bioaccessibility decreased which could reduce blood cholesterol	Blood cholesterol- and glucose-lowering capabilities	[77]

Table 2. *Cont.*

Ingredient Added	Active Ingredient	Substitution Range	Impact on Pasta	Predicted Health Benefits from Data Presented	Reference
Carrot leaf meal and Oregano leaf meal	alpha-linolenic acid, omega-3 fatty acids	0, 5, 10 of each and combinations	Increased AO, and omega-3 fatty acid content from as little as 5% but pasta with higher cooking loss, shorter optimum cooking time, reduced weight increase but all formulations were acceptable by sensory the best being 10% oregano and carrot leaf meal	Unknown	[78]
Soy okra soybean by-product	protein, lipid, dietary fibre isoflavones, phytosterols, coumestans, lignans, phytates, and saponins	0, 10, 20, 30, 40	Increasing soy okra flour reduced pasta optimum cooking time, increased cooking loss and altered taste, texture and colour tolerating only 10%. However, AO and total phenolic contents increased and predicted GI (IVSD) decreased	Lower GI, TPA presumed benefits	[79]
Mushroom powder (white button, shiitake and porcini)	proteins, acidic polysaccharides, dietary fibre and antioxidants	0, 5, 10, 15	mushroom powder increased pasta cooking loss and cooked firmness The addition of shiitake mushroom powder resulted in pasta with the highest firmness and tensile strength	unknown	[80]
Tomato peel	Antioxidants, carotenoids, DF	0, 10, 15	Detrimental effect on pasta such as colour, break resistance, high firmness, reduced cooking loss, inferior sensory taste and overall quality at 10% and higher. However, by adding CMC or gums could negate some of these effects on sensory. Nutritionally tomato peel enhanced b-carotene, lycopene and TDF	to scavenge reactive oxygen species and protect against degenerative diseases like cancer and cardiovascular diseases	[20]
Onion powder	Flavonoids, Quercetin, Proteins, saponins and phenolic components	0, 5, 10, 15	Onion powder up to 10% does not affects sensory characteristics and provides 2.2 mg/100 g of quercetin	Unknown	[81]

Table 3. Database search history showing database, search term and number of hits.

Web of Science (2000–2021)									
Pasta and human health	Glycaemic index and pasta	Cardiovascular disease and durum pasta	Diabetes and durum pasta	Obesity and durum pasta	Insulin and durum pasta	Dietary fibre and durum pasta	Cholesterol and durum pasta		
27	19	8	8	5	8	53	4		
PubMed (2000–2021)									
Pasta and health	CVD and durum wheat pasta	diabetes and durum wheat pasta	obesity and durum wheat pasta	weight gain and durum wheat pasta	cancer and durum wheat pasta	insulin and durum wheat pasta	cholesterol and durum wheat pasta	dietary fibre and durum wheat pasta	dietary fibre and durum wheat pasta and health
707	6	5	5	3	3	15	8	62	20
Cochrane Registry									
durum pasta and obesity	durum pasta and CVD	durum pasta and weight gain	durum pasta and cancer	durum pasta and insulin	durum pasta and dietary fibre and health	durum pasta and cholesterol			
0	1	2	1	6	0	1			

2. Results and Discussion

2.1. Health Based Evidence for Functional Pasta

Different approaches can be used to enhance the nutritional and potentially health promoting properties of pasta. Consuming pasta simply as a wholegrain or wholemeal product which has been available commercially for many years is probably the simplest way. Wholemeal pasta is when bran has been added back to the semolina, while wholegrain pasta refers to "the intact, cracked, ground or flaked caryopsis, whose anatomical parts, endosperm, bran and germ are found in the same quantity as present in the intact original grain" [82]. Since the bran and germ contain many biologically active compounds such as vitamins, minerals, essential fatty acids, amino acids and many phytochemicals, they have been linked to reducing the risk of many lifestyle diseases [83]. However, consumer preference is for pasta made from refined semolina or flour due to better taste, appearance and texture despite fewer health benefits compared to wholemeal/wholegrain counterparts. Other approaches to enhance the nutritional properties of pasta include the addition of specific ingredients or combination of ingredients to provide specific functionalities based on knowledge about their function in isolation or from research studies in other foods. To enhance the consumption of pasta with added health benefits researchers, industry and relevant agencies need to overcome some of the barriers to their uptake such as improving the sensory quality, processing issues (cooking time), availability and media mixed messages.

Studies where health promoting ingredients have been added to pasta and evidence of a health affect using animal, human clinical studies or in vivo measurements are discussed. These are divided into the major disease risk categories.

2.1.1. Hypoglycaemic Effects

Lowering the absorption of carbohydrate into the blood stream from the intestine has been shown to reduce the risk of developing metabolic disease and type II diabetes mellitus (2-DM) while lowering insulin demand caused by eating slowly absorbed carbohydrates less likely to induce insulin resistance in healthy people [84]. It has been demonstrated in vitro and in vivo that durum wheat pasta made from a high amylose durum wheat (with elevated resistant starch), at least above ~50%, reduces the postprandial glycaemic response (PPGR) compared to regular durum wheat pasta with amylose 25–30% [7]. Food structure plays an important role in determining a foods glycaemic response. Both the compact structure of pasta and the presence of the gluten network which surrounds the starch granules together interferes with α-amylase breakdown of the starch is thought to be the mechanism for this effect [85–87]. Similarly in noodles [88] fed 12 healthy subjects noodles with amylose contents ranging from 15–45% obtained by blending high amylose wheat flour and showed a reduction in the PPGR in the 45% amylose noodles compared to the low amylose, 15% noodles but no difference to the 19.6% amylose noodles. This is supported by earlier studies substituting semolina for high amylose (>75%) maize flour in pasta with significantly lower PPGR and postprandial insulin levels [89].

Other ingredients added to pasta have also shown a reduction in the glycaemic response and some studies are discussed. Lupin (*Lupinus albus*) flour contains a protein called Y-conglutin, shown to decrease glycaemia in humans and an extract enriched in this protein was added to pasta (125 mg of pure protein in 100 g of pasta) which was fed to hyperglycaemic rats as uncooked food for three weeks. The protein enriched pasta with Y-conglutin led to a decrease in food intake, and a reduction in glycaemia [90]. Authors noted that their results could have been affected by the lower carbohydrate content in the lupin meal with respect to the control and that the pasta was not cooked before feeding as the stability of the Y-conglutin protein could have been affected. Goñi and Valentín-Gamazo (2003) [91] fed 12 healthy subjects test meals of durum spaghetti and spaghetti containing 25% chickpea flour and the latter had significantly lower GI than the regular spaghetti as well as increasing the mineral, fat and indigestible content of the pasta. Authors suggested this was due to the presence in the chickpea flour of non-starch polysaccharides resistant to enzymic digestion.

Soluble dietary fibres such as β-glucan, guar gum, psyllium and alginate can reduce elevations in postprandial glucose [92,93] because of their viscosity properties which adjusts the rate of gastric emptying. The insulin response of 11 healthy males fed a high fibre pasta made from 40% barley flour high in β-glucan was compared to regular wheat flour pasta. Carbohydrate was more slowly absorbed from the high fibre pasta with a reduced insulin response [94]. Some of these components are already present in certain foods (β-glucan rich sources are oats and barley; Plantago ovata plant for psyllium) and efforts to isolate these fibre components for use as supplemental dietary fibre in functional food design is attractive. To be certain of the effectiveness of the active ingredient, food processing and the form of food consumption (cooked, steamed, etc.) may modify the food structure, ingredient stability and fibre viscosity and potentially impact any proposed health claims. Thus, food manufacturing process may or may not preserve the beneficial properties of the added ingredient and should be considered.

Pasta made from debranned durum wheat flour, enriched in polyphenols and with added barley β-glucan and *Bacillus coagulans* GBI-30, 6086 (probiotic) had good cooking quality with high content of bound ferulic acid compared to control pasta. The probiotic strain remained viable during the pasta-making and cooking processes. However, the PPGR measured in healthy volunteers was no different to control pasta [95]. Frost et al. (2003) [96] included soluble fibre psyllium into pasta to see the impact of a viscous fibre fed to 10 subjects. While there was no effect on gastric emptying or the incremental area under the curve for glucagon-like peptide 1 compared with the control pasta, the added polyunsaturated fat (30 g) and sodium propionate (3 g) in the pasta recipe did alter these parameters which could reduce the risk of diabetes and improve coronary risk factor profiles. Authors suggested the combined high-fat meal with psyllium-enriched pasta may affect the intestinal milieu, affecting carbohydrate digestion and glucose uptake from the small intestine with slower rates of gastric emptying [96]. The addition of fat to a food can reduce glucose response to carbohydrate.

Evidence for efficacy of soluble fibres on PPGR in other foods is extensive [97] but there are issues with their application particularly with regards to sensory acceptance, due to the requirement for relatively large quantities necessary to confer the intended health benefit. To maximize the bioavailability and physiological effects of soluble DF in relation to PPGR, functional food design and assessing processing effects is needed. For example, during extrusion there are forces and heat developed that can reduce the soluble fibre molecular weight, reducing viscosity and effectiveness on PPGR [98]. The most effective soluble fibre from clinical studies in attenuating the PPGR when consumed with a high carbohydrate food like pasta seems to be β-glucan provided it undergoes minimal processing [99]. This efficacy can be diminished with food processing for example Bourdon et al. (1999) [94] found no effect on PPGR when β-glucan was added to pasta because the food structure was not altered by the food processing. More research is needed to develop food manufacturing procedures that minimise disruptions to pasta structure and the resulting viscosity. While a positive effect on PPGR in clinical studies is desirable, longer clinical trials are needed to establish a link between attenuation of blood glycaemia and a reduction in incidence of lifestyle diseases related to PPGR.

Taha and Wasif (1996) [100] fed diabetic rats a diet consisting of semolina only pasta, or wholemeal pasta or wholemeal pasta supplemented with 12% soy flour and 3% methionine for 28 days. They showed that the latter pasta diet lowered total glycerides and cholesterol, and within 10 d, it lowered the PPGR compared to rats fed only semolina or wholemeal pasta, which was maintained at a lower level over the study period. Using wholegrain pasta as a control, pasta containing barley β-glucans and *Bacillus coagulans* BC30, 6086 were fed to healthy overweight or obese volunteers (n = 41) for a 12-week intervention study. The study found that a daily serving of symbiotic whole-grain pasta reduced glycaemia (plasma high-sensitivity C-reactive protein) and plasma LDL/HDL cholesterol ratio [101].

Recently, a review of the GI of 74 pasta products consisting of refined and wholewheat pasta made from durum semolina or white wheat flour, together with pasta made with added egg or legumes or vegetable or algae or other ingredients were described. This database of pasta GI studies (minimum 10 subjects) show a large variability with GI ranging from 18 to 93. Most pasta products had low to medium GI with the median value of 52.5, which is low GI < 55 by definition [102]. The variability within each group reflects the different processing methods for manufacturing, and different subject groups and laboratories conducting the GI test, but, overall, the review concludes that pasta is generally a low GI food. Details on the influence of pasta processing on starch digestion is discussed elsewhere [103].

2.1.2. Hypocholesterolemic Properties and Beneficial Effects on Cardiovascular Disease (CVD)

Attempts to reduce the risk of CVD with diet are varied and aim to prevent the move towards use of drugs which impart their own risks. Recent guidelines recommend consumption of functional foods with evidence from epidemiological studies indicating adequate consumption of whole-wheat or wholegrain foods is associated with reduced CVD risk [104,105]. Favari et al. (2020) [106] fed 41 subjects daily for 12 weeks a whole-wheat pasta (control) and a new innovative whole-wheat pasta enriched in barley β-glucans (2.3 g/100 g) and supplemented with spores of *Bacillus coagulans* GBI-30, 6086 (10^8–10^9 CFU/100 g). They showed improvement in serum cholesterol efflux capacity in overweight/obese participants, indicating the potential of a functional food to improve athero-protective high-density lipoprotein cholesterol function. Patients with hypercholesterolemia fed a soy-germ-enriched pasta containing isoflavone aglycons displayed improved serum lipid markers of cardiovascular risk [107]. A similar study in patients with T2D [108] showed the same soy-germ-enriched pasta significantly reduced blood pressure, and oxidative stress thought to be due to the high antioxidant capacity of the isoflavones in soy protein [109].

Use of non-live bacterial cells (paraprobiotics defined as inactivated microbial cells or cell fractions) as alternative to probiotics decreases risks in certain individuals and avoids need to use dairy foods as a delivery vehicle. Since pasta is processed and consumed after heat treatments, use of paraprobiotics has an advantage over probiotics. Almada et al. (2021) [110] investigated the effects of consumption of wheat-durum pasta with added *Bifidobacterium animalis* inactivated by gamma-irradiation on the health and gut microbiota of rats. Durum wheat pasta with added *B. animalis* was prepared, cooked and dried and the ground material fed to rats for 15 days. This pasta was found to reduce the serum glucose and total cholesterol levels in healthy rats compared to a standard control (non-pasta) and changed the gut microbiota. Pasta can be an effective vehicle to deliver this paraprobiotic.

A common pre-biotic, inulin (a fructan carbohydrate) has been shown to reduce serum triglycerides that might help reduce the development of the metabolic syndrome. Inulin (Raftline HP = Gel) was incorporated into pasta (11%) and together with regular 100% semolina control pasta fed to 22 healthy males in two 5 week feeding periods in a crossover design. Inulin enriched pasta improved lipid (reduced triglycerides and increased HDL-cholesterol) and glucose metabolism (lower fasting glucose and haemoglobin A1c) and delayed gastric emptying. Delayed gastric emptying could be caused by colonic fermentation of the inulin leading to short chain fatty acid production inhibiting gastric emptying [24]. Slowing the gastric emptying can also decrease glucose absorption of foods, reducing PPGR. Indeed, improved metabolic control in the group treated with inulin-enriched pasta was observed. This level of inulin addition (11%) from other studies has been shown to have a minimal impact on traditional pasta quality measures depending on the degree of polymerisation of the inulin used [53]. No side effects on the gastrointestinal tract were found in the study [24].

Opuntia ficus-indica (prickly pear) is an important source of vitamins C, B1, B2, A, and E and minerals such as potassium, calcium, magnesium, and phosphorus. Durum wheat pasta was supplemented with 3% *Opuntia* and fed to 49 people with metabolic syndrome for 4 weeks. Improved atherogenic benefits were obtained such as reduced waist circumference, plasma glucose and triglycerides indicating beneficial effects of this extract [111].

In a randomised controlled trial consisting of meals of regular pasta (control) or pasta with 40% sprouted chickpea flour fed to 22 participants, a higher AO content and brachial artery flow-mediated dilation was achieved eating the functional pasta indicating potential benefits to cardiovascular health [112].

2.1.3. Antihypertensive Effects

In a recent study Valdez-Meza et al. (2019) [113] prepared pasta at different protein contents with amaranth protein concentrate and an amaranth hydrolysate to evaluate antihypertensive properties in rats compared to regular pasta. The antihypertensive amaranth activity of the hydrolysate was maintained after incorporation in the pasta and after pasta ingestion, reducing blood pressure in the rats, confirming bioavailability. These additives reduced the sensory desirability of the pasta as assessed by 30 untrained panellists compared to regular pasta. Hydrolysis of amaranth proteins with microbial alcalase can release ACE-1 inhibitory peptides that can reduce the activity of angiotensin-1-converting enzyme which is involved in the pathogenesis of hypertension. The presence of these proteins in pasta was evaluated as a vehicle for consumption of these proteins in a food matrix. Pasta was supplemented with an alcalase-treated amaranth protein concentrate and compared to regular pasta. This ingredient negatively impacted the overall acceptability but antihypertensive measures in rats indicated reduced blood pressure [113].

2.1.4. Oxidative Stress and Aging Effects

Oxidative stress is a condition where there is an imbalance between the generation of free radicals, such as reactive oxygen/nitrogen species, and the antioxidant defences (endogenous antioxidants glutathione, catalase and superoxide dismutase). Lack of dietary intake of foods rich in antioxidants, such as polyphenols, can play a role in the development and progression of many chronic diseases, such as CVD [114–116] and diabetes. While some information exists on the level of AO in pasta enriched in various ingredients [4,117] impacts on the AO status in vivo is lacking for most functional pasta studies. Epidemiological studies have shown an inverse association between the consumption of polyphenolic-rich foods and the risk of chronic diseases associated with oxidative stress [118]. Khan et al., 2014 [119] fed cooked pasta containing red wholegrain sorghum flour (30% *w*/*w*) to 20 healthy subjects and found elevated levels of polyphenols, antioxidant capacity and superoxide dismutase activity in their blood compared to 100% semolina pasta control thus improving the antioxidant status. This level of incorporation was found to be acceptable to consumers [119].

Laus et al., 2016 [120] fed 7 healthy subjects pasta bran enriched in lipophilic antioxidants or bran enriched in phenolics compared to non-supplemented pasta control. These pastas were similar in sensory score to control pasta. Lipophilic pasta improved the antioxidant status of the serum similar to a wheat AO rich commercial dietary supplement called Lisosan G while the phenolic antioxidants enriched pasta effected serum AO status. There were no differences in the AO status of the pasta extracts by in vitro assay.

Pasta enriched with tartary buckwheat (*Fagopyrum tataricum* Gaertn.) sprouts (30%) was characterised by a high quercetin content and antioxidant activity. When fed to rats for six weeks, the rats exhibited a significant decrease in DNA damage (38%) and more efficient DNA repair (84%) compared to rats fed with commercial pasta [121,122]. Pasta enriched with 6% β-glucan can lower oxidative stress in people based on a longitudinal study that lasted 30 days [123].

Healthy diets have been linked to delaying the onset of aging disabilities and pathologies. Cactus pear extract was added to pasta (3% *w*/*w*) and fed to healthy human subjects for 30 days which led to decreased glycaemic and anti-inflammatory responses with putative effect on the aging process and related metabolic disorders [124].

2.1.5. Other Effects

Weight gain and obesity are critical societal issues facing many communities worldwide and the push for foods that are less energy dense and promote satiety is strong. Pulse flours are higher in protein than cereals and contain slowly digestible and resistant starch. They also provide a better amino acid balance with higher levels of cereal deficient lysine and threonine. Up to 35% faba bean flour has been incorporated into pasta but can reduce pasta quality thought to be related to structural impacts [63]. These and possibly the presence of α-amylase inhibitors may explain the slowing of starch digestion and a lower postprandial glycaemic response. Faba bean flour and protein concentrate were added to pasta (25% dwb) and fed to 15 human subjects and compared to a durum semolina pasta. Pasta with faba bean added had reduced postprandial blood glucose response and improved satiety with acceptable sensory liking for the faba bean flour pasta [125]. Greffeuille et al. (2015) [126] over a two and a half month period fed 15 healthy subjects cooked durum wheat pasta dried at a low temperature (control), and pasta enriched with 35% faba bean dried at either a low or very high temperature and the GI was determined and visual analogue scale (degree of fullness). Inclusion of 35% faba bean flour in pasta increased resistant starch content but had no effect on starch digestion extent in vitro or the in vivo GI, despite disruption to the pasta structure. Using a high-temperature drying cycle during pasta manufacture but with no impact on pasta GI did improve its global digestive comfort and led to a decrease in appetite after eating.

A recent review and meta-analysis of randomised controlled trials of pasta consumption in adults showed a significant reduction in body weight gain and body mass index compared with higher GI dietary patterns, dispelling the myth that a carbohydrate staple food such as pasta is a cause of the obesity epidemic [127].

In a study by Costabile et al. (2018) [128] a randomized, controlled, crossover trial (14 subjects) consumed whole grain (with 13% higher TDF) instead of refined wheat pasta and this improved appetite control but did not influence acute energy balance. After the wholemeal pasta, the desire to eat and the sensation of hunger were lower (−16%, $p = 0.04$ and −23%, $p = 0.004$, respectively) and satiety was higher (+13%; $p = 0.08$) compared with the control pasta. After consumption of wholemeal pasta, the blood glucose and triglyceride levels increased compared to control pasta. Insulin response at 30 min ($p < 0.05$) and ghrelin at 60 min ($p = 0.03$) were lower and PYY (anorexigenic gastrointestinal hormone Peptide YY) levels higher (AUC = +44%, $p = 0.001$) in subjects that ate the wholemeal compared to the refined wheat pasta.

Fibres can be used to reduce digestion and absorption in the human small intestine and thus reduce the daily caloric intake [16]. Typical DF used in pasta are legume fibre, wheat bran insoluble fibre, inulin, psyllium fibre, olive powder, psyllium seed husk, oat β-glucans, *Lentinus edodes* β-glucans, resistant starch, common bean flour, and some non-starch polysaccharides such as locust bean gum, xanthan gum, guar gum, and pectin (Table 2).

2.2. *Pasta with Added Functional Ingredients with No Direct Evidence of Health Benefits*

There is a plethora of reports on adding ingredients into a pasta formula without any evidence of health effects from animal or human studies [2–4,6,13–19]. Rather than repeat the approach taken in these reviews, a summary of studies where durum only semolina has been substituted in part with an ingredient are listed in a table grouped into arbitrary categories: barley components, dietary fibre, fish products, herbs, inulin, legumes, oat, proteins, resistant starch, soy and vegetable (Table 2). For each of the listed studies (n = 60), information on the ingredient added to the pasta, the amounts, the likely active ingredient(s), the reported impact on the pasta quality and the authors predicted health benefits from the data are presented (Table 2).

The majority of the studies reviewed in the previous section focus mostly on evidence for reducing the PPGR in humans, a few on CVD risk reduction and blood pressure brought about by a range of pasta supplemented ingredients (β-glucan, soluble fibres, high amylose flours, inulin) and other flours from other crop species (chickpeas, sorghum, pseudocereals, faba bean, lupin). No other health conditions seemed to have been looked at with functional pasta to date that meet the search criteria and inclusion restrictions. For studies where barley fractions or commercial β-glucan has been included in the pasta recipe, the in vitro studies show a reduction in starch digestion extent [36–38,41] that compares with the in vivo data from clinical studies [43,92–94]. Amounts less than 20% are generally effective in reducing in vitro starch digestion however there are impacts on the pasta quality especially above 10% if using β-glucan extract or commercial sources (e.g., Barley Balance®) but this depends on the β-glucan and if used in combination with other ingredients like vital wheat gluten or gums that can overcome some limitations of β-glucan [44]. Although Peressini et al. (2020) [43] confirmed differences in sensory attributes between Barley Balance® enriched (15%) pasta samples and control pasta, these differences were not judged detrimental for the overall quality.

Inulin (a prebiotic) addition to pasta has been shown in laboratory studies to lower starch digestion up to ~5% inclusion [53] and at higher levels > 7.5% [54] backed up by clinical studies showing 11% inulin pasta slows the gastric emptying causing a decrease in PPGR [24]. Clinical studies adding flours from chickpea [91,129], faba bean [125,126], lupin [92] and red lentil flour [130] to pasta show reductions in GI. Laboratory studies with these flour additions to pasta support reduction in GI [61] while similar in vitro studies in faba bean are lacking and only one study looked at pasta fortified with lupin protein isolate (17%) showing a reduction in in vitro starch digestion [66]. Soy has also been shown in clinical studies when consumed with pasta to show benefits such as reduced GI [100] and reduced blood pressure [107] while only Kamble et al. (2019) [79] study using soy-okra provided evidence of a reduction in in vitro GI. Wholemeal pasta has been shown to have a low GI from a survey by Di Pede et al. (2021) [102] ranging from 35–65, while studies examining the in vitro starch digestion of wholemeal pasta are rare. One study found no effect on the in vitro starch digestion in pasta prepared with fine bran 10–30% [27]. Vegetables have been added to pasta for many years with two studies showing low pasta GI with added vegetable pulps [131] supported by the in vitro studies [77,79]. The other ingredients added to pasta listed in Table 2 seem not to have been evaluated in human clinical trials (gums, debranning fractions, wheat embryo, herbs, protein extracts) so more evaluation is needed. As always, cost of human trials can be prohibitive as well as obtaining ethics approval. Also, there is a need to look at other health indicators besides those discussed in this review such as ingredients added to pasta that can demonstrate benefits to mental health, slowing aging, improving the microbiome health.

Overall, it seems many of the in vitro claims are met by the in vivo results although the level of affect in vivo could be higher than the in vitro studies suggest. For example, raising the amylose content of pasta to mid-40s% while significantly increasing the in vitro starch digestion extent had no significant impact on the GI in a 10 subject glucose tolerance test [7]. The in vitro studies provide a guide to the likely impact in the human but claims for health benefits require proof from the human feeding trials and longer-term data to provide good evidence for using a functional pasta for health benefits. Much more research is needed in this area. Interactions between active compounds and protein matrix while understood for some ingredients like bran, inulin, soluble fibre and resistant starch [6] are not understood for many novel approaches proposed in Table 1. Only a few studies have considered synergism or interactions between individual compounds affecting pasta product quality. While a health benefit is sought after in the many studies discussed, a very important consideration is consumer acceptance of the functional pasta. Many of the studies listed in Table 2 evaluate the technological quality of the resultant pasta with a range of instrumental, cooking procedures and colour as well as the important sensory analysis, most often using a trained panel mostly limited to 10 people. More expensive and time-consuming consumer panels involving many people are needed to give an indication of the market acceptance of the product since taste, appearance, smell and texture are important to consumers. However, gender, race, country of origin can affect peoples perceived acceptance of functional foods [132]. Generally, these studies are not done for specific functional foods but more generally for food categories like wholegrain foods [132]. Another important consideration in the manufacture and consumption of functional pasta is the storage stability. Dried pasta made from 100% semolina typically has a water activity in the range 0.3–0.5 [133] and if stored in sealed containers typically lasts at least 2 or more years, often well passed the shelf-life given by the manufacturer, which is often conservative. However, the composition of the pasta with added ingredients needs to be considered. For example, any egg products used in the manufacture of the dried pasta may not be as stable for such a length because of their high content of lipids, the pasta can turn rancid. Discolouration or off-odours are good indicators of spoilage. Fresh pasta has a much shorter shelf life of 2–3 days with refrigeration because of the high water activity (0.92–0.99) [133] and will deteriorate rapidly if not stored properly. Various chemicals and natural antimicrobials can be used to extend shelf-life [6]. There are limited studies on the storage stability of pasta and impacts on pasta nutritional value. One example is the use of modified packaging using high-density polyethylene and biaxially oriented polypropylene films were compared with the former providing a longer shelf life for multigrain pasta [134]. Another investigated lipid oxidation in spaghetti enriched in long chain *n*-3 polyunsaturated fatty acids with functional spaghetti having a shelf life comparable to control pasta [135]. Thus, it is important to include storage stability studies in functional food design to ensure no deterioration in the functional value occurs with storage.

3. Conclusions

Pasta is a popular food and has already been shown to be a good method to incorporate increased nutritional or functional compounds. Care is needed to ensure good technological quality in pasta with a substituted ingredient and consumer acceptability at an affordable price. Therefore, the manufacturer of such products must be profitable, and a ready supply of the desired ingredient be assured before a manufacturer prepares such products as well as a likely market. Interactions between active compounds and protein matrix while understood for some ingredients like bran, inulin, soluble fibre and resistant starch, are not understood for many novel approaches proposed in Table 1. Only a few studies have considered synergism or interactions between individual compounds affecting pasta product quality. However, legislation in many countries require proof before a health claim can be made on a food, such as low GI, cholesterol lowering, heart safe etc. For this reason, more research is needed to evaluate the most promising functional pasta with human clinical trials to validate the actual health benefit. Health claims together with good taste, texture and appearance at an acceptable price will help drive consumer demand for such foods.

Funding: This research received no external funding.

Institutional Review Board Statement: Not applicable.

Informed Consent Statement: Not applicable.

Data Availability Statement: The data presented in this study are available on request from the corresponding author. The data are not publicly available due to institutional privacy requirements.

Acknowledgments: NSW Department of Primary Industries, Australia.

Conflicts of Interest: The author declares no conflict of interest.

Abbreviations

GI	glycaemic index
IVSD	in vitro starch digestion
TPA	total phenolic acids
AO	antioxidant
TDF	total dietary fibre
DF	dietary fibre
CMC	carboxymethylcellulose
XG	xanthan gum
EAA	essential amino acids
ACE	angiotensin 1-converting enzyme
RS	resistant starch
CMC	carboxymethylcellulose
GG	guar gum

References

1. Sissons, M. Pasta. In *Encyclopedia of Food Grains*, 2nd ed.; Wrigley, C., Corke, H., Seetharaman, K., Faubion, J., Eds.; Academic Press: Oxford, UK, 2016; pp. 79–89.
2. Bustos, M.C.; Perez, G.T.; Leon, A.E. Structure and quality of pasta enriched with functional ingredients. *RSC Adv.* **2015**, *5*, 30780–30792. [CrossRef]
3. Mercier, S.; Moresoli, C.; Mondor, M.; Villeneuve, S.; Marcos, B. A Meta-analysis of enriched pasta: What are the effects of enrichment and process specifications on the quality attributes of pasta? *Compr. Rev. Food Sci. Food Saf.* **2016**, *15*, 685–704. [CrossRef]
4. Wahanik, A.L.; Chang, Y.K.; Clerici, P.S.; Teresa, M. How to make pastas healthier? *Food Rev. Int.* **2018**, *34*, 23–30. [CrossRef]
5. Food and Nutrition Board of the National Academy of Sciences. Available online: https://www.nationalacademies.org/fnb/food-and-nutrition-board (accessed on 2 December 2021).
6. Li, M.; Zhu, K.-X.; Guo, X.-N.; Brijs, K.; Zhou, H.-M. Natural additives in wheat-based pasta and noodle products: Opportunities for enhanced nutritional and functional properties. *Compr. Rev. Food Sci. Food Saf.* **2014**, *13*, 347–357. [CrossRef]
7. Sissons, M.; Sestili, F.; Botticella, E.; Masci, S.; Lafiandra, D. Can manipulation of durum wheat amylose content reduce the glycaemic index of spaghetti? *Foods* **2020**, *9*, 693. [CrossRef]
8. Yu, S.; Tian, L. Breeding major cereal grains through the lens of nutrition sensitivity. *Mol. Plant* **2018**, *11*, 23–30. [CrossRef]
9. Gebruers, K.; Dornez, E.; Boros, D.; Fraś, A.; Dynkowska, W.; Bedő, Z.; Rakszegi, M.; Delcour, J.A.; Courtin, C.M. Variation in the content of dietary fiber and components thereof in wheats in the HEALTHGRAIN diversity screen. *J. Agric. Food Chem.* **2008**, *56*, 9740–9749. [CrossRef]
10. Pirzadah, T.B.; Malik, B. Pseudocereals as super foods of 21st century: Recent technological interventions. *J. Agric. Food Res.* **2020**, *2*, 100052. [CrossRef]
11. Mefleh, M.; Conte, P.; Fadda, C.; Giunta, F.; Piga, A.; Hassounc, G.; Motzoa, R. From ancient to old and modern durum wheat varieties: Interaction among cultivar traits, management, and technological quality. *J. Sci. Food Agric.* **2018**, *99*, 2059–2067. [CrossRef]
12. Onwulata, C.I. Encapsulation of new active ingredients. *Annu. Rev. Food Sci. Technol.* **2012**, *3*, 183–202. [CrossRef]
13. Bianchi, F.; Tolve, R.; Rainero, G.; Bordiga, M.; Brennan, C.S.; Simonato, B. Technological, nutritional and sensory properties of pasta fortified with agro-industrial by-products: A review. *Int. J. Food Sci. Technol.* **2021**, *56*, 4356–4366. [CrossRef]
14. Brennan, C.S.; Tudorica, C.M. Evaluation of potential mechanisms by which dietary fibre additions reduce the predicted glycaemic index of fresh pastas. *Int. J. Food Sci. Technol.* **2008**, *43*, 2151–2162. [CrossRef]
15. Fuad, T.; Prabhasankar, P. Role of ingredients in pasta product quality: A review on recent developments. *Crit. Rev. Food Sci. Nutr.* **2010**, *50*, 787–798. [CrossRef]
16. Krishnan, M.; Prabhasankar, P. Health based pasta: Redefining the concept of the next generation convenience food. *Crit. Rev. Food Sci. Nutr.* **2012**, *52*, 9–20. [CrossRef]
17. Marconi, E.; Messia, M.C. Pasta made from non-traditional raw materials: Technological and nutritional aspects. In *Durum Wheat Chemistry and Technology 2012*, 2nd ed.; Sissons, M., Abecassis, J., Marchylo, B., Carcea, M., Eds.; AACC International Press: St Paul, MN, USA, 2012; pp. 85–124.
18. Nilusha, R.A.T.; Jayasinghe, J.M.J.K.; Perera, O.D.A.N.; Perera, P.I.P. Development of Pasta Products with Nonconventional Ingredients and Their Effect on Selected Quality Characteristics: A Brief Overview. *Int. J. Food Sci.* **2019**, *2019*, 6750726. [CrossRef]
19. Sissons, M.; Fellows, C.M. Sensory, technological and health aspects of adding fibre to wheat-based pasta. In *Wheat and Rice in Disease Prevention and Health*; Watson, R.R., Preedy, V.R., Zibaldi, S., Eds.; Elsevier: Amsterdam, The Netherlands, 2014; pp. 211–226.
20. Padalino, L.; Conte, A.; Lecce, L.; Likyova, D.; Sicari, V.; Pellicano, T.M.; Poiana, M.; Del Nobile, M.A. Functional pasta with tomato by-product as a source of antioxidant compounds and dietary fibre. *Czech J. Food Sci.* **2017**, *35*, 48–56. [CrossRef]

21. Marti, A.; Pagani, M.A. What can play the role of gluten in gluten free pasta? *Trends Food Sci. Technol.* **2013**, *31*, 63–71. [CrossRef]
22. Torres, A.; Frias, J.; Granito, M.; Guerra, M.; Vidal-Valverde, C. Chemical, biological and sensory evaluation of pasta products supplemented with alpha-galactoside-free lupin flours. *J. Sci. Food Agric.* **2007**, *87*, 74–81. [CrossRef]
23. Gallegos-Infante, J.A.; Rocha-Guzman, N.E.; Gonzalez-Laredo, R.F.; Ochoa-Martínez, L.A.; Corzo, N.; Bello-Perez, L.A.; Medina-Torres, L.; Peralta-Alvarez, L.E. Quality of spaghetti pasta containing Mexican common bean flour (*Phaseolus vulgaris* L.). *Food Chem.* **2010**, *119*, 1544–1549. [CrossRef]
24. Russo, F.; Riezzo, G.; Chiloiro, M.; De Michele, G.; Chimienti, G.; Marconi, E.; D'Attoma, B.; Linsalata, M.; Clemente, C. Metabolic effects of a diet with inulin-enriched pasta in healthy young volunteers. *Curr. Pharm. Des.* **2010**, *16*, 825–831. [CrossRef]
25. Delvecchio, L.N.; Pasqualone, A. Production trials of fresh pasta enriched with phenolic compounds extracted from wheat by KOH-induced hydrolysis. *Prog. Nutr.* **2012**, *14*, 247–251.
26. Martin-Esparza, M.E.; Raga, A.; Gonzalez-Martinez, C.; Albors, A. Micronised bran-enriched fresh egg tagliatelle: Significance of gums addition on pasta technological features. *Food Sci. Technol. Int.* **2018**, *24*, 309–320. [CrossRef]
27. Aravind, N.; Sissons, M.J.; Egan, N.E.; Fellows, C.M. Effect of insoluble dietary fibre addition on technological, sensory, and structural properties of durum wheat spaghetti. *Food Chem.* **2012**, *130*, 299–309. [CrossRef]
28. Kaur, G.; Sharma, S.; Nagi, H.P.S.; Dar, B.N. Functional properties of pasta enriched with variable cereal brans. *J. Food Sci. Technol.-Mysore* **2012**, *49*, 467–474. [CrossRef]
29. Bagdi, A.; Szabo, F.; Gere, A.; Kókai, Z.; Sipos, L.; Tömösközi, S. Effect of aleurone-rich flour on composition, cooking, textural and sensory properties of pasta. *LWT—Food Sci. Technol.* **2014**, *59*, 996–1002. [CrossRef]
30. Marinelli, V.; Padalino, L.; Conte, A.; Del Nobile, M.A.; Briviba, K. Red grape marc flour as food ingredient in durum wheat spaghetti: Nutritional evaluation and bioaccessibility of bioactive compounds. *Food Sci. Technol. Res.* **2018**, *24*, 1093–1100. [CrossRef]
31. Desai, A.S.; Brennan, M.A.; Guo, X.; Zeng, X.-A.; Brennan, C.S. Fish protein and lipid interactions on the digestibility and bioavailability of starch and protein from durum wheat pasta. *Molecules* **2019**, *24*, 839. [CrossRef]
32. Kadam, S.U.; Prabhasankar, P. Evaluation of cooking, microstructure, texture and sensory quality characteristics of shrimp meat-based pasta. *J. Texture Stud.* **2012**, *43*, 268–274. [CrossRef]
33. Boudalia, S.; Gueroui, Y.; Boumaza, B.; Bousbia, A.; Benada, M.; Leksir, C.; Mezroua, E.Y.; Zemmouchi, K.R.; Saoud, A.; Chemmam, M. Evaluation of Physicochemical Properties and sensory qualities of pasta enriched with freeze-dried sweet whey. *Sci. Agric. Bohem.* **2020**, *51*, 75–85. [CrossRef]
34. Zen, C.K.; Tiepo, C.B.V.; da Silva, R.V.; Reinehr, C.O.; Gutkoski, L.C.; Oro, T.; Colla, L.M. Development of functional pasta with microencapsulated *Spirulina*: Technological and sensorial effects. *J. Sci. Food Agric.* **2018**, *100*, 2018–2026. [CrossRef]
35. Cárdenas-Hernández, A.; Beta, T.; Loarca-Piña, G.; CastanoTostado, E.; Nieto-Barrera, J.O.; Mendoza, S. Improved functional properties of pasta: Enrichment with amaranth seed flour and dried amaranth leaves. *J. Cereal. Sci.* **2016**, *72*, 84–90. [CrossRef]
36. Aravind, N.; Sissons, M.J.; Egan, N.; Fellows, C.M.; Blazek, J.; Gilbert, E.P. Effect of β-glucan on technological, sensory, and structural properties of durum wheat pasta. *Cereal. Chem.* **2012**, *89*, 84–93. [CrossRef]
37. Chillo, S.; Ranawana, D.V.; Henry, C.J.K. Effect of two barley beta-glucan concentrates on in vitro glycaemic impact and cooking quality of spaghetti. *LWT—Food Sci. Technol.* **2011**, *44*, 940–948. [CrossRef]
38. Cleary, L.; Brennan, C. The influence of a (1 -> 3)(1 -> 4)-beta-D-glucan rich fraction from barley on the physico-chemical properties and in vitro reducing sugars release of durum wheat pasta. *Int. J. Food Sci. Technol.* **2006**, *41*, 910–918. [CrossRef]
39. Knuckles, B.E.; Hudson, C.A.; Chiu, M.M.; Sayre, R.N. Effect of beta-glucan barley fractions in high-fibre bread and pasta. *Cereal. Foods World* **1997**, *42*, 94–97.
40. Marconi, E.; Graziano, M.; Cubadda, R. Composition and utilization of barley pearling by-products for making functional pastas rich in dietary fiber and beta-glucans. *Cereal. Chem.* **2000**, *77*, 133–139. [CrossRef]
41. Montalbano, A.; Tesoriere, L.; Diana, P.; Barraja, P.; Carbone, A.; Spano, V.; Parrino, B.; Attanzio, A.; Livrea, M.A.; Cascioferro, S.; et al. Quality characteristics and in vitro digestibility study of barley flour enriched ditalini pasta. *LWT—Food Sci. Technol.* **2016**, *72*, 223–228. [CrossRef]
42. Verardo, V.; Gomez-Caravaca, A.M.; Messia, M.C.; Marconi, E.; Caboni, M.F. Development of functional spaghetti enriched in bioactive compounds using barley coarse fraction obtained by air classification. *J. Agric. Food Chem.* **2011**, *59*, 9127–9134. [CrossRef]
43. Peressini, D.; Cavarape, A.; Brennan, M.A.; Gao, J.; Brennan, C.S. Viscoelastic properties of durum wheat doughs enriched with soluble dietary fibres in relation to pasta-making performance and glycaemic response of spaghetti. *Food Hydrocoll.* **2020**, *102*, 105613. [CrossRef]
44. Krawęcka, A.; Sobota, A.; Sykut-Domańska, E. Physicochemical, sensory, and cooking qualities of pasta enriched with oat beta-glucans, xanthan gum, and vital gluten. *Foods* **2020**, *9*, 1412. [CrossRef]
45. Aravind, N.; Sissons, M.J.; Fellows, C.M. Effect of soluble fibre (guar gum and carboxymethylcellulose) addition on technological, sensory and structural properties of durum wheat spaghetti. *Food Chem.* **2012**, *131*, 893–900. [CrossRef]
46. Ciccoritti, R.; Taddei, F.; Nicoletti, I.; Gazza, L.; Corradini, D.; D'Egidio, M.G.; Martini, D. Use of bran fractions and debranned kernels for the development of pasta with high nutritional and healthy potential. *Food Chem.* **2017**, *225*, 77–86. [CrossRef]
47. Foschia, M.; Peressini, D.; Sensidoni, A.; Brennan, M.A.; Brennan, C.S. Synergistic effect of different dietary fibres in pasta on in vitro starch digestion? *Food Chem.* **2015**, *172*, 245–250. [CrossRef]

48. Levent, H.; Koyuncu, M.; Bilgicli, N.; Adiguzel, E.; Dedeoglu, M. Improvement of chemical properties of noodle and pasta using dephytinized cereal brans. *LWT—Food Sci. Technol.* **2020**, *128*, 109470. [CrossRef]
49. Manthey, F.A.; Schorno, A.L. Physical and cooking quality of spaghetti made from whole wheat durum. *Cereal. Chem.* **2002**, *79*, 504–510. [CrossRef]
50. Piwińska, M.; Wyrwisz, J.; Wierzbicka, A. Effect of micronization of high-fiber oat powder and vacuum-drying on pasta quality. *J. Food* **2016**, *14*, 433–439. [CrossRef]
51. Calanche, J.; Beltran, H.; Marquina, P.; Roncales, P.; Beltran, J.A. Eating fish in another way: Development of functional pasta with added concentrates of farmed sea bass (*Dicentrarchus labrax*). *Cereal. Chem.* **2019**, *96*, 856–865. [CrossRef]
52. Sato, R.; de Lima Cilli, L.P.; de Oliveira, B.E.; Vieira Maciel, V.B.; Venturini, A.C.; Pedroso Yoshida, C.M. Nutritional improvement of pasta with *Pereskia aculeata* Miller: A non-conventional edible vegetable. *Food Sci. Technol.* **2019**, *39*, 28–34. [CrossRef]
53. Aravind, N.; Sissons, M.J.; Fellows, C.M.; Blazek, J.; Gilbert, E.P. Effect of inulin soluble dietary fibre addition on technological, sensory, and structural properties of durum wheat spaghetti. *Food Chem.* **2012**, *132*, 993–1002. [CrossRef]
54. Brennan, C.S.; Kuri, V.; Tudorica, C.M. Inulin-enriched pasta: Effects on textural properties and starch degradation. *Food Chem.* **2004**, *86*, 189–193. [CrossRef]
55. Sabanis, D.; Makri, E.; Doxastakis, G. Effect of durum flour enrichment with chickpea flour on the characteristics of dough and lasagne. *J. Sci. Food Agric.* **2006**, *86*, 1939–1944. [CrossRef]
56. Wood, J.A. Texture, processing and organoleptic properties of chickpea-fortified spaghetti with insights to the underlying mechanisms of traditional durum pasta quality. *J. Cereal. Sci.* **2009**, *49*, 128–133. [CrossRef]
57. Chillo, S.; Monro, J.A.; Mishra, S.; Henry, C.J. Effect of incorporating legume flour into semolina spaghetti on its cooking quality and glycaemic impact measured in vitro. *Int. J. Food Sci. Nutr.* **2010**, *61*, 149–160. [CrossRef]
58. De Pasquale, I.; Verni, M.; Verardo, V.; Gomez-Caravaca, A.M.; Rizzello, C.G. Nutritional and functional advantages of the use of fermented black chickpea flour for semolina-pasta fortification. *Foods* **2021**, *10*, 182. [CrossRef]
59. Di Stefano, V.; Pagliaro, A.; Del Nobile, M.A.; Conte, A.; Melilli, M.G. Lentil fortified spaghetti: Technological properties and nutritional characterization. *Foods* **2021**, *10*, 4. [CrossRef]
60. Laleg, K.; Barron, C.; Cordelle, S.; Schlich, P.; Walrand, S.; Micard, V. How the structure, nutritional and sensory attributes of pasta made from legume flour is affected by the proportion of legume protein. *LWT—Food Sci. Technol.* **2017**, *79*, 471–478. [CrossRef]
61. Osorio-Diaz, P.; Agama-Acevedo, E.; Mendoza-Vinalay, M.; Tovar, J.; Bello-Perez, L.A. Pasta added with chickpea flour: Chemical composition, in vitro starch digestibility and predicted glycemic index. *Cienc. Y Tecnol. Aliment.* **2008**, *6*, 6–12. [CrossRef]
62. Shreenithee, C.R.; Prabhasankar, P. Effect of different shapes on the quality, microstructure, sensory and nutritional characteristics of yellow pea flour incorporated pasta. *Food Meas.* **2013**, *7*, 166–176. [CrossRef]
63. Petitot, M.; Boyer, L.; Minier, C.; Micard, V. Fortification of pasta with split pea and faba bean flours: Pasta processing and quality evaluation. *Food Res. Int.* **2010**, *43*, 634–641. [CrossRef]
64. Wu, Y.V.; Hareland, G.A.; Warner, K. Protein enriched spaghetti fortified with corn gluten meal. *J. Agric. Food Chem.* **2001**, *49*, 3906–3910. [CrossRef]
65. Jayasena, V.; Nasar-Abbas, S.M. Development and quality evaluation of high-protein and high-dietary-fiber pasta using lupin flour. *J. Texture Stud.* **2012**, *43*, 153–163. [CrossRef]
66. Cutillo, S.; Farahnaky, A.; Marcotuli, I.; Gadaleta, A.; Sissons, M. In vitro starch digestion and technological properties of spaghetti fortified with lupin protein isolate. *Int. J. Food Sci. Technol.* **2021**, *56*, 3567–3577. [CrossRef]
67. Alzuwaid, N.T.; Sissons, M.; Laddomada, B.; Fellows, C.M. Nutritional and functional properties of durum wheat bran protein concentrate. *Cereal. Chem.* **2020**, *97*, 304–315. [CrossRef]
68. Alzuwaid, N.T.; Pleming, D.; Fellows, C.M.; Sissons, M. Fortification of durum wheat spaghetti and common wheat bread with wheat bran protein concentrate-impacts on nutrition and technological properties. *Food Chem.* **2021**, *334*, 1–9. [CrossRef]
69. Alzuwaid, N.T.; Laddomada, B.; Fellows, C.M.; Egan, N.; Sissons, M. Supplementation of durum wheat spaghetti with wheat bran protein concentrate: Impacts on phytochemical profile and starch digestion. *Cereal. Chem.* **2021**, *98*, 642–650. [CrossRef]
70. Jayawardena, S.R.; Morton, J.D.; Brennan, C.S.; Bekhit, A.E.-D.A. Utilisation of beef lung protein powder as a functional ingredient to enhance protein and iron content of fresh pasta. *Int. J. Food Sci. Technol.* **2019**, *54*, 610–618. [CrossRef]
71. Sadeghi, M.; Bhagya, S. Quality characterization of pasta enriched with mustard protein isolate. *J. Food Sci.* **2008**, *73*, S229–S237. [CrossRef]
72. Segura-Campos, M.R.; Garcia-Rodriguez, K.; Carlos Ruiz-Ruiz, J.; Chel-Guerrero, L.; Betancur-Ancona, D. Effect of incorporation of hard-to-cook bean (*Phaseolus vulgaris* L.) protein hydrolysate on physical properties and starch and dietary fiber components of semolina pasta. *J. Food Process. Preserv.* **2015**, *39*, 1159–1165. [CrossRef]
73. Gelencser, T.; Gal, V.; Salgo, A. Effects of applied process on the in vitro digestibility and resistant starch content of pasta products. *Food Bioprocess Technol.* **2010**, *3*, 491–497. [CrossRef]
74. Aravind, N.; Sissons, M.J.; Fellows, C.M.; Blazek, J.; Gilbert, E.P. Optimisation of resistant starch II and III levels in durum wheat pasta to reduce in vitro digestibility while maintaining processing and sensory characteristics. *Food Chem.* **2013**, *136*, 1100–1109. [CrossRef]
75. Hernandez-Nava, R.G.; Berrios, J.D.J.; Pan, J.; Osorio-Diaz, P.; Bello-Perez, L.A. Development and characterization of spaghetti with high resistant starch content supplemented with banana starch. *Food Sci. Technol. Int.* **2009**, *15*, 73–78. [CrossRef]

76. Sun-Waterhouse, D.; Jin, D.; Waterhouse, G.I.N. Effect of adding elderberry juice concentrate on the quality attributes, polyphenol contents and antioxidant activity of three fibre-enriched pastas. *Food Res. Int.* **2013**, *54*, 781–789. [CrossRef]
77. Attanzio, A.; Diana, P.; Barraja, P.; Carbone, A.; Spano, V.; Parrino, B.; Cascioferro, S.M.; Allegra, M.; Cirrincione, G.; Tesoriere, L.; et al. Quality, functional and sensory evaluation of pasta fortified with extracts from Opuntia ficus-indica cladodes. *J. Sci. Food Agric.* **2019**, *99*, 4242–4247. [CrossRef]
78. Boroski, M.; de Aguiar, A.C.; Boeing, J.S.; Rotta, E.M.; Wibby, C.L.; Bonafé, E.G.; de Souza, N.E.; Visentainer, J.V. En-hancement of pasta antioxidant activity with oregano and carrot leaf. *Food Chem.* **2011**, *125*, 696–700. [CrossRef]
79. Kamble, D.B.; Singh, R.; Rani, S.; Pratap, D. Physicochemical properties, in vitro digestibility and structural attributes of okara-enriched functional pasta. *J. Food Process Preserv.* **2019**, *43*, e14232. [CrossRef]
80. Lu, X.K.; Brennan, M.A.; Serventi, L.; Mason, S.; Brennan, C.S. How the inclusion of mushroom powder can affect the physicochemical characteristics of pasta. *Int. J. Food Sci. Technol.* **2016**, *51*, 2433–2439. [CrossRef]
81. Rajeswari, G.; Susanna, S.; Prabhasankarn, P.; Venkateswara Rao, G. Influence of onion powder and its hydrocolloid blends on pasta dough, pasting, microstructure, cooking and sensory characteristics. *Food Biosci.* **2013**, *4*, 13–20. [CrossRef]
82. AACC International. *Approved Methods of Analysis*, 11th ed.; AACC International: St Paul, MN, USA, 2010.
83. Allai, F.M.; Azad, Z.R.A.A.; Gul, K.; Dar, B.N. Wholegrains: A review on the amino acid profile, mineral content, physicochemical, bioactive composition and health benefits. *Int. J. Food Sci. Technol.* **2021**. [CrossRef]
84. Krawęcka, A.; Sobota, A.; Sykut-Domańska, E. Functional cereal products in the diet for type 2 diabetes patients. *Int. J. Food Sci.* **2019**, *2019*, 4012450. [CrossRef]
85. Jenkins, D.J.A.; Wolever, T.M.S.; Jenkins, A.L.; Lee, R.; Wong, G.S.; Josse, R. Glycemic response to wheat products: Reduced response to pasta but no effect of fiber. *Diabetes Care* **1983**, *6*, 155–159. [CrossRef]
86. Colonna, P.; Barry, J.-L.; Cloarec, D.; Bornet, F.; Gouilloud, S.; Galmiche, J.-P. Enzymic susceptibility of starch from pasta. *J. Cereal Sci.* **1990**, *11*, 59–70. [CrossRef]
87. Zou, W.; Sissons, M.; Gidley, M.J.; Gilbert, R.G.; Warren, F.J. Combined techniques for characterising pasta structure reveals how the gluten network slows enzymic digestion rate. *Food Chem.* **2015**, *188*, 559–568. [CrossRef]
88. Ang, K.; Bourgy, C.; Fenton, H.; Regina, A.; Newberry, M.; Diepeveen, D.; Lafiandra, D.; Grafenauer, S.; Hunt, W.; Solah, V. Noodles made from high amylose wheat flour attenuate postprandial glycaemia in healthy adults. *Nutrients* **2020**, *12*, 2171. [CrossRef]
89. Hospers, J.J.; van Amelsvoort, J.M.M.; Westrate, J.A. Amylose-amylopectin ratio in pasta affects PPBG and insulin responses and satiety in males. *J. Food Sci.* **1994**, *59*, 1144–1149. [CrossRef]
90. Capraro, J.; Magni, C.; Scarafoni, A.; Caramanico, R.; Rossi, F.; Morlacchini, M.; Duranti, M. Pasta supplemented with isolated lupin protein fractions reduces body weight gain and food intake of rats and decreases plasma glucose concentration upon glucose overload trial. *Food Funct.* **2014**, *5*, 375–380. [CrossRef]
91. Goñi, L.; Valentín-Gamazo, C. Chickpea flour ingredient slows glycemic response to pasta in healthy volunteers. *Food Chem.* **2003**, *81*, 511–515. [CrossRef]
92. Brownlee, I.A. The physiological roles of dietary fibre. *Food Hydrocoll.* **2011**, *25*, 238–250. [CrossRef]
93. Yokoyama, W.M.; Hudson, C.A.; Knuckles, B.E. Effect of barley ß-glucan in durum wheat pasta on human glycemic response. *Cereal. Chem.* **1997**, *74*, 293–296. [CrossRef]
94. Bourdon, I.; Yokoyama, W.; Davis, P.; Hudson, C.; Backus, R.; Richter, D.; Schneeman, B.O. Postprandial lipid, glucose, insulin, and cholecystokinin responses in men fed barley pasta enriched with beta-glucan. *Am. J. Clin. Nutr.* **1999**, *69*, 55–63. [CrossRef]
95. Fares, C.; Menga, V.; Martina, A.; Pellegrini, N.; Scazzina, F.; Torriani, S. Nutritional profile and cooking quality of a new functional pasta naturally enriched in phenolic acids, added with b-glucan and *Bacillus coagulans* GBI-30,6086. *J. Cereal. Sci.* **2015**, *65*, 260–266. [CrossRef]
96. Frost, G.; Brynes, A.; Dhillo, W.; Bloom, S.; McBurney, M. The effects of fiber enrichment of pasta and fat content on gastric emptying, GLP-1, glucose, and insulin responses to a meal. *Eur. J. Clin. Nutr.* **2003**, *57*, 293. [CrossRef]
97. Cassidy, Y.M.; McSorley, E.M.; Allsopp, P.J. Effect of soluble dietary fibre on postprandial blood glucose response and its potential as a functional food ingredient. *J. Funct. Foods* **2018**, *46*, 423–439. [CrossRef]
98. Regand, A.; Tosh, S.M.; Wolever, T.M.; Wood, P.J. Physicochemical properties of β-glucan in differently processed oat foods influence glycemic response. *J. Agric. Food Chem.* **2009**, *57*, 8831–8838. [CrossRef]
99. Wood, P.J.; Braaten, J.T.; Fraser, W.S.; Riedel, D.; Postel, M. Comparisons of viscous properties of oat and guar gum and effects of these and oat bran on glycemic index. *J. Agric. Food Chem.* **1990**, *38*, 753–757. [CrossRef]
100. Taha, S.; Wasif, M. Hypoglycemic effect and protein nutritive quality of soy and methionine-supplemented whole durum pasta products. *Food/Nahrung* **1996**, *40*, 281–287. [CrossRef]
101. Angelino, D.; Martina, A.; Rosi, A.; Veronesi, L.; Antonini, M.; Mennella, I.; Vitaglione, P.; Grioni, S.; Brighenti, F.; Zavaroni, I.; et al. Glucose- and lipid-related biomarkers are affected in healthy obese or hyperglycemic adults consuming a whole-grain pasta enriched in prebiotics and probiotics: A 12-week randomized controlled trial. *J. Nutr.* **2019**, *149*, 1714–1723. [CrossRef]
102. Di Pede, G.; Dodi, R.; Scarpa, C.; Brighenti, F.; Dall'Asta, M.; Scazzina, F. Glycemic index values of pasta products: An Overview. *Foods* **2021**, *10*, 2541. [CrossRef]
103. Petitot, M.; Abecassis, J.; Micard, V. Structuring of pasta components during processing: Impact on starch and protein digestibility and allergenicity. *Trends. Food Sci. Technol.* **2009**, *20*, 521–532. [CrossRef]

104. Catapano, A.L.; Graham, I.; De Backer, G.; Wiklund, O.; Chapman, M.J.; Drexel, H.; Hoes, A.W.; Jennings, C.S.; Landmesser, U.; Pedersen, T.R.; et al. ESC/EAS guidelines for the management of dyslipidaemias. *Eur. Heart J.* **2016**, *37*, 2999–3058. [CrossRef]
105. Brunner, E.J.; Thorogood, M.; Rees, K.; Hewitt, G. Dietary advice for reducing cardiovascular risk. *Cochrane Database Syst. Rev.* **2005**, *19*, CD002128.
106. Favari, E.; Angelino, D.; Cipollari, E.; Adorni, M.P.; Zimetti, F.; Bernini, F.; Ronda, N.; Pellegrini, N. Functional pasta consumption in healthy volunteers modulates ABCG1-mediated cholesterol efflux capacity of HDL. *Nutr. Metab. Cardiovasc. Dis.* **2020**, *30*, 1768–1776. [CrossRef]
107. Clerici, C.; Setchell, K.D.R.; Battezzati, P.M.; Pirro, M.; Giuliano, V.; Asciutti, S.; Castellani, D.; Nardi, E.; Sabatino, G.; Orlandi, S.; et al. Pasta naturally enriched with isoflavone aglycons from soy germ reduces serum lipids and improves markers of cardiovascular risk. *J. Nutr.* **2007**, *137*, 2270–2278. [CrossRef]
108. Clerici, C.; Nardi, E.; Battezzati, P.M.; Asciutti, S.; Castellani, D.; Corazzi, N.; Giuliano, V.; Gizzi, S.; Perriello, G.; Di Matteo, G.; et al. Novel soy germ pasta improves endothelial function, blood pressure, and oxidative stress in patients with type 2 diabetes. *Diabetes Care* **2011**, *34*, 1946–1948. [CrossRef]
109. Ruiz-Larrea, M.B.; Mohan, A.R.; Paganga, G.; Miller, N.J.; Bolwell, G.P.; Rice-Evans, C.A. Antioxidant activity of phytoestrogenic isoflavones. *Free Radic. Res.* **1997**, *26*, 63–70. [CrossRef]
110. Almada, C.N.; Almada-Erix, C.N.; Costa, W.K.A.; Silva, J.; Cabral, L.; Noronha, M.F.; Goncalves, A.E.S.S.; dos Santos, A.; Lollo, P.C.; Magnani, M.; et al. Wheat-durum pasta added of inactivated *Bifidobacterium animalis* decreases glucose and total cholesterol levels and modulates gut microbiota in healthy rats. *Int. J. Food Sci. Nutr.* **2021**, *72*, 781–793. [CrossRef]
111. Giglio, R.V.; Carruba, G.; Cicero, A.F.G.; Banach, M.; Patti, A.M.; Nikolic, D.; Cocciadiferro, L.; Zarcone, M.; Montalto, G.; Stoian, A.P.; et al. Pasta supplemented with opuntia ficus-indica extract improves metabolic parameters and reduces atherogenic small dense low-density lipoproteins in patients with risk factors for the metabolic syndrome: A four-week intervention study. *Metabolites* **2020**, *26*, 428. [CrossRef]
112. Bruno, J.A.; Feldman, C.H.; Konas, D.W.; Kerrihard, A.L.; Matthews, E.L. Incorporating sprouted chickpea flour in pasta increases brachial artery flow-mediated dilation. *Physiol. Int.* **2019**, *16*, 207–212. [CrossRef]
113. Vernaza, M.G.; Biasutti, E.; Schmiele, M.; Jaekel, L.Z.; Bannwart, A.; Chang, Y.K. Effect of supplementation of wheat flour with resistant starch and monoglycerides in pasta dried at high temperatures. *Int. J. Food Sci. Technol.* **2012**, *47*, 1302–1312. [CrossRef]
114. Strobel, N.A.; Fassett, R.G.; Marsh, S.A.; Coombes, J.S. Oxidative stress biomarkers as predictors of cardiovascular disease. *Int. J. Cardiol.* **2011**, *147*, 191–201. [CrossRef]
115. Giacco, F.; Brownlee, M. Oxidative stress and diabetic complications. *Circ. Res.* **2010**, *107*, 1058–1070. [CrossRef]
116. Scalbert, A.; Williamson, G. Dietary intake and bioavailability of polyphenols. *J. Nutr.* **2000**, *130*, 2073Se85S. [CrossRef]
117. Camelo-Mendez, G.A.; Ferruzzi, M.G.; Gonzalez-Aguilar, G.A.; Bello-Perez, L.A. Carbohydrate and phytochemical digestibility in pasta. *Food Eng. Rev.* **2016**, *8*, 76–89. [CrossRef]
118. Arts, I.C.W.; Hollman, P.C. Polyphenols and disease risk in epidemiologic studies. *Am. J. Clin. Nutr.* **2005**, *81* (Suppl. S1), 317S–325S. [CrossRef]
119. Khan, I.; Yousif, A.M.; Johnson, S.K.; Gamlath, S. Effect of sorghum flour addition on in vitro starch digestibility, cooking quality and consumer acceptability of durum wheat pasta. *J. Food Sci.* **2014**, *79*, S1560–S1567. [CrossRef]
120. Laus, M.N.; Soccio, M.; Alfarano, M.; Pasqualone, A.; Lenucci, M.S.; Miceli, G.D.; Pastore, D. Different effectiveness of two pastas supplemented with either lipophilic or hydrophilic/phenolic antioxidants in affecting serum as evaluated by the novel Antioxidant/Oxidant Balance approach. *Food Chem.* **2016**, *221*, 278–288. [CrossRef]
121. Meschini, R.; Filippi, S.; Molinari, R.; Costantini, L.; Bonafaccia, G.; Merendino, N. Pasta containing tartary buckwheat sprouts prevents DNA damage in spontaneously hypertensive rats. *Int. J. Food Sci. Nutr.* **2015**, *66*, 574–578. [CrossRef]
122. Merendino, N.; Molinari, R.; Costantini, L.; Mazzuccato, A.; Pucci, A.; Bonafaccia, F.; Esti, M.; Brunella Ceccantoni, B.; Papeschi, C.; Bonafaccia, G. A new "functional" pasta containing tartary buckwheat sprouts as an ingredient improves the oxidative status and normalizes some blood pressure parameters in spontaneously hypertensive rats. *Food Funct.* **2014**, *5*, 1017–1026. [CrossRef]
123. Barera, A.; Buscemi, S.; Monastero, R.; Caruso, C.; Caldarella, R.; Ciaccio, M.; Vasto, S. β-glucans: Ex vivo inflammatory and oxidative stress results after pasta intake. *Immun. Ageing* **2016**, *13*, 14. [CrossRef]
124. Aiello, A.; Di Bona, D.; Candore, G.; Carru, C.; Zinellu, A.; Di Miceli, G.; Nicosia, A.; Gambino, C.M.; Ruisi, P.; Caruso, C.; et al. Targeting Aging with Functional Food: Pasta with Opuntia Single-Arm Pilot Study. *Rejuvenation Res.* **2018**, *21*, 249–256. [CrossRef]
125. Chan, C.K.Y.; Fabek, H.; Mollard, R.C.; Jones, P.J.H.; Tulbek, M.C.; Chibbar, R.N.; Gangola, M.P.; Ramadoss, B.R.; Sanchez-Hernandez, D.; Anderson, G.H. Faba bean protein flours added to pasta reduce post-ingestion glycaemia, and increase satiety, protein content and quality. *Food Funct.* **2019**, *10*, 7476–7488. [CrossRef]
126. Greffeuille, V.; Marsset-Baglieri, A.; Molinari, N.; Cassan, D.; Sutra, T.; Avignon, A.; Micard, V. Enrichment of pasta with faba bean does not impact glycemic or insulin response but can enhance satiety feeling and digestive comfort when dried at very high temperature. *Food Funct.* **2015**, *6*, 2996–3005. [CrossRef]
127. Chiavaroli, L.; Kendall, C.W.C.; Braunstein, C.R.; Blanco Mejia, S.; Leiter, L.A.; Jenkins, D.J.A.; Sievenpiper, J.L. Effect of pasta in the context of low-glycaemic index dietary patterns on body weight and markers of adiposity: A systematic review and meta-analysis of randomised controlled trials in adults. *BMJ Open* **2018**, *8*, e019438. [CrossRef]

128. Costabile, G.; Griffo, E.; Cipriano, P.; Vetrani, C.; Vitale, M.; Mamone, G.; Rivellese, A.A.; Riccardi, G.; Giacco, R. Subjective satiety and plasma PYY concentration after wholemeal pasta. *Appetite* **2018**, *125*, 172–181 . [CrossRef]
129. Turco, I.; Bacchetti, T.; Morresi, C.; Padalino, L.; Ferretti, G. Polyphenols and the glycaemic index of legume pasta. *Food Funct.* **2019**, *10*, 5931–5938. [CrossRef]
130. Fujiwara, N.; Hall, C.; Jenkins, A.L. Development of low glycemic index (GI) foods by incorporating pulse ingredients into cereal-based products: Use of in vitro screening and in vivo methodologies. *Cereal. Chem.* **2017**, *94*, 110–116. [CrossRef]
131. Scazzina, F.; Dall'Asta, M.; Casiraghi, M.C.; Sieri, S.; Del Rio, D.; Pellegrini, N.; Brighenti, F. Glycemic index and glycemic load of commercial Italian foods. *Nutr. Metab. Cardiovasc. Dis.* **2016**, *26*, 419–429. [CrossRef]
132. Deana, M.; Shepherda, R.; Arvolab, A.; Vassalloc, M.; Winkelmannd, M.; Claupeind, E.; Lähteenmäki, L.; Raatsa, M.M.; Saba, A. Consumer perceptions of healthy cereal products and production methods. *J. Cereal. Sci.* **2007**, *46*, 188–196. [CrossRef]
133. Schmidt, S.J.; Fontana, A.J. Water Activity Values of Select Food Ingredients and Products. Available online: https://onlinelibrary.wiley.com/doi/pdf/10.1002/9780470376454.app5 (accessed on 15 December 2021).
134. Kamble, D.B.; Singh, R.; Pal Kaur, B.; Rani, S. Storage stability and shelf life prediction of multigrain pasta under dif-ferent packaging material and storage conditions. *J. Food Process. Preserv.* **2020**, *44*, e14585. [CrossRef]
135. Verardo, V.; Ferioli, F.; Riciputi, Y.; Iafelice, G.; Marconi, E.; Caboni, M.F. Evaluation of lipid oxidation in spaghetti pasta enriched with long chain n−3 polyunsaturated fatty acids under different storage conditions. *Food Chem.* **2009**, *114*, 472–477. [CrossRef]

foods MDPI

Review

Glycemic Index Values of Pasta Products: An Overview

Giuseppe Di Pede [1], Rossella Dodi [1], Cecilia Scarpa [1], Furio Brighenti [1], Margherita Dall'Asta [2] and Francesca Scazzina [1,*]

[1] Department of Food and Drug, University of Parma, Via Volturno 39, 43124 Parma, Italy; giuseppe.dipede@unipr.it (G.D.P.); rossella.dodi@unipr.it (R.D.); cecilia.scarpa@studenti.unipr.it (C.S.); furio.brighenti@unipr.it (F.B.)

[2] Department of Animal Science, Food and Nutrition (DiANA), Università Cattolica del Sacro Cuore, Via Parmense 84, 29122 Piacenza, Italy; margherita.dallasta@unicatt.it

* Correspondence: francesca.scazzina@unipr.it; Tel.: +39-0521-903913

Abstract: Durum wheat pasta is considered a low-glycemic index (GI) food. In recent years, the interest in developing enriched pasta has increased. Since both the formulation and processing technologies may affect the GI, this study aimed to investigate the GI values of pasta products (pp) reported in the literature until 2020. GI values of pp analyzed following the ISO guidelines were included in this survey. A total of 95 pp were identified and, according to their formulation, classified into 10 categories (n, mean GI): category n 1: 100% refined wheat (35, 55); category n 2: 100% whole wheat (6, 52); category n 3: other cereal-based products (8, 52); category n 4: containing egg (5, 52); category n 5: gluten free (11, 60); category n 6: containing legumes (9, 46); category n 7: noodles and vermicelli (9, 56); category n 8: containing vegetable or algae (6, 51); category n 9: containing other ingredients (5, 37); category n 10: stuffed (1, 58). Overall, pasta is confirmed to be a medium–low-GI food, even if a high variability among or within each category emerged. The formulation of enriched pp able to elicit a controlled glycemic response could represent a strategy to improve the nutritional value of pasta.

Keywords: pasta; glycemic index; enrichment; carbohydrate; ingredient; database; formulation; quality

Citation: Di Pede, G.; Dodi, R.; Scarpa, C.; Brighenti, F.; Dall'Asta, M.; Scazzina, F. Glycemic Index Values of Pasta Products: An Overview. *Foods* **2021**, *10*, 2541. https://doi.org/10.3390/foods10112541

Academic Editor: Mike Sissons

Received: 28 August 2021
Accepted: 19 October 2021
Published: 22 October 2021

1. Introduction

Cereals, tubers and pulses are the main dietary sources of carbohydrates within the human diet [1], which are well known as the main dietary components affecting postprandial blood glucose levels [2–4]. The glycemic index (GI), proposed by Jenkins [5], is a tool for quantifying the relative rise in blood glucose level after consuming a carbohydrate-containing food. The GI is defined as the incremental area under the two-hour blood glucose response curve (IAUC) after ingestion of a food with a certain amount of available carbohydrates, expressed as a percentage of the IAUC after consumption of a standard meal in an iso-glucidic portion [5,6]. Pasta, a traditional food item within the Italian diet, is now globally consumed, becoming an important source of complex carbohydrates (i.e., starch) in many countries [7,8]. Since durum wheat pasta is produced by mixing semolina with water and with energy input [9], its nutritional properties are prevalently linked to its matrix structure formed during the extrusion and drying processes [10–12]. As a consequence of this technological process, the microstructure of pasta is compact and relatively dense, limiting the hydrolysis of internal starch granules, which explains its richness in slow digestible starch and its reduced enzymatic susceptibility during digestion [9,12]. Postprandial studies conducted in both healthy and diabetic volunteers confirmed that durum wheat pasta induced a lower postprandial glucose response than other wheat-based products (i.e., bread) by virtue of its compact dense physical structure (dried pasta) and the network of gluten surrounding the starch granules [13–16]. On the other hand, refined wheat pasta is significantly lower in fiber and micronutrients (i.e., minerals and vitamins) with respect to whole grain pasta [9], and it is well known that the biological value of wheat proteins is low due to the deficiency in some essential amino acids, such as lysine and threonine [17]. Due to the importance and role of pasta as one of the main staple foods in the human diet, the interest in developing enriched pasta with high nutritional values has grown [18–22]. To achieve this goal, different approaches have been developed, as pasta could be used as dietary carrier of macronutrients, vitamins, minerals and/or phytochemicals by adding legumes, flour from vegetables/marine foods, and flour of refined or whole cereals different from wheat [19,20,23–25] (Figure 1).

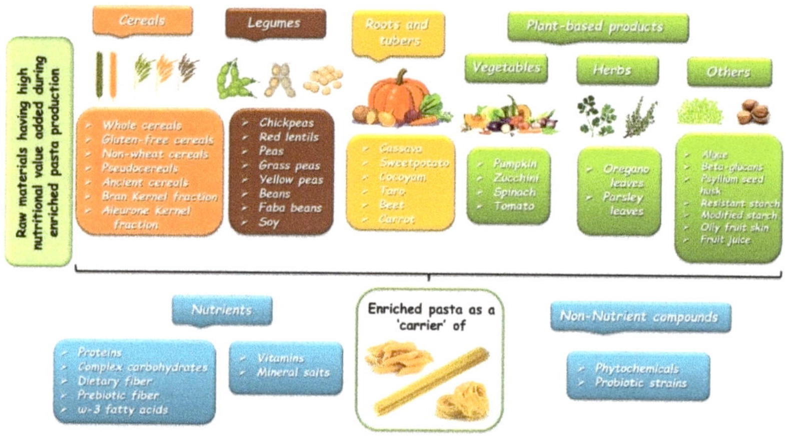

Figure 1. Raw materials commonly employed to produce enriched pasta product at high nutritional value.

Within the context of enriched pasta, functional ingredients can be added, as functional food consumption has increased in recent years [26,27]. Their consumption, by virtue of their physiologically active components, should provide health benefits beyond basic nutrition [28]. Since pasta formulation could affect the glycemic response after consumption, and therefore, its GI, beyond the processing method [29–31], a large number of human intervention studies have investigated the GI of enriched pasta products [18,21,32–34]. Thus, since the GI represents one of the most important parameters considered for evaluating the quality of dietary carbohydrates, this study aimed to gather the GI values of pasta products (pp) published in the literature until 2020.

2. Materials and Methods

2.1. Data Collection

A literature search to collect data on GI values of pp published in the literature without any restrictions was performed in December 2020 by using Pubmed, Scopus, Web of Science and Science Direct. Keywords used for data collection were: "glyc(a)emic index" AND pasta. Taking into consideration the ISO guidelines [35] for GI determination, exclusion criteria for data collection were as follows: (I) GI values obtained in the context of mixed meals or with the addition of any condiments; (II) GI values obtained using a sample size of less than ten subjects, and/or unhealthy subjects; (III) GI values calculated by using a standard meal other than glucose solution or white bread; (IV) GI values calculated considering IAUCs obtained before or after two postprandial hours following pasta consumption; (V) human intervention studies not specifying the number of subjects enrolled and/or the standard meal used; (VI) GI values calculated using in vitro models (i.e., estimated GI).

2.2. Database Development

Data on (i) pasta characteristics (types and formulation), (ii) GI values (mean value and data distribution expressed as standard deviation (SD) or standard error of the mean (SEM)), and (iii) experimental protocol for GI measurement (blood sample, sample size, standard meal, available carbohydrate (Av. CHO)/portion in grams, and place of analysis) were collected from research papers that met the inclusion criteria. According to their formulation, pp were classified into ten categories: (category n 1) 100% refined wheat; (category n 2) 100% whole wheat; (category n 3) other cereal-based products; (category n 4) containing egg; (category n 5) gluten free (GF); (category n 6) containing legumes; (category n 7) noodles and vermicelli; (category n 8) containing vegetable or algae; (category n 9) containing other ingredients; (category n 10) stuffed. Furthermore, pp within the same category were further subdivided into 'Low' GI ($0 \geq GI \leq 55$), 'Medium' GI ($55 > GI \leq 70$), and 'High' GI ($70 > GI \leq 100$) [5,35].

2.3. Data Analysis

The normality of data distribution within each category was verified through the Kolmogorov–Smirnov test, and GI values for the 10 categories of pp were expressed as the mean. The number of items at low, medium and high GI was provided as a percentage value with respect to the total number of pp within each category (data distribution). The statistical analysis was carried out using SPSS software (IBM SPSS Statistics, Version 25.0, IBM Corp., Chicago, IL, USA).

3. Results

3.1. GI Data

GI values of 95 pp were gathered from 28 research articles and are reported in Table 1. Category n 1 (100% refined wheat) was the largest group, including 35 items, among which six values were collected for 100% whole wheat pasta (category n 2), eight for other cereal-based products (category n 3), five for egg pasta (category n 4), 11 for GF (category n 5), nine for products containing legume (category n 6), nine for noodles and vermicelli (category n 7), six for pasta containing vegetable or algae (category n 8), five for items containing other ingredients (category n 9), and only one for stuffed pasta (category n 10). As reported in Figure 2, the GI of pp belonging to the same category are highly variable. Low-GI pastas were present in all the investigated categories, with the only exception of category n 10 (stuffed pp). No data on medium GI food items were recovered for products containing egg and containing other ingredients (categories n 4 and n 9, respectively). Conversely, high GI pastas fell within the 100% refined wheat pasta (category n 1), other cereal-based products (category n 3), GF pasta (category n 5) categories, and within products containing legumes (category n 6).

According to the GI classification rank (http://www.glycemicindex.com, accessed on 20 July 2021), pp belonging to categories n 1 (100% refined wheat), n 2 (100% whole wheat), n 3 (other cereal-based products), n 4 (containing egg), n 6 (containing legumes), n 8 (containing vegetable or algae), and n 9 (containing other ingredient) can be classified as low-GI foods. Items belonging to categories n 5 (gluten free), n 7 (noodles and vermicelli) and n 10 (stuffed) had a medium GI.

Table 1. Pasta product characteristics, glycemic index and experimental protocol data.

Pasta Product Characteristics		GI Data		Experimental Protocol Data					
Types	Formulation	Mean Value	Data Distribution	Blood Sample Type	Sample Size	Standard Meal	Av. CHO (g)/Portion	Place of Analysis	Ref.
Category n 1: 100% refined wheat									
Low GI									
-spaghetti, dried at high temperature (80 °C) *	durum wheat (*var. Duilio*) flour	32.6	6.1 º	capillary	10	G	50.0	Italy	[36]
-spaghetti §	durum wheat flour	33.0	6.0 º	capillary	10	G	50.0	Italy	[37]
-spaghetti *	white wheat flour	36.4	35.8 **	venous	12	G	50.0	IS	[38]
-spaghetti *	white wheat flour	42.1	10.8 **	capillary	10	G	50.0	IS	[38]
-spaghetti *	white wheat flour	43.8	9.2 **	capillary	10	G	50.0	IS	[38]
-spaghetti *	durum wheat flour	44.0	7.0 º	capillary	13	G	50.0	Italy	[39]
-spaghetti * (CT: 15 min)	white flour	44.0	3.0 º	n.a.	10	G	48.0	Australia	[40]
-spaghetti *	white wheat flour	44.1	19.8 **	capillary	10	G	50.0	IS	[38]
-penne §	durum wheat flour	47.0	4.0 º	capillary	10	G	50.0	Italy	[37]
-spaghetti *	semolina flour	47.0	n.a.	capillary	12	G	50.0	Iran	[41]
-spaghetti, dry *	durum wheat (*var. Svevo*) flour	48.0	4.0 º	n.a.	10	G	50.0	Italy	[42]
-penne §	durum wheat flour	50.0	7.0 º	capillary	10	G	50.0	Italy	[37]
-spaghetti §	durum wheat flour	50.0	9.0 º	capillary	10	G	50.0	Italy	[37]
-spaghetti §	durum wheat flour	51.0	9.0 º	capillary	10	G	50.0	Italy	[37]
-spaghetti, dry *	durum wheat (*var. Svevo*) flour	52.0	3.0 º	n.a.	10	G	50.0	Italy	[42]
-spaghetti, dry § (CT: 8 min)	durum wheat semolina	52.0	4.0 º	capillary	10	G	50.0	Italy	[43]
-spaghetti, dried at low temperature (55 °C) § (CT:10 min)	durum wheat flour	52.3	7.0 º	capillary	15	G	50.0	France	[44]
-pasta #	durum wheat semolina flour	52.5	8.4 º	capillary	15	G	50.0	Italy	[21]
-short penne §	durum wheat flour	53.0	5.0 º	capillary	10	G	50.0	Italy	[37]
-fusilli, dry # (CT: 10 min)	durum wheat flour	54.0	11.0 º	capillary	10	G	50.0	UK	[45]
-spaghetti #	durum wheat flour	54.9	n.a.	n.a.	10	WB	50.0	Italy	[46]
Medium GI									
-spaghetti * (CT: 15 min)	100% durum wheat semolina	58.0	6.8 º	capillary	10	WB	50.0	Sweden	[47]
-small penne §	durum wheat flour	59.0	11.0 º	capillary	10	G	50.0	Italy	[37]
-spaghetti *, infused	common wheat (*var. Nol*) flour	60.0	n.a.	capillary	12	G	50.0	Iran	[41]
-macaroni *	wheat flour	61.0	5.0 º	n.a.	10	WB	50.0	Canada	[48]
-fusilli * (CT: 10 min)	durum wheat semolina	61.0	9.0 º	n.a.	10	G	50.0	UK	[49]
-white spaghetti * and stored	white wheat flour	62.0	5.0 º	capillary	10	WB	50.0	Brasile	[50]
-spaghetti * and infused	semolina flour	63.0	n.a.	capillary	12	G	50.0	Iran	[41]

Table 1. *Cont.*

Pasta Product Characteristics		GI Data		Experimental Protocol Data					
Types	**Formulation**	**Mean Value**	**Data Distribution**	**Blood Sample Type**	**Sample Size**	**Standard Meal**	**Av. CHO (g)/Portion**	**Place of Analysis**	**Ref.**
Category n 1: 100% refined wheat									
Medium GI									
-white spaghetti and stored *	white wheat flour	64.0	7.0 º	capillary	10	WB	50.0	Brasile	[50]
-spaghetti *, infused	common wheat (*var. Nol*) flour	68.0	n.a.	capillary	12	G	50.0	Iran	[41]
-spaghetti *	white wheat flour	70.0	10.0 º	n.a.	12	WB	44.0	Australia	[40]
High GI									
-pasta *	wheat flour	72.6	n.a.	capillary	10	WB	n.a.	Indonesia	[18]
-spaghetti *	wheat refined flour	72.8	5.0 º	capillary	12	G	n.a.	Spain	[19]
-pasta, fresh # (CT: 20 min)	semolina flour	78.0	8.0 º	capillary	10	WB	50.0	Canada	[51]
-spaghetti *	white wheat flour	83.6	9.6 º	capillary	19	G	50.0	Canada	[33]
Category n 2: 100% whole wheat									
Low GI									
-spaghetti §	whole-meal durum wheat flour	35.0	3.0 º	capillary	10	G	50.0	Italy	[37]
-short penne §	whole-meal durum wheat flour	48.0	9.0 º	capillary	10	G	50.0	Italy	[37]
-spaghetti §	whole-meal durum wheat flour	53.0	10.0 º	capillary	10	G	50.0	Italy	[37]
-spaghetti §	whole-meal durum wheat flour	55.0	10.0 º	capillary	10	G	50.0	Italy	[37]
-fusilli, dry # (CT: 10 min)	whole wheat flour	55.0	8.0 º	capillary	10	G	50.0	UK	[45]
Medium GI									
-spaghetti *	whole wheat flour	65.0	n.a.	n.a.	10	WB	40.0	Canada	[40]
Category n 3: other cereal-based products									
Low GI									
-spaghetti, dried at high temperature (80 °C) *	emmer wheat flour (emmer genotype 399)	18.1	2.6 º	capillary	10	G	50.0	Italy	[36]
-spaghetti, dried at high temperature (80 °C) *	emmer wheat flour (emmer genotype 257)	30.5	4.7 º	capillary	10	G	50.0	Italy	[36]
-spaghetti #	Kamut® (*T. polonicum*) flour	41.6	n.a.	n.a.	10	WB	50.0	Italy	[46]

Table 1. *Cont.*

Pasta Product Characteristics		GI Data		Experimental Protocol Data					
Types	Formulation	Mean Value	Data Distribution	Blood Sample Type	Sample Size	Standard Meal	Av. CHO (g)/Portion	Place of Analysis	Ref.
Category n 3: other cereal-based products									
Medium GI									
-spaghetti #	spelt (*T. dicoccum*) flour	56.5	n.a.	n.a.	10	WB	50.0	Italy	[46]
-pasta fresh # (CT: 5 min)	Celebrity barley cultivar (white pearled) flour	58.0	4.0 º	capillary	10	WB	50.0	Canada	[51]
-pasta fresh # (CT: 5 min)	AC Parkhill barley cultivar (white pearled) flour	64.0	4.0 º	capillary	10	WB	50.0	Canada	[51]
High GI									
-pasta fresh # (CT: 5 min)	Celebrity barley cultivar (whole grain) flour	71.0	6.0 º	capillary	10	WB	50.0	Canada	[51]
-pasta fresh # (CT: 5 min)	AC Parkhill barley cultivar (whole grain) flour	73.0	7.0 º	capillary	10	WB	50.0	Canada	[51]
Category n 4: containing egg									
Low GI									
-fettuccine *	egg pasta	47.0	n.a.	venous	14	G	46.0	NZ	[52]
-tagliatelle §	durum wheat flour, eggs	51.0	7.0 º	capillary	10	G	50.0	Italy	[37]
-lasagne, dry # (CT: 10 min)	egg pasta	53.0	9.0 º	capillary	10	G	50.0	UK	[45]
-tagliatelle *	egg pasta	54.0	5.0 º	capillary	10	G	50.0	UK	[53]
-tagliatelle, dry §	durum wheat flour, eggs	55.0	4.0 º	capillary	10	G	50.0	Italy	[37]
Category n 5: gluten free									
Low GI									
-penne, dry §	corn flour, millet flour, sugar cane syrup	48.1	n.a.	capillary	10	G	50.0	Italy	[34]
-spaghetti	rice and high amylose maize flour	51.0	5.0 º	n.a.	10	G	49.0	Australia	[40]
-pasta	rice flour	51.0	n.a.	n.a.	10	G	47.0	Australia	[40]
-fusilli, dry §	100% corn flour, water	54.4	n.a.	capillary	10	G	50.0	Italy	[34]

Table 1. *Cont.*

Pasta Product Characteristics		GI Data		Experimental Protocol Data					
Types	**Formulation**	**Mean Value**	**Data Distribution**	**Blood Sample Type**	**Sample Size**	**Standard Meal**	**Av. CHO (g)/Portion**	**Place of Analysis**	**Ref.**
Category n 5: gluten free									
Medium GI									
-tagliatelle, fresh §	rice, corn and chickpea flour, eggs (20%), egg white, water	59.6	n.a.	capillary	10	G	50.0	Italy	[34]
-tortellini, fresh §	rice, corn and chickpea flour, eggs (20%), egg white, water, stuffed with pork meat	60.6	n.a.	capillary	10	G	50.0	Italy	[34]
-pasta macaroni, dry #	parboiled rice flour	61.0	n.a.	capillary	10	G	40.0	Italy	[54]
-pasta, macaroni dry #	parboiled rice flour	65.0	n.a.	capillary	10	G	40.0	Italy	[54]
-vermicelli *	finger millet flour, defatted soy, resistant maltodextrin	65.5	5.5 º	capillary	16	G	50.0	India	[55]
High GI									
-macaroni, dry #	rice flour	71.0	n.a.	capillary	10	G	40.0	Italy	[54]
-pasta *	corn flour	78.0	10.0 º	n.a.	10	G	42.0	Australia	[40]
Category n 6: containing legumes									
Low GI									
-pasta #	60% grass pea flour, 40% chickpea flour	20.0	7.6 º	capillary	15	G	50.0	Italy	[21]
-pasta #	100% red lentil flour	22.3	6.9 º	capillary	15	G	50.0	Italy	[21]
-pasta #	100% pea flour	23.3	6.7 º	capillary	15	G	50.0	Italy	[21]
-spaghetti, dried at low temperature (55 °C) § (CT: 10.5 min)	35% faba bean flour, durum wheat semolina	41.9	5.7 º	capillary	15	G	50.0	France	[44]
-spaghetti	soy flour	47.0	7.4 º	capillary	10	G	25.0	Australia	[32]
-spaghetti, dried at high temperature (90 °C) § (CT: 13.5 min)	35% faba bean flour, durum wheat semolina	49.4	6.8 º	capillary	15	G	50.0	France	[44]
-macaroni *	50% red lentil flour	55.0	8.0 º	n.a.	10	WB	50.0	Canada	[48]

Table 1. *Cont.*

Pasta Product Characteristics		GI Data		Experimental Protocol Data					
Types	**Formulation**	**Mean Value**	**Data Distribution**	**Blood Sample Type**	**Sample Size**	**Standard Meal**	**Av. CHO (g)/Portion**	**Place of Analysis**	**Ref.**
Category n 6: containing legumes									
Medium GI									
-spaghetti * (CT: 10 min)	75% durum wheat flour, 25% chickpea flour	58.9	6.0 º	capillary	12	G	n.a.	Spain	[19]
High GI									
-spaghetti *	30% whole yellow pea flour, white durum wheat flour	93.3	9.4 º	capillary	19	G	50.0	Canada	[33]
Category n 7: noodles and vermicelli									
Low GI									
-noodles, dry *	wheat flour	46.0	5.8 º	venous	10	G	50.0	China	[56]
-noodle, dried	wheat	46.0	2.0 º	n.a.	10	G	42.0	China	[40]
-noodles, instant 'two-minute'	n.a.	48.0	n.a.	venous	15	G	26.0	NZ	[52]
-noodles, instant, all flavors	n.a.	52.0	5.0 º	n.a.	10	G	22.0	Australia	[40]
Medium GI									
-Jianxi vermicelli * (CT: 8 min)	rice flour	56.0	7.0 º	capillary	10	G	50.0	HK	[57]
-Sau tao Beijing noodles * (CT: 3 min)	wheat flour, salt, tapioca starch	61.0	5.0 º	capillary	10	G	50.0	HK	[57]
-noodles, reheated (CT: 5 min)	udon pasta, plain	62.0	8.0 º	n.a.	10	G	48.0	Australia	[40]
-Sau tao chicken-flavored Sichuan spicy noodles * (CT: 3 min)	wheat flour, salt	65.0	4.0 º	capillary	10	G	50.0	HK	[57]
-Taiwan vermicelli * (CT: 2 min)	rice, maize starch	68.0	12.0 º	capillary	10	G	50.0	HK	[57]

Table 1. *Cont.*

Pasta Product Characteristics		GI Data		Experimental Protocol Data					
Types	**Formulation**	**Mean Value**	**Data Distribution**	**Blood Sample Type**	**Sample Size**	**Standard Meal**	**Av. CHO (g)/Portion**	**Place of Analysis**	**Ref.**
Category n 8: containing vegetable or algae									
Low GI									
-small farfalle §	durum wheat flour, carrot and pumpkin pulps	44.0	5.0 º	capillary	10	G	50.0	Italy	[37]
-pasta, dry *	wheat flour, algae (*eucheuma cottonii*) flour (21%), eggs, cooking oil	44.4	n.a.	capillary	10	WB	N/A	Indonesia	[18]
-small pipe §	durum wheat flour, tomato and carrot pulps	47.0	7.0 º	capillary	10	G	50.0	Italy	[37]
-small penne §	durum wheat flour, zucchini and spinach pulps	48.0	5.0 º	capillary	10	G	50.0	Italy	[37]
Medium GI									
-pasta, dry *	wheat flour, algae (*eucheuma cottonii*) flour (14%), eggs, cooking oil	56.3	n.a.	capillary	10	WB	n.a.	Indonesia	[18]
-pasta, dry *	wheat flour, algae (*eucheuma cottonii*) flour (7%), eggs, cooking oil	66.4	n.a.	capillary	10	WB	n.a.	Indonesia	[18]
Category n 9: containing other ingredients									
Low GI									
-pasta *	protein enriched	28.0	1.0 º	n.a.	10	G	49.0	Australia	[40]
-spaghetti, dry § (CT: 8.5 min)	85% durum wheat semolina, 15% Barley Balance®	33.0	5.0 º	capillary	10	G	50.0	Italy	[43]
-spaghetti, dry § (CT: 8 min)	85% durum wheat semolina, 7.5% Barley Balance®, 7.5% psyllium seed husk	35.0	3.0 º	capillary	10	G	50.0	Italy	[43]
-spaghetti, dry *	durum wheat (*var Svevo*, line SBEIIa) flour, 58% amylose, 7.36% RS	38.0	3.0 º	n.a.	10	G	50.0	Italy	[42]
-spaghetti, dry *	durum wheat (*var Svevo* line SSIIa) flour, 44% amylose, 2.06% RS	49.0	3.0 º	n.a.	10	G	50.0	Italy	[42]
Category n 10: stuffed									
Medium GI									
-ravioli, fresh §	durum wheat flour, stuffed with calf meat	58.0	7.0 º	capillary	10	G	50.0	Italy	[37]

GI = glycemic index; º = data distribution is expressed as standard error of mean (SEM); ** = data distribution is expressed as standard deviation (SD); G = glucose solution; WB = white bread; Av. CHO = available carbohydrates; n.a. = not available; UK = United Kingdom; NZ = New Zealand; HK: Hong Kong; IS = interlaboratory study: the study was performed in Canada, Italia, Australia, Sweden, New Zealand, West Indies and South Africa; GF = gluten free; § = boiled in salted water; # = boiled in unsalted water; * = boiled in water; CT = cooking time; RS = resistant starch; Var = variety.

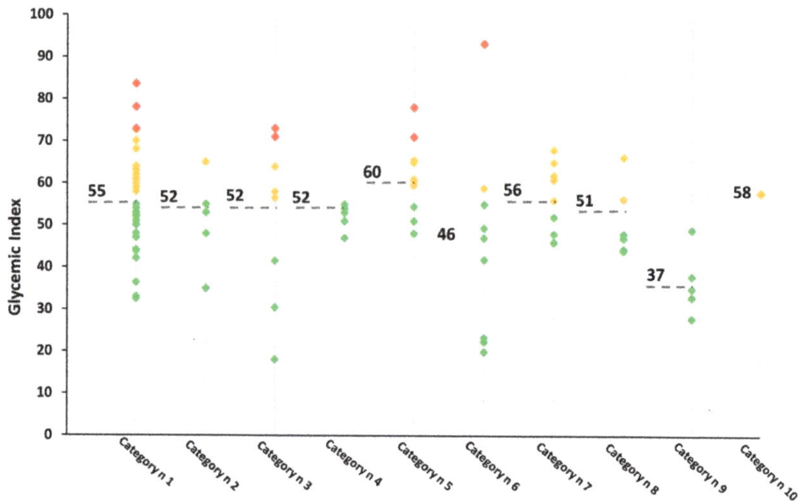

Figure 2. GI values of the 10 categories of pp analyzed. Red diamonds correspond to high GI pp; orange diamonds to medium GI pp; green diamonds to low-GI pp. Values reported in the figure correspond to the mean value for each category. Note: category n 1: 100% refined wheat; category n 2: 100% whole wheat; category n 3: other cereal-based products; category n 4: containing egg; category n 5: gluten free; category n 6: containing legumes; category n 7: noodles and vermicelli; category n 8: containing vegetable or algae; category n 9: containing other ingredients; category n 10: stuffed.

3.2. Formulations

Flours from barley and emmer were the main flours employed to produce pp with other cereals (category n 3), followed by spelt and Kamut® flours. GF items (category n 5) were formulated using GF cereal flours (rice, corn, and millet) and adding legumes (chickpea, soy), or modified starches (high amylose or resistant maltodextrin). Among the items containing legumes (category n 6), only three were formulated with 100% legume flour (red lentil, pea, and soy), while the remaining products were produced through a combination of legume (faba bean, chickpea, and whole yellow pea) and durum wheat flour, or by mixing different legume flours (i.e., grass pea and chickpea flours). Flours from wheat, rice, corn, or tubers (i.e., tapioca) were raw materials used for the formulation of noodles and vermicelli (category n 7). Pulps from carrot, pumpkin, tomato, zucchini and spinach were used for pasta containing vegetable formulations, while only one algae flour type (Eucheuma cottonii), added at different percentages (7%, 14%, and 21%), was used for pasta containing algae production (category n 8). Items containing protein, starchy ingredients (amylose and resistant starch) or fiber (Barley Balance®, psyllium seed husk) were included in category n 9 (containing other ingredients).

3.3. Experimental Protocol Data

A total of 71 GI values (equal to 75% of the total GI values) were obtained from capillary blood with respect to venous blood (used for 4% of the total GI values), and in the remaining studies, this information was not available. A total of 74 GI values (78% of the total GI values) were calculated with a sample size of 10 subjects. Glucose solution as a standard meal was used for the determination of 76 GI values (80% of the total GI values). For 73 GI values, the amount of available carbohydrates (Av. CHO) contained for each portion of pasta was 50.0 g, while for 16 GI values, the Av. CHO content in pasta portion size ranged from 22.0 g to 49.0 g; no data were available for the six remaining products. Italy was the place of analyses for 42 GI values (equal to 44% of the total GI values), while a great heterogeneity emerged for the remaining items.

4. Discussion

This study aimed to develop a database of GI values of pp based on the collection of the data recently reported in the literature. To the best of our knowledge, this is the first database specifically designed for reporting all GI data on pp, even if several databases on GI values, calculated either in healthy or diabetic patients, of a wide range of food items, have been proposed [37,40,45,57–60]. High-GI foods elicit higher postprandial glycemic responses, which have been associated with several chronic diseases, among which type 2 diabetes [61,62], cancer [2,63], and cardiovascular diseases [3,61] are the most relevant. Hence, since low-GI food consumption was associated with weight reduction and decreased incidence of several pathological conditions [3,4,6,61,64,65], the adherence to low-GI dietary patterns is strongly recommended by several national guidelines aiming at cardiovascular disease and diabetes prevention worldwide [66–70]. The present work confirmed that the GI of refined wheat pasta is low, even if a relevant variability was observed among GI values belonging to category n 1 (100% refined wheat). Indeed, among GI values gathered for category n 1, 60% of them were low (*n* = 21), followed by items of medium and high GI (29% (*n* = 10) and 11% (*n* = 4) for 100% refined wheat pastas at medium and high GI, respectively). The physical structure of the gluten matrix, formed by durum wheat starch and wheat proteins, is the main intrinsic factor supposed to explain the lower glycemic response of 100% refined wheat pasta products with respect to other products prepared with refined wheat [10–13,71]. In fact, it is well established that wheat pasta may elicit a lower postprandial glycemic response compared with bread or potatoes in both healthy and diabetic subjects [11,13,14,72,73]. The presence of high-GI pp among those belonging to category n 1 could have been probably linked to a different area of production [18,19,33,51], which reflects a certain heterogeneity in both pasta formulation and processing technology. The 100% whole wheat items (category n 2) had prevalently low GI, confirming the tendency of wheat fiber to positively modulate postprandial glycemic excursions [74]. It seems that the overall concept of the low GI of durum wheat pasta should be contextualized with the raw materials (common or durum wheat, refined or whole wheat), their origin, and the technological process used to produce it, rather than with the experimental conditions (i.e., sample size, characteristics and dietary patterns of the enrolled subjects, and inter-day variability) applied throughout the study. Despite pp belonging to category n 3 were classified as low-GI foods, it should be noted that pp formulated with whole barley flours resulted in high GI, probably due to a weaker food structure by virtue of the higher amount of insoluble fiber in whole barley [51]. Further human intervention studies are needed to fully clarify the influence of using other cereals (both in their refined and whole version) on the GI of pasta. It is well known that food formulation, as well as processing technologies, has been recognized as the most important factors affecting the GI of food products [29–31,75]. In the present work, enriched pp were classified into seven categories, reflecting the high variety of raw materials employed throughout the technological processing to enrich them. Nowadays, several food production/formulation strategies are implemented to enrich pasta by improving its nutritional [20,76–78], technological [79–82] and sensorial attributes [83–87]. Moreover, both nutritional and health claims could be obtained following food enrichment [88], positively affecting consumer choices [89–92]. Egg pp samples (category n 4) had a low GI by virtue of egg macronutrients, such as protein and lipids, which may mediate a reduction in the glycemic excursion [93]. Considering all the samples included in the enriched pp categories (from categories n 4 to n 10), 29 items (equal to 63% of the total enriched pp) were categorized as low GI, while the remaining 14 and 3 pp were medium and high GI, respectively (equal to 30% and 7% of the total enriched pp for those at medium and high GI, respectively). Based on these results, it is clear that enriched pasta also tends to maintain a food matrix able to make starch poorly accessible to the enzymatic activity within the gastro-intestinal tract. On the other hand, it should be considered that some raw materials added for pasta enrichment might negatively influence its GI. Among pp belonging to categories n 5 and n 7 (GF, noodles and vermicelli, respectively), a high heterogeneity in GI values for items formulated from the same starchy source (i.e., rice and corn) emerged. In this case, the absence of further details concerning both the composition and the technological processes employed for both GF and noodle and vermicelli production limits any exhaustive conclusions on the link between a product's characteristics and its GI. Furthermore, 78% of the total legume pp (category n 6) were categorized as low-GI items. Legumes are low-GI components of the Mediterranean diet by virtue of their nutritional properties (i.e., richness in protein and low digestible starch) [94–96]. Similarly, 67% of pp belonging to category n 8 (containing vegetable or algae) were also low GI. If vegetables are cooked or dressed with healthful oils, they could be considered important low-GI foods within our diet [97]. On the other hand, algae are recognized for their capacity to modulate glycemic response possibly thanks to the richness in bioactive compounds able to modulate glucose absorption and disposal [98]. As reported in Table 1, it should be presumed that both soluble fiber and modified starches or protein did not affect the food matrix structure and, consequently, carbohydrate bioavailability of pastas. Indeed, 100% of the items belonging to category n 9 were low GI. Dietary fiber, hydrocolloids, resistant starches and proteins have been shown to be able to slow the carbohydrate digestion rate [99,100]. Especially for other cereal-based items (category n 3), for GF pp (category n 5), and for those containing legumes (category n 6), GI values belonging to the same category were highly variable, reflecting the putative role of food properties [29,31,101], technological processing methods [14,15,20,54,102] and cooking time [12,31,103,104] in affecting carbohydrate bioavailability for pp, which could appear similar. Furthermore, since GI data for similar pp were presented as mean values and were collected from different human studies, the possible inter-individual variability in carbohydrate metabolism should also be taken into account [31,101,105,106]. The same factors may explain the variability observed among items belonging to different categories, which were not comparable. Similar pp (i.e., in terms of type, size, and shape) have different GI, since they could have been formulated by different brands or food factories and by means of several different raw materials (i.e., non-local flours) or a variety of technological methods. This variability could be greater for foods prepared to be sold in different national markets, given that the same product could be formulated depending on the country in which it will be commercialized [19]. We collected pp without any condiment added to avoid any confounding factors, since their role in modifying the glycemic excursion was clearly demonstrated [39,107–109]. Finally, both data on pasta formulation or regarding the experimental protocol employed for GI measurement were not always available, representing a limitation of the present study and proving the need for well-designed studies. The lack of data for some categories limits the conclusions for a clear relation between pasta formulation and GI value.

5. Conclusions and Future Perspectives

Overall, pasta is confirmed to be a medium–low-GI food. The present database would be a useful tool for pasta producers to formulate enriched pp with a high nutritional value. In fact, pasta with a high nutritional value and a low GI should be the industrial target, also keeping in mind specific consumer categories (e.g., celiac disease or type 2 diabetes patients). The observed variability for GI values of pp belonging to the same category, and to different categories, proves the inevitable role of formulation in influencing the GI of pasta, one of the most consumed starchy foods in our diet. Further human intervention studies are needed to obtain a clearer picture of this relationship.

Author Contributions: Conceptualization, G.D.P. and F.S.; methodology, G.D.P., R.D. and C.S.; investigation, G.D.P., M.D. and F.S.; data curation, G.D.P., R.D., C.S. and M.D.; writing—original draft preparation, G.D.P. and R.D.; writing—review and editing, G.D.P., F.B., M.D. and F.S.; supervision, M.D. and F.S. All authors have read and agreed to the published version of the manuscript.

Funding: This research received no external funding.

Conflicts of Interest: The authors declare no conflict of interest.

References

1. FAO. *Food Outlook-Biannual Report on Global Food Markets;* FAO: Rome, Italy, 2017; ISBN 9789251097823.
2. Blaak, E.E.; Antoine, J.-M.; Benton, D.; Björck, I.; Bozzetto, L.; Brouns, F.; Diamant, M.; Dye, L.; Hulshof, T.; Holst, J.J.; et al. Impact of Postprandial Glycaemia on Health and Prevention of Disease. *Obes. Rev.* **2012**, *13*, 923–984. [CrossRef]
3. Ludwig, D.S. the Glycemic Index: Physiological Mechanisms Relating to Obesity, Diabetes, and Cardiovascular Disease. *J. Am. Med. Assoc.* **2002**, *287*, 2414–2423. [CrossRef]
4. Ludwig, D.S.; Hu, F.B.; Tappy, L.; Brand-Miller, J. Dietary Carbohydrates: Role of Quality and Quantity in Chronic Disease. *BMJ* **2018**, *361*, k2340. [CrossRef]
5. Jenkins, D.J.; Wolever, T.M.; Taylor, R.H.; Barker, H.; Fielden, H.; Baldwin, J.M.; Bowling, A.C.; Newman, H.C.; Jenkins, A.L.; Goff, D. V Glycemic Index of Foods: A Physiological Basis for Carbohydrate Exchange. *Am. J. Clin. Nutr.* **1981**, *34*, 362–366. [CrossRef]
6. Willett, W.; Manson, J.; Liu, S. Glycemic Index, Glycemic Load, and Risk of Type 2 Diabetes. *Am. J. Clin. Nutr.* **2002**, *76*, 274S–280S. [CrossRef]
7. Mastorakou, D.; Rabaeus, M.; Salen, P.; Pounis, G.; De, L.M. *Mediterranean Diet: A Health-Protective Dietary Pattern for Modern Times;* Elsevier Inc.: Amsterdam, The Netherlands, 2019; ISBN 9780128145562.
8. International Pasta Organisation (IPO). *The World Pasta Industry Status Report;* International Pasta Organization: Rome, Italy, 2019. Available online: Http://Www.Internationalpasta.Org (accessed on 25 April 2021).
9. Giacco, R.; Vitale, M.; Riccardi, G. Pasta: Role in Diet. In *Encyclopedia of Food and Health;* Elsevier Inc.: Amsterdam, The Netherlands, 2016; pp. 242–245; ISBN 9780123849533.
10. Colonna, P.; Barry, J.-L.; Cloarec, D.; Bornet, F.; Gouilloud, S.; Galmiche, J.-P. Enzymic Susceptibility of Starch From Pasta. *J. Cereal Sci.* **1990**, *11*, 59–70. [CrossRef]
11. Granfeldt, Y.; Björck, I. Glycemic Response to Starch in Pasta: A Study of Mechanisms of Limited Enzyme Availability. *J. Cereal Sci.* **1991**, *14*, 47–61. [CrossRef]
12. Petitot, M.; Abecassis, J.; Micard, V. Structuring of Pasta Components during Processing: Impact on Starch and Protein Digestibility and Allergenicity. *Trends Food Sci. Technol.* **2009**, *20*, 521–532. [CrossRef]
13. Granfeldt, Y.; Bjorck, I.; Hagander, B. on the Importance of Processing Conditions, Product Thickness and Egg Addition for the Glycaemic and Hormonal Responses to Pasta: A Comparison with Bread Made from "Pasta Ingredients". *Eur. J. Clin. Nutr.* **1991**, *45*, 489–499.
14. Jenkins, D.J.A.; Wolever, T.M.S.; Jenkins, A.L.; Lee, R.; Wong, G.S.; Josse, R. Glycemic Response to Wheat Products: Reduced Response to Pasta But No Effect of Fiber. *Diabetes Care* **1983**, *6*, 155–159. [CrossRef]
15. Wolever, T.M.S.; Jenkins, D.J.A.; Kalmusky, J.; Giordano, C.; Giudici, S.; Thompson, L.U.; Wong, G.S.; Josse, R.G. Glycemic Response to Pasta: Effect of Surface Area, Degree of Cooking, and Protein Enrichment. *Diabetes Care* **1986**, *9*, 401–404. [CrossRef]
16. Zou, W.; Sissons, M.; Gidley, M.J.; Gilbert, R.G.; Warren, F.J. Combined Techniques for Characterising Pasta Structure Reveals How the Gluten Network Slows Enzymic Digestion Rate. *Food Chem.* **2015**, *188*, 559–568. [CrossRef]
17. Abdel-Aal, E.S.M.; Hucl, P. Amino Acid Composition and in Vitro Protein Digestibility of Selected Ancient Wheats and their End Products. *J. Food Compos. Anal.* **2002**, *15*, 737–747. [CrossRef]
18. Firdaus, M.; Yahya; Nugraha, G.R.H.; Utari, D.D. Fortification of Seaweed (Eucheuma Cottonii) Flour on Nutrition, Iodine, and Glycemic Index of Pasta. In *The IOP Conference Series: Earth and Environmental Science;* Institute of Physics Publishing: Bristol, UK, 2017; Volume 89.
19. Goñi, I.; Valentín-Gamazo, C. Chickpea Flour Ingredient Slows Glycemic Response to Pasta in Healthy Volunteers. *Food Chem.* **2003**, *81*, 511–515. [CrossRef]
20. Oliviero, T.; Fogliano, V. Food Design Strategies to Increase Vegetable Intake: The Case of Vegetable Enriched Pasta. *Trends Food Sci. Technol.* **2016**, *51*, 58–64. [CrossRef]
21. Turco, I.; Bacchetti, T.; Morresi, C.; Padalino, L.; Ferretti, G. Polyphenols and the Glycaemic Index of Legume Pasta. *Food Funct.* **2019**, *10*, 5931–5938. [CrossRef]
22. Wahanik, A.L.; Chang, Y.K.; Clerici, M.T.P.S. How to Make Pastas Healthier? *Food Rev. Int.* **2018**, *34*, 52–69. [CrossRef]
23. Aravind, N.; Sissons, M.; Fellows, C.M. Effect of Soluble Fibre (Guar Gum and Carboxymethylcellulose) Addition on Technological, Sensory and Structural Properties of Durum Wheat Spaghetti. *Food Chem.* **2012**, *131*, 893–900. [CrossRef]
24. Gallegos-Infante, J.-A.; Bello-Perez, L.A.; Rocha-Guzman, N.E.; Gonzalez-Laredo, R.F.; Avila-ontiveros, M. Effect of the Addition of Common Bean (*Phaseolus Vulgaris* L.) Flour on the in Vitro Digestibility of Starch and Undigestible Carbohydrates in Spaghetti. *J. Food Sci.* **2010**, *75*, 151–156. [CrossRef]
25. Khan, I.; Yousif, A.; Johnson, S.K.; Gamlath, S. Effect of Sorghum Flour Addition on Resistant Starch Content, Phenolic Profile and Antioxidant Capacity of Durum Wheat Pasta. *Food Res. Int.* **2013**, *54*, 578–586. [CrossRef]
26. Bogue, J.; Collins, O.; Troy, A.J. Market Analysis and Concept Development of Functional Foods. In *Developing New Functional Food and Nutraceutical Products;* Elsevier Inc.: Amsterdam, The Netherlands, 2017; pp. 29–45; ISBN 9780128027790.
27. Karelakis, C.; Zevgitis, P.; Galanopoulos, K.; Mattas, K. Consumer Trends and Attitudes to Functional Foods. *J. Int. Food Agribus. Mark.* **2020**, *32*, 266–294. [CrossRef]
28. Chavez-Gonzalez, M.L.; Buenrostro-Figueroa, J.J.; Aguilar, C.N. *Handbook of Research on Food Science and Technology: Volume 3: Functional Foods and Nutraceuticals;* Apple Academic Press: Palm Bay, FL, USA, 2021; ISBN 9781774635308.

29. Björck, I.; Granfeldt, Y.; Liljeberg, H.; Tovar, J.; Asp, N.G. Food Properties Affecting the Digestion and Absorption of Carbohydrates. *Am. J. Clin. Nutr.* **1994**, *59*, 699s–705s. [CrossRef]
30. Brand, J.C.; Nicholson, P.L.; Thorburn, A.W.; Truswell, A.S.; Brand, C.; Nicholson, L. Food Processing and the Glycemic Index. *Am. J. Clin. Nutr.* **1985**, *42*, 1192–1196. [CrossRef]
31. Englyst, K.N.; Englyst, H.N. Carbohydrate Bioavailability. *Br. J. Nutr.* **2005**, *94*, 1–11. [CrossRef]
32. Blair, R.M.; Henley, E.C.; Tabor, A. Soy Foods Have Low Glycemic and Insulin Response Indices in Normal Weight Subjects. *Nutr. J.* **2006**, *5*, 35. [CrossRef]
33. Marinangeli, C.P.F.; Kassis, A.N.; Jones, P.J.H. Glycemic Responses and Sensory Characteristics of Whole Yellow Pea Flour Added to Novel Functional Foods. *J. Food Sci.* **2009**, *74*, S385–S389. [CrossRef]
34. Scazzina, F.; Dall'asta, M.; Pellegrini, N.; Brighenti, F. Glycaemic Index of Some Commercial Gluten-Free Foods. *Eur. J. Nutr.* **2015**, *54*, 1021–1026. [CrossRef]
35. International Standards Organisation. *Food Products Determination of the Glycaemic Index (Gi) and Recommendation for Food Classification*; International Standards Organisation: Geneva, Switzerland, 2010; ISO 26642.
36. Fares, C.; Codianni, P.; Nigro, F.; Platani, C.; Scazzina, F.; Pellegrini, N. Processing and Cooking Effects on Chemical, Nutritional and Functional Properties of Pasta Obtained From Selected Emmer Genotypes. *J. Sci. Food Agric.* **2008**, *88*, 2435–2444. [CrossRef]
37. Scazzina, F.; Dall'asta, M.; Casiraghi, M.C.; Sieri, S.; Del Rio, D.; Pellegrini, N.; Brighenti, F. Glycemic Index and Glycemic Load of Commercial Italian Foods. *Nutr. Metab. Cardiovasc. Dis.* **2016**, *26*, 419–429. [CrossRef]
38. Wolever, T.M.S.; Vorster, H.H.; Björck, I.; Brand-Miller, J.; Brighenti, F.; Mann, J.I.; Ramdath, D.D.; Granfeldt, Y.; Holt, S.; Perry, T.L.; et al. Determination of the Glycaemic Index of Foods: Interlaboratory Study. *Eur. J. Clin. Nutr.* **2003**, *57*, 475–482. [CrossRef]
39. Chiavaroli, L.; Di Pede, G.; Dall'asta, M.; Cossu, M.; Francinelli, V.; Goldoni, M.; Scazzina, F.; Brighenti, F. the Importance of Glycemic Index on Post-Prandial Glycaemia in the Context of Mixed Meals: A Randomized Controlled Trial on Pasta and Rice. *Nutr. Metab. Cardiovasc. Dis.* **2020**, *31*, 615–625. [CrossRef]
40. Atkinson, F.S.; Foster-Powell, K.; Brand-Miller, J.C. International Tables of Glycemic Index and Glycemic Load Values: 2008. *Diabetes Care* **2008**, *31*, 2281–2283. [CrossRef]
41. Hajifaraji, M.; Rezvani, V.; Yaghoobi, A.S.; Morteza, H.; Maddah, M. Glycemic Indices of Three Commonly Consumed Foods: A Clinical Trial in Iranian Healthy Adults. *Med. J. Nutr. Metab.* **2012**, *5*, 253–257. [CrossRef]
42. Sissons, M.; Sestili, F.; Botticella, E.; Masci, S.; Lafiandra, D. Can Manipulation of Durum Wheat Amylose Content Reduce the Glycaemic Index of Spaghetti? *Foods* **2020**, *9*, 693. [CrossRef]
43. Peressini, D.; Cavarape, A.; Brennan, M.A.; Gao, J.; Brennan, C.S. Viscoelastic Properties of Durum Wheat Doughs Enriched with Soluble Dietary Fibres in Relation to Pasta-Making Performance and Glycaemic Response of Spaghetti. *Food Hydrocoll.* **2020**, *102*, 105613. [CrossRef]
44. Greffeuille, V.; Marsset-Baglieri, A.; Molinari, N.; Cassan, D.; Sutra, T.; Avignon, A.; Micard, V. Enrichment of Pasta with Faba Bean Does Not Impact Glycemic or Insulin Response But Can Enhance Satiety Feeling and Digestive Comfort When Dried at Very High Temperature. *Food Funct.* **2015**, *6*, 2996–3005. [CrossRef]
45. Henry, C.J.K.; Lightowler, H.J.; Strik, C.M.; Renton, H.; Hails, S. Glycaemic Index and Glycaemic Load Values of Commercially Available Products in the Uk. *Br. J. Nutr.* **2005**, *94*, 922–930. [CrossRef]
46. Simonato, B.; Curioni, A.; Pasini, G. Digestibility of Pasta Made with Three Wheat Types: A Preliminary Study. *Food Chem.* **2015**, *174*, 219–225. [CrossRef]
47. Granfeldt, Y.; Wu, X.; Björck, I. Determination of Glycaemic Index; Some Methodological Aspects Related to the Analysis of Carbohydrate Load and Characteristics of the Previous Evening Meal. *Eur. J. Clin. Nutr.* **2006**, *60*, 104–112. [CrossRef]
48. Fujiwara, N.; Hall, C.; Jenkins, A.L. Development of Low Glycemic Index (Gi) Foods By Incorporating Pulse Ingredients Into Cereal-Based Products: Use of in Vitro Screening and in Vivo Methodologies. *Cereal Chem. J.* **2017**, *94*, 110–116. [CrossRef]
49. Henry, C.J.K.K.; Lightowler, H.J.; Kendall, F.L.; Storey, M. the Impact of the Addition of toppings/Fillings on the Glycaemic Response to Commonly Consumed Carbohydrate Foods. *Eur. J. Clin. Nutr.* **2006**, *60*, 763–769. [CrossRef]
50. Carreira, M.C.; Lajolo, F.M.; De Menezes, E.W. Glycemic Index: Effect of Food Storage Under Low Temperature. *Braz. Arch. Biol. Technol.* **2004**, *47*, 569–574. [CrossRef]
51. Aldughpassi, A.; Abdel-Aal, E.-S.M.; Wolever, T.M.S. The Journal of Nutrition Nutrient Physiology, Metabolism, and Nutrient-Nutrient Interactions Barley Cultivar, Kernel Composition, and Processing Affect the Glycemic Index 1-3. *J. Nutr.* **2012**, *142*, 1666–1671. [CrossRef]
52. Perry, T.; Mann, J.; Mehalski, K.; Gayya, C.; Wilson, J.; Thompson, C. Glycaemic Index of New Zealand Foods. *N. Z. Med. J.* **2000**, *113*, 140–142.
53. Aston, L.M.; Gambell, J.M.; Lee, D.M.; Bryant, S.P.; Jebb, S.A. Determination of the Glycaemic Index of Various Staple Carbohydrate-Rich Foods in the UK Diet. *Eur. J. Clin. Nutr.* **2008**, *62*, 279–285. [CrossRef]
54. Barbiroli, A.; Bonomi, F.; Casiraghi, M.C.; Iametti, S.; Pagani, M.A.; Marti, A. Process Conditions Affect Starch Structure and Its Interactions with Proteins in Rice Pasta. *Carbohydr. Polym.* **2013**, *92*, 1865–1872. [CrossRef]
55. Shobana, S.; Selvi, R.P.; Kavitha, V.; Gayathri, N.; Geetha, G.; Gayathri, R.; Vijayalakshmi, P.; Balasubramaniam, K.K.G.; Ruchi, V.; Sudha, V.; et al. Development and Evaluation of Nutritional, Sensory and Glycemic Properties of Finger Millet (*Eleusine Coracana* L.) Based Food Products. *Asia Pac. J. Clin. Nutr.* **2018**, *27*, 84–91. [CrossRef]

56. Yang, Y.X.; Wang, H.W.; Cui, H.M.; Wang, Y.; Yu, L.D.; Xiang, S.X.; Zhou, S.Y. Glycemic Index of Cereals and Tubers Produced in China. *World J. Gastroenterol.* **2006**, *12*, 3430–3433. [CrossRef]
57. Lok, K.Y.; Chan, R.; Chan, D.; Li, L.; Leung, G.; Woo, J.; Lightowler, H.J.; Henry, C.J.K. Glycaemic Index and Glycaemic Load Values of a Selection of Popular Foods Consumed in Hong Kong. *Br. J. Nutr.* **2010**, *103*, 556–560. [CrossRef]
58. Foster-Powell, K.; Holt, S.H.; Brand-Miller, J.C. International Table of Glycemic Index and Glycemic Load Values: 2002. *Am. J. Clin. Nutr.* **2002**, *76*, 5–56. [CrossRef]
59. Foster-Powell, K.; Miller, J.B. International Tables of Glycemic Index. *Am. J. Clin. Nutr.* **1995**, *62*, 871s–890s. [CrossRef]
60. Atkinson, F.S.; Brand-Miller, J.C.; Foster-Powell, K.; Buyken, A.E.; Goletzke, J. International Tables of Glycemic Index and Glycemic Load Values 2021: A Systematic Review. *Am. J. Clin. Nutr.* **2021**, *31*, 2281–2283. [CrossRef]
61. Augustin, L.S.A.; Kendall, C.W.C.; Jenkins, D.J.A.; Willett, W.C.; Astrup, A. Nutrition, Metabolism & Cardiovascular Diseases Glycemic Index, Glycemic Load and Glycemic Response: An International Scientific Consensus Summit From the International Carbohydrate Quality Consortium (ICQC). *Nutr. Metab. Cardiovasc. Dis.* **2015**, *25*, 795–815. [CrossRef]
62. Cox, D.J.; Fang, K.; Mccall, A.L.; Conaway, M.R.; Banton, T.A.; Moncrief, M.A.; Diamond, A.M.; Taylor, A.G. Behavioral Strategies to Lower Postprandial Glucose in Those with Type 2 Diabetes May Also Lower Risk of Coronary Heart Disease. *Diabetes Ther.* **2019**, *10*, 277–281. [CrossRef]
63. Sieri, S.; Agnoli, C.; Pala, V.; Grioni, S.; Brighenti, F.; Pellegrini, N.; Masala, G.; Palli, D.; Mattiello, A.; Panico, S.; et al. Dietary Glycemic Index, Glycemic Load, and Cancer Risk: Results From the Epic-Italy Study. *Sci. Rep.* **2017**, *7*, 9757. [CrossRef]
64. Mirrahimi, A.; De Souza, R.J.; Chiavaroli, L.; Sievenpiper, J.L.; Beyene, J.; Hanley, A.J.; Augustin, L.S.A.; Kendall, C.W.C.; Jenkins, D.J.A. Associations of Glycemic Index and Load with Coronary Heart Disease Events: A Systematic Review and Meta-Analysis of Prospective Cohorts. *J. Am. Heart Assoc.* **2012**, *1*, e000752. [CrossRef]
65. Wolever, T.M.; Jenkins, D.J.; Jenkins, A.L.; Josse, R.G. the Glycemic Index: Methodology and Clinical Implications. *Am. J. Clin. Nutr.* **1991**, *54*, 846–854. [CrossRef]
66. Arnett, D.K.; Blumenthal, R.S.; Albert, M.A.; Buroker, A.B.; Goldberger, Z.D.; Hahn, E.J.; Himmelfarb, C.D.; Khera, A.; Lloyd-Jones, D.; Mcevoy, J.W.; et al. 2019 Acc/Aha Guideline on the Primary Prevention of Cardiovascular Disease: Executive Summary: A Report of the American College of Cardiology/American Heart Association Task Force on Clinical Practice Guidelines. *Circulation* **2019**, *140*, E563–E595. [CrossRef]
67. Piepoli, M.F.; Hoes, A.W.; Agewall, S.; Albus, C.; Brotons, C.; Catapano, A.L.; Cooney, M.T.; Corrà, U.; Cosyns, B.; Deaton, C.; et al. 2016 European Guidelines on Cardiovascular Disease Prevention in Clinical Practice. *Eur. Heart J.* **2016**, *37*, 2315–2381. [CrossRef]
68. Davies, M.J.; D'alessio, D.A.; Fradkin, J.; Kernan, W.N.; Mathieu, C.; Mingrone, G.; Rossing, P.; Tsapas, A.; Wexler, D.J.; Buse, J.B. Management of Hyperglycemia in Type 2 Diabetes, 2018. A Consensus Report By the American Diabetes Association (ADA) and the European Association for the Study of Diabetes (EASD). *Diabetes Care* **2018**, *41*, 2669–2701. [CrossRef]
69. Evert, A.B.; Boucher, J.L.; Cypress, M.; Dunbar, S.A.; Franz, M.J.; Mayer-Davis, E.J.; Neumiller, J.J.; Nwankwo, R.; Verdi, C.L.; Urbanski, P.; et al. Nutrition therapy Recommendations for the Management of Adults with Diabetes. *Diabetes Care* **2014**, *37*, S120–S143. [CrossRef]
70. Sievenpiper, J.L.; Chan, C.B.; Dworatzek, P.D.; Med, C.F.; Williams Med, S.L. 2018 Clinical Practice Guidelines Nutrition therapy Diabetes Canada Clinical Practice Guidelines Expert Committee. *Can. J Diabetes* **2018**, *42*, S64–S79. [CrossRef]
71. Huang, M.; Li, J.; Ha, M.A.; Riccardi, G.; Liu, S. A Systematic Review on the Relations between Pasta Consumption and Cardio-Metabolic Risk Factors. *Nutr. Metab. Cardiovasc. Dis.* **2017**, *27*, 939–948. [CrossRef]
72. Kristensen, M.; Jensen, M.G.; Riboldi, G.; Petronio, M.; Bügel, S.; Toubro, S.; Tetens, I.; Astrup, A. Wholegrain Vs. Refined Wheat Bread and Pasta. Effect on Postprandial Glycemia, Appetite, and Subsequent Ad Libitum Energy Intake in Young Healthy Adults. *Appetite* **2010**, *54*, 163–169. [CrossRef]
73. Riccardi, G.; Clemente, G.; Giacco, R. Glycemic Index of Local Foods and Diets: The Mediterranean Experience. *Nutr. Rev.* **2003**, *61*, 56–60. [CrossRef] [PubMed]
74. Anderson, J.W.; Baird, P.; Davis, R.H.; Ferreri, S.; Knudtson, M.; Koraym, A.; Waters, V.; Williams, C.L. Health Benefits of Dietary Fiber. *Nutr. Rev.* **2009**, *67*, 188–205. [CrossRef] [PubMed]
75. Bahado-Singh, P.S.; Wheatley, A.O.; Ahmad, M.H.; St A Morrison, E.Y.; Asemota, H.N. Food Processing Methods Influence the Glycaemic Indices of Some Commonly Eaten West Indian Carbohydrate-Rich Foods. *Br. J. Nutr.* **2020**, *96*, 476–481. [CrossRef]
76. Betoret, E.; Betoret, N.; Vidal, D.; Fito, P. Functional Foods Development: Trends and Technologies. *Trends Food Sci. Technol.* **2011**, *22*, 498–508. [CrossRef]
77. Fogliano, V.; Vitaglione, P. Functional Foods: Planning and Development. *Mol. Nutr. Food Res.* **2005**, *49*, 256–262. [CrossRef] [PubMed]
78. Foschia, M.; Horstmann, S.W.; Arendt, E.K.; Zannini, E. Legumes as Functional Ingredients in Gluten-Free Bakery and Pasta Products. *Annu. Rev. Food Sci. Technol.* **2017**, *8*, 75–96. [CrossRef]
79. Bustos, M.C.; Pérez, G.T.; León, A.E. Effect of Four Types of Dietary Fiber on the Technological Quality of Pasta. *Food Sci. Technol. Int.* **2011**, *17*, 213–221. [CrossRef]
80. Bustos, M.C.; Perez, G.T.; Leon, A.E. Structure and Quality of Pasta Enriched with Functional Ingredients. *Rsc Adv.* **2015**, *5*, 30780–30792. [CrossRef]
81. Bustos, M.C.; Paesani, C.; Quiroga, F.; León, A.E. Technological and Sensorial Quality of Berry-Enriched Pasta. *Cereal Chem.* **2019**, *96*, 967–976. [CrossRef]

82. Zen, C.K.; Tiepo, C.B.V.; Silva, R.V.; Reinehr, C.O.; Gutkoski, L.C.; Oro, T.; Colla, L.M. Development of Functional Pasta with Microencapsulated *Spirulina*: Technological and Sensorial Effects. *J. Sci. Food Agric.* **2020**, *100*, 2018–2026. [CrossRef] [PubMed]
83. Biernacka, B.; Dziki, D.; Gawlik-Dziki, U.; Różyło, R.; Siastała, M. Physical, Sensorial, and Antioxidant Properties of Common Wheat Pasta Enriched with Carob Fiber. *LWT-Food Sci. Technol.* **2017**, *77*, 186–192. [CrossRef]
84. El-Sohaimy, S.; Brennan, M.; Darwish, A.M.G.; Brennan, C. Physicochemical, Texture and Sensorial Evaluation of Pasta Enriched with Chickpea Flour and Protein Isolate. *Ann. Agric. Sci.* **2020**, *65*, 28–34. [CrossRef]
85. Fois, S.; Campus, M.; Piu, P.P.; Siliani, S.; Sanna, M.; Roggio, T.; Catzeddu, P. Fresh Pasta Manufactured with Fermented Whole Wheat Semolina: Physicochemical, Sensorial, and Nutritional Properties. *Foods* **2019**, *8*, 422. [CrossRef] [PubMed]
86. Giménez, M.A.; González, R.J.; Wagner, J.; Torres, R.; Lobo, M.O.; Samman, N.C. Effect of Extrusion Conditions on Physicochemical and Sensorial Properties of Corn-Broad Beans (Vicia Faba) Spaghetti Type Pasta. *Food Chem.* **2013**, *136*, 538–545. [CrossRef] [PubMed]
87. Panghal, A.; Kaur, R.; Janghu, S.; Sharma, P.P.; Sharma, P.P.; Chhikara, N. Nutritional, Phytochemical, Functional and Sensorial Attributes of Syzygium Cumini L. Pulp Incorporated Pasta. *Food Chem.* **2019**, *289*, 723–728. [CrossRef]
88. Regolamento (Ce), N. 1924/2006 Del Parlamento Europeo E Del Consiglio Del 20 Dicembre 2006 Relativo Alle Indicazioni Nutrizionali E Sulla Salute Fornite Sui Prodotti Alimentari. Available online: https://Eur-Lex.Europa.Eu/Legal-Content/It/Txt/Pdf/?Uri=Celex:32006r1924&From=It (accessed on 25 April 2021).
89. Aschemann-Witzel, J.; Hamm, U. Do Consumers Prefer Foods with Nutrition and Health Claims? Results of a Purchase Simulation. *J. Mark. Commun.* **2010**, *16*, 47–58. [CrossRef]
90. Lemken, D.; Knigge, M.; Meyerding, S.; Spiller, A. the Value of Environmental and Health Claims on New Legume Products: A Non-Hypothetical online Auction. *Sustainability* **2017**, *9*, 1340. [CrossRef]
91. Sogari, G.; Li, J.; Lefebvre, M.; Menozzi, D.; Pellegrini, N.; Cirelli, M.; Gómez, M.I.; Mora, C. The Influence of Health Messages in Nudging Consumption of Whole Grain Pasta. *Nutrients* **2019**, *11*, 2993. [CrossRef] [PubMed]
92. Van Trijp, H.C.M.; Van Der Lans, I.A. Consumer Perceptions of Nutrition and Health Claims. *Appetite* **2007**, *48*, 305–324. [CrossRef] [PubMed]
93. Pelletier, X.; Thouvenot, P.; Belbraouet, S.; Chayvialle, J.A.; Hanesse, B.; Mayeux, D.; Debry, G. Effect of Egg Consumption in Healthy Volunteers: Influence of Yolk, White or Whole-Egg on Gastric Emptying and on Glycemic and Hormonal Responses. *Ann. Nutr. Metab.* **1996**, *40*, 109–115. [CrossRef]
94. Singh, M.; Manickavasagan, A.; Shobana, S.; Mohan, V. Glycemic Index of Pulses and Pulse-Based Products: A Review. *Crit. Rev. Food Sci. Nutr.* **2020**, *61*, 1567–1588. [CrossRef]
95. Trinidad, T.P.; Mallillin, A.C.; Loyola, A.S.; Sagum, R.S.; Encabo, R.R. The Potential Health Benefits of Legumes as a Good Source of Dietary Fibre. *Br. J. Nutr.* **2010**, *103*, 569–574. [CrossRef]
96. Venn, B.J.; Mann, J.I. Cereal Grains, Legumes and Diabetes. *Eur. J. Clin. Nutr.* **2004**, *58*, 1443–1461. [CrossRef]
97. Ludwig, D.S. Dietary Glycemic Index and Obesity. *J. Nutr.* **2000**, *130*, 280s–283s. [CrossRef]
98. Parada, J.; Pérez-Correa, J.R.; Pérez-Jiménez, J. Design of Low Glycemic Response Foods Using Polyphenols from Seaweed. *J. Funct. Foods* **2019**, *56*, 33–39. [CrossRef]
99. Belorio, M.; Gómez, M. Psyllium: A Useful Functional Ingredient in Food Systems. *Crit. Rev. Food Sci. Nutr.* **2020**, *21*, 1–12. [CrossRef] [PubMed]
100. Capriles, V.D.; Arêas, J.A.G. Approaches to Reduce the Glycemic Response of Gluten-Free Products: In Vivo and in Vitro Studies. *Food Funct.* **2016**, *7*, 1266–1272. [CrossRef]
101. Dall'asta, M.; Del Rio, D.; Tappy, L.; Potì, F.; Agostoni, C.; Brighenti, F. Critical and Emerging topics in Dietary Carbohydrates and Health. *Int. J. Food Sci. Nutr.* **2020**, *71*, 286–295. [CrossRef]
102. Padalino, L.; Conte, A.; Del Nobile, M. Overview on the General Approaches to Improve Gluten-Free Pasta and Bread. *Foods* **2016**, *5*, 87. [CrossRef] [PubMed]
103. Englyst, K.N.; Vinoy, S.; Englyst, H.N.; Lang, V. Glycaemic Index of Cereal Products Explained By their Content of Rapidly and Slowly Available Glucose. *Br. J. Nutr.* **2003**, *89*, 329–339. [CrossRef] [PubMed]
104. Ross, S.W.; Brand, J.C.; Thorburn, A.W.; Truswell, S. Glycemic Index of Processed Wheat Products. *Am. J. Clin. Nutr.* **1987**, *46*, 631–635. [CrossRef] [PubMed]
105. Vega-López, S.; Ausman, L.M.; Griffith, J.L.; Lichtenstein, A.H. Interindividual Variability and Intra-Individual Reproducibility of Glycemic Index Values for Commercial White Bread. *Diabetes Care* **2007**, *30*, 1412–1417. [CrossRef] [PubMed]
106. Vrolix, R.; Mensink, R.P. Variability of the Glycemic Response to Single Food Products in Healthy Subjects. *Contemp. Clin. Trials* **2010**, *31*, 5–11. [CrossRef] [PubMed]
107. Collier, G.; Mclean, A.; O'dea, K. Effect of Co-Ingestion of Fat on the Metabolic Responses to Slowly and Rapidly Absorbed Carbohydrates. *Diabetologia* **1984**, *26*, 50–54. [CrossRef]
108. Moghaddam, E.; Vogt, J.A.; Wolever, T.M.S. The Effects of Fat and Protein on Glycemic Responses in Nondiabetic Humans Vary with Waist Circumference, Fasting Plasma Insulin, and Dietary Fiber Intake. *J. Nutr.* **2006**, *136*, 2506–2511. [CrossRef]
109. Sun, L.; Tan, K.W.J.; Lim, J.Z.; Magkos, F.; Henry, C.J. Dietary Fat and Carbohydrate Quality Have Independent Effects on Postprandial Glucose and Lipid Responses. *Eur. J. Nutr.* **2018**, *57*, 243–250. [CrossRef]

foods

MDPI

Article

Role of Hydrocolloids in the Structure, Cooking, and Nutritional Properties of Fiber-Enriched, Fresh Egg Pasta Based on Tiger Nut Flour and Durum Wheat Semolina

Maria Eugenia Martín-Esparza [1,*], Maria Dolores Raigón [2], Maria Dolores García-Martínez [3] and Ana Albors [1]

1 Food Technology Department, Research Institute of Food Engineering for Development, Universitat Politècnica de València, 46022 Valencia, Spain; analbors@tal.upv.es
2 Department of Chemistry, Universitat Politècnica de València, 46022 Valencia, Spain; mdraigon@qim.upv.es
3 Institute for the Preservation and Improvement of Valencian Agro-Diversity, Universitat Politècnica de València, 46022 Valencia, Spain; magarma8@qim.upv.es
* Correspondence: eesparza@tal.upv.es; Tel.: +34-963879834

Abstract: The aim of this work concerns the manufacturing process of fresh egg tagliatelle labeled as a "source of fiber" based on tiger nut flour and wheat semolina. An attempt to improve the quality attributes and cooking properties of the obtained product was made by means of structuring agents. More specifically, a combination of three hydrocolloids (carboximethylcellulose, CMC; xanthan gum, XG; and locust bean gum, LBG) was tested. A Box–Behnken design with randomized response surface methodology was used to determine a suitable combination of these gums to achieve fewer cooking losses, higher water gain and swelling index values, and better texture characteristics before and after cooking. Positive effects on textural characteristics were observed when incorporating XG into the pasta formulation. Cooking and fiber loss also significantly diminished with the XG-CMC combination over 0.8%. No significant effect was found for the other evaluated parameters. A synergistic interaction between LBG and XG was only significant for the water absorption index. The cooked pasta was considered a source of fiber in all cases.

Keywords: dietary fiber; hydrocolloids; food texture; cooking properties

Citation: Martín-Esparza, M.E.; Raigón, M.D.; García-Martínez, M.D.; Albors, A. Role of Hydrocolloids in the Structure, Cooking, and Nutritional Properties of Fiber-Enriched, Fresh Egg Pasta Based on Tiger Nut Flour and Durum Wheat Semolina. *Foods* **2021**, *10*, 2510. https://doi.org/10.3390/foods10102510

Academic Editor: Mike Sissons

Received: 7 September 2021
Accepted: 18 October 2021
Published: 19 October 2021

1. Introduction

Edible, sweet, brown-colored tiger nut (*Cyperus esculentus* L.) tubers are widely cultivated in Spain, Burkina Faso, Mali, Niger, and Nigeria [1]. Although it is underutilized in many countries in the world, tiger nut is an important crop in Spain [2], where it is used to produce a milky beverage and has also been employed for animal feed. This tuber is rich in carbohydrates, lipids, fiber, some minerals (potassium, phosphorus, calcium), and vitamins E and C [3]. It is also rich in lipids with a fatty acid profile, similarly to olive and hazelnut oils. The large amount of fiber content (8–15 g/100 g) and omega-6 fatty acids confers this tuber healthy properties [3,4] and plays a key role in the prevention of certain diseases, such as coronary heart disease, colon cancer, diabetes, and obesity [5]. For this reason, several scientific studies on tiger nuts have been conducted. They have focused mainly on the qualitative and quantitative assessments of their nutritional properties and also on the utilization of these components for industrial food purposes. Tiger nut flour (TNF) can be obtained by directly milling clean tubers followed by sieving to achieve the desired homogeneous particle size. This flour has been assessed to produce bakery products [6] and fresh or dry egg pastas [7,8] and for preparing gluten-free (GF) noodles [1] or GF bread with good baking and nutritional characteristics [9,10].

Pasta is a staple food thanks to its simple preparation, variety, versatility, sensory characteristics, and low price [11,12]. Fresh pasta has gained a market share in the last few years. The global fresh pasta market was valued at 1004.6 million USD in 2020 and is expected to grow at a 2.0% CAGR during 2021–2026 [13]. Europe not only purchases the most fresh pasta and consumed 411.49 million tons in 2018 [13] but is also the fastest growing area given its traditional cultural inheritance and good business environment [13]. Adding TNF to the pasta formula is an interesting option for increasing dietary fiber intake, which remains a challenge. Endeavors have been made by several authors to improve nutritional pasta properties, which include pea, oat, teff, quinoa, maize, soy, and amaranth as other plant source flours, mostly to enrich proteins in GF products, dietary fiber, or antioxidants [14–21]. Pasta quality, as affected by an increase in soluble and insoluble fibers, vitamins, and minerals, has been studied by other authors [22–24]. The glycemic index can be lowered by including dietary fiber, which additionally may offer other health benefits [22,24–26].

Durum wheat proteins can form a continuous viscoelastic network when flour is mixed with water during pasta production. The resulting dough may present optimal properties in mixing and extrusion steps [27], which lead to a final product with better strength and stability. The structure-forming protein in flour (gluten) is also important to reach a correct pasta behavior during cooking, mainly represented by low cooking losses and "al dente" pasta texture. When using GF flours (i.e., tiger nut), lack of gluten must be counteracted by employing ingredients that help to overcome loss of extensibility and elasticity. The literature points out that substances that swell in water (i.e., hydrocolloids), can be utilized to mimic viscoelastic gluten properties by improving acceptability, structural mouthfeel, and shelf life [28]. Hydrocolloids' film-forming properties can also act as a lubricant in batters and help prevent damage on other formulation ingredients caused by mixing, especially starch granules [29]. The structure of these hydrophilic molecules is variable (linear, branched, with/without chain flexibility) and may interfere with gluten development in relation to their chemical structure. Previously, research has reported that adding hydrocolloids can lead to strong intermolecular hydrogen bonding between the OH groups of gluten proteins and polysaccharide [30–32]. Carboxymethylcellulose (CMC), a derivative of cellulose, is a widespread thickening agent employed to modify the viscosity of some food matrices like cake mixes, dairy products, and jellies [16]. Adding CMC (soluble fiber) to cereal-based food has beneficial effects on fasting plasma cholesterol and blood glucose regulation [33]. Non-starch polysaccharides, such as locust bean gum (LBG) and xanthan gum (XG), strongly affect pasta viscoelastic properties. They can be utilized to improve not only its elastic texture but also the mouthfeel and firmness of end products [18]. As far as the authors know, no research is available about evaluating these hydrocolloids being employed to develop fresh egg pasta based on durum wheat semolina (DWS) and TNF.

As reported by [34], response surface methodology (RSM) is a statistical technique that employs quantitative data acquired from suitable experimental designs to establish and simultaneously solve multivariate equations [35]. This tool is effective in optimizing complex processes and has many applications in several food operations [36–40]. By employing tiger nut as a potential source of food nutrients (emphasis is especially placed on the quantity of fiber), the work reported here was done to search the optimum CMC–XG–LBG combination to obtain a high quality "source of fiber", namely (>3%) tiger nut-based fresh pasta. The most relevant technological pasta properties (consistency, firmness, elasticity, color attributes, cooking loss, water absorption index, swelling index) were assessed. The results may offer a basis for developing fresh tagliatelle using tiger nut-DWS blends with the desired quality and enriched fiber values.

2. Materials and Methods

2.1. Raw Materials

Commercial TNF and DWS (65% extraction) were respectively supplied by Tigernuts Traders S.L. (L'Eliana, Valencia, Spain) and Harinas Villamayor S.A. (Huesca, Spain). Avícola Llombai S.A. (Llombai, Valencia, Spain) supplied the pasteurized liquid egg (LE). Hydrocolloids (carboxymethylcellulose CMC-3500-4000 cps, locust bean gum LBG-2800 cps, and xanthan gum XG-1400 cps) were, respectively, supplied by Quimica Amtex, S.A. (Mexico City, Mexico), Lbg Sicilia Srl. (Ragusa, Italy) and Shandong Fufeng Fermentation Co., Ltd. (Linyi, China). The same batch formed by the above materials was employed in all the experiments. Raw materials were examined for moisture content, protein, fat, crude fiber, and ash according to AACC methods 44–40.01, 46–10.01, 30–20.01, 32–10.01, and 08–01.01 [41]. The proximate chemical composition of both flours, LE, and hydrocolloids (suppliers sent the data) are summarized in Table 1.

Table 1. Proximate chemical composition of: tiger nut flour (TNF); durum wheat semolina (DWS); liquid egg (LE) (g/100 g); hydrocolloids (CMC, XG, LBG). The mean values of three replicates are provided (standard deviation) for TNF, DWS, and LE.

	DWS	TNF	LE	CMC *	XG *	LBG *
Water	13.67 (0.03)	8.83 (0.05)	79 (2)	10	15	12
Protein	13.2 (0.7)	4.95 (0.07)	11.4 (0.6)	-	-	-
Fat	0.90 (0.05)	25.07 (0.02)	7.83 (0.98)	-	-	-
Ash	1.71 (0.07)	2.05 (0.04)	0.60 (0.04)	-	13	1
Dietary Fiber	10.00 (0.02)	15.85 (0.03)	-	90	72	87
DC **	60.54 (0.02)	43.25 (0.03)	0.91 (0.02)	-	-	-

* Suppliers provided the data. ** Digestible carbohydrates were calculated by the difference.

2.2. Experimental Design

The effect of the different factor combinations (three independent variables: CMC, XG, LBG) on the various response variables (randomized RSM) was evaluated by a Box–Behnken design with a quadratic model. The included variables were: nutritional losses during cooking (%P, protein; %F, fat; %M, minerals; %CF, crude fiber; and %DC, digestible carbohydrates), cooking loss (%CL), swelling index (%SI), water absorption index (WAI), CIEL*a*b* color coordinates, chrome (C^*_{ab}), tone (h^*_{ab}), firmness (F), consistency (A), and elasticity (S_i). Both the experimental design and statistical analysis were carried out by version 16.1.17 of the Statgraphics® Centurion XVI statistical software (StatPoint Technologies, Inc., Warrenton, VA, USA, 2011). Both the upper and lower limits of the factor levels were selected after contemplating the preliminary trials (data not provided). Their range went from 0 to 0.8% *w*/*w* (coded as 0 = 0%, 1 = 0.4%, 2 = 0.8%). Fifteen trials were run with three replicates of the central point (Table 2). A multiple regression analysis was followed to assess the significance of the linear, quadratic and interactive effects of factors (CMC, XG, LBG amounts) on the response variables. These parameters were measured in both the uncooked (subscript o) and cooked (subscript c) pasta samples. A second-order polynomial equation describes the regression model (Equation (1)), and every response variable (Y) is associated with the obtained linear (β_i), quadratic (β_{ii}), and interactive (β_{ij}) regression coefficients, i.e., to the relative weight of every analyzed effect (G_1-CMC, G_2-XG, and G_3-LBG, alone or combined). Constant β_o represents the response if no gum was taken into account.

$$Y = \beta_o + \sum_{i=1}^{3} \beta_i \cdot G_i + \sum_{i=1}^{3} \beta_{ii} \cdot G_i^2 + \sum_{i=1}^{3} \sum_{j>1}^{3} \beta_{ij} \cdot G_i \cdot G_j \qquad (1)$$

To better visualize the overall trends, 3-dimensional graphs were employed for the models. Non-significant terms were not included in the model equations to obtain these plots. All the formulations were performed in duplicate.

The basic pasta dough formulation was achieved by mixing tap water (16% *w/w*), DWS (71% *w/w*), and pasteurized LE (13% *w/w*). The quantity of added water was adjusted in earlier tests to obtain dough that was easy to handle and process. TNF was included in recipes at the 42.6% DWS replacement level (*w/w*). This gave a product with a fiber content of about 4%, which was labeled as "source of fiber" (>3 g dietary fiber/100 g food) according to the Nutritional Requirements for Dietary Fiber Foods [42]. The chemical composition of the raw materials was considered to estimate fiber content.

Table 2. Response variable values for the various pasta formulations that correspond to the experimental design in accordance to the levels of the three gum concentrations (CMC, carboxymethylcellulose; LBG, locust bean gum; XG, xanthan gum).

	Factors *			Response Variable **																					
Run Order	CMC	XG	LBG	Y_{Fo}	Y_{Sio}	Y_{Ao}	Y_{Fc}	Y_{Sic}	Y_{Ac}	Y_{WI}	$Y_{\%CL}$	$Y_{\%SI}$	Y_{L^*o}	Y_{a^*o}	Y_{b^*o}	$Y_{C^*ab,o}$	$Y_{h^*ab,o}$	Y_{L^*c}	Y_{a^*c}	Y_{b^*c}	$Y_{C^*ab,c}$	$Y_{h^*ab,c}$	%P	%L	%A	%FL
1	+1	+1	0	6.15	0.48	18.38	1.72	0.21	5.89	1.39	2.424	1.69	55.89	4.82	19.37	19.96	1.33	62.00	2.06	13.67	13.83	1.42	−0.13	−0.16	−0.15	−0.54
2	−1	+1	0	19.59	1.41	30.64	1.98	0.24	6.75	1.10	6.64	1.28	55.69	5.09	18.58	19.26	1.30	62.92	2.03	14.07	14.22	1.43	−0.13	−0.06	−0.22	−0.50
3	0	−1	+1	14.01	1.13	21.69	1.23	0.18	4.44	1.18	8.11	1.36	56.09	5.32	20.56	21.24	1.32	61.42	2.55	14.94	15.16	1.40	−0.09	−0.002	−0.15	−0.51
4	+1	−1	0	14.89	1.34	22.34	1.27	0.19	4.99	1.17	–	1.68	56.25	4.93	20.17	20.76	1.33	56.72	1.95	12.08	12.24	1.41	−0.10	0.03	−0.08	−0.49
5	0	0	0	5.29	0.36	10.68	1.47	0.24	5.03	1.26	6.84	1.49	55.36	4.79	20.21	20.77	1.34	61.18	2.23	13.71	13.89	1.41	−0.11	−0.04	−0.14	−0.46
6	0	+1	+1	7.62	0.57	22.68	1.87	0.21	6.26	1.34	5.61	1.69	56.23	4.67	19.40	19.96	1.33	62.86	1.50	12.75	12.84	1.45	−0.10	−0.08	−0.17	−0.54
7	0	0	0	5.30	0.09	10.76	1.47	0.22	4.75	1.23	7.08	1.52	56.43	4.94	20.86	21.43	1.34	60.71	2.39	14.52	14.72	1.41	−0.06	−0.19	−0.18	−0.50
8	−1	0	−1	12.87	0.63	15.59	1.62	0.27	5.77	1.20	7.67	1.37	56.61	5.26	19.73	20.42	1.31	61.55	2.16	13.22	13.39	1.41	−0.16	−0.12	−0.18	−0.39
9	−1	−1	0	9.07	0.34	11.40	1.50	0.20	5.71	1.15	6.95	0.94	54.63	4.87	20.25	20.83	1.34	61.17	2.47	14.44	14.65	1.40	−0.11	−0.06	−0.18	−0.40
10	0	+1	−1	9.89	1.35	21.26	1.92	0.25	7.32	1.16	6.25	1.39	58.24	4.37	18.39	18.90	1.34	62.98	2.12	13.86	14.03	1.42	−0.10	−0.14	−0.13	−0.51
11	+1	0	−1	7.78	0.87	18.05	1.57	0.22	6.20	1.31	7.12	1.59	59.15	3.91	17.81	18.23	1.36	61.69	2.35	14.28	14.48	1.41	−0.13	0.07	−0.04	−0.53
12	0	−1	−1	11.60	0.67	11.84	1.28	0.21	4.77	1.15	–	1.30	56.65	5.13	20.29	20.93	1.32	64.55	1.97	14.24	14.38	1.43	−0.09	−0.13	−0.24	−0.43
13	0	0	+1	6.75	0.59	17.71	1.54	0.22	5.25	1.14	6.60	1.48	53.94	5.69	20.97	21.73	1.31	60.38	2.05	13.37	13.53	1.42	−0.10	−0.15	−0.06	−0.57
14	+1	0	+1	6.34	0.45	12.68	1.39	0.19	4.27	1.22	5.60	1.52	56.06	4.91	20.39	20.97	1.33	60.73	1.96	13.53	13.68	1.43	−0.12	−0.17	0.0005	−0.53
15	0	0	0	5.67	0.40	11.82	1.49	0.24	4.71	1.18	7.23	1.57	53.84	5.55	21.26	21.97	1.32	60.96	2.08	13.77	13.92	1.42	−0.11	−0.09	−0.13	−0.52

* Factors CMC, XG, and LBG stand for carboxymethylcellulose; xanthan gum; locust bean gum; −1 = 0% *w/w*; 0 = 0.4% *w/w*; +1 = 0.8% *w/w*;.** Response variables Y_{Fo}, Y_{Fc}, Y_{Sio}, Y_{Sic}, Y_{Ao}, Y_{Ac}, Y_{WI}, $Y_{\%CL}$, $Y_{\%SI}$, Y_{L^*o}, Y_{a^*o}, Y_{b^*o}, $Y_{C^*ab,o}$, $Y_{h^*ab,o}$, Y_{L^*c}, Y_{a^*c}, Y_{b^*c}, $Y_{C^*ab,c}$, $Y_{h^*ab,c}$ stand for elasticity (S_i), firmness (F), consistency (A), water absorption index (WI), cooking loss (%CL), swelling index (%SI), lightness (L*), redness (color coordinate a*), yellowness (color coordinate b*), chrome (C^*_{ab}), hue angle (h^*_{ab}), losses of proteins (%P), lipids (%L), minerals (%A), and fiber (%FL), respectively. Subscripts *o* and *c* refer to the uncooked and cooked pasta samples, respectively.

2.3. Pasta Preparation

After weighing (0.001 g accuracy, PFB 300-3, Kern & Sohn GmbH, Balingen), both the liquid (egg/water) and dry (DWS/TNF/gums) ingredients were premixed in an electric cooking device (Thermomix TM-31, Vorwerk Spain M.S.L., S.C., Madrid, Spain) at medium speed, mixing egg and water for 15 s, adding gums to be mixed for 40 s, and then incorporating DWS/TNF powders and kneading for 45 s more yielded a uniform blend. The blends were kneaded in the same cooking device for 2.5 min and then placed inside plastic bags for 20 min for sample relaxation purposes. Then, tagliatelle was made with a pasta-making device (Simplex SP150, Imperia, Italy) coupled to a specific motor (A2500, Imperia, Italy). Dough was laminated by passing it between rollers 5 times before gradually narrowing the gap between rollers to make 1-mm-thick sheets, which were cut into 4-mm-wide tagliatelle. Tagliatelle was left to stand for 10 min to prevent stickiness before cooking began. A temperature of 20 °C was maintained while preparing and analyzing dough. Tagliatelle samples were made to be immediately tested for their mass, dimensions (volume), water content, mechanical properties, and color attributes (see the analysis explained below). There were three replicates (5 for mechanical properties) per pasta formulation.

2.4. Pasta Cooking

The cooking trial for each pasta formulation was done in triplicate. Cooked pasta was prepared by boiling 25 g of 7-cm-long samples in 300 mL of deionized water. Water volume was left at 90% of its initial volume by adding boiling water and covering flasks to prevent loss of evaporation. At 4 min (optimal cooking time for 100% DWS fresh egg tagliatelle according to the AACC method 16–50 [41]), pasta was removed from flasks before quickly stopping the cooking process by adding 50 mL of cold deionized water. Next, pasta samples were drained for 2 min, weighed (0.001 g accuracy, PFB 300-3, Kern & Sohn GmbH, Balingen, Germany), and evaluated for their water absorption index (WI), cooking losses (%CL), swelling index (%SI, volume changes), proximate chemical composition, mechanical properties, and color attributes (analysis explained below).

2.5. Proximate Chemical Composition of Both Cooked and Uncooked Pasta Samples

Cooked tagliatelle was analyzed for its water content, crude fiber, protein, ash, and fat according to AACC methods 44–40.01, 46–10.01, 30–20.01, 32–10.01, and 08–01.01 [41]. Digestible carbohydrates were calculated by difference. There were three replicates per formulation. Moisture content was immediately analyzed after cooking; for the other chemical measurements, cooked pasta was freeze-dried (Telstar, Lyoalfa-6, Azbil, Spain) for 24 h at 0.1 mbar and stored at room temperature in sealed polyethylene bags until further analyses. The proximate chemical composition of the raw materials was employed to calculate that of the uncooked pasta samples to know the corresponding percentage losses caused by cooking.

2.6. Pasta Technological Properties

The water absorption index (WAI, g/g) was calculated from both mass gain and increased water content after cooking. Cooking loss (quantity of solid substance lost to cooking water; %CL) was determined by the AACC-approved method 16–50 [41], with some modifications. After cooking, both the cooking and rinse waters were collected and left in an aluminum container to be evaporated to dryness by two steps: placing in an air oven at 100 °C to reduce 2/3 volume and freeze-drying (Telstar, Lyoalfa-6, Azbil, Spain). The residue was weighed and indicated as a percentage of starting material. There were three replicates per formulation. Tagliatelle dimensions (thickness, length, width) were taken using a caliper (PCE-DCP 200N, PCE Ibérica S.L., Albacete, Spain).

Pasta swelling (%SI) was expressed as the relative volume changes between the cooked and uncooked samples. There were three replicates per formulation.

Tagliatelle color measurements were taken over the surface reflectance spectra obtained by a spectrocolorimeter (Minolta CM-3600D) from 400 to 700 nm (iluminant D65, 10° standard observer) on a white background. Determinations were made for all the pasta formulations in triplicate both before and after cooking (0 and 4 min). The CIEL*a*b* color coordinates L* (lightness), a* (redness-greenness), and b* (yellowness-blueness) were obtained from the reflectance spectra, and the results were expressed in terms of chromatic magnitudes: color saturation ($C^*_{ab} = \sqrt{a^{*2} + b^{*2}}$) and hue angle ($h^*_{ab} = arctg\frac{b^*}{a^*}$).

A Texture Analyzer (TA.XT2, Stable Micro Systems, Godalming, UK), coupled to a PC with data acquisition and version 1.22 of the Texture Exponent software (Stable Micro Systems), was employed to determine the mechanical properties. Tests were run in accordance with the AACC Method 16–50 [41]. Five 7-cm-long adjacent strands were cut using the A/LKB-F cutting probe at 0.17 mm/s until total sample deformation was achieved. A 5-kg load cell was employed. At least five replicates for the uncooked and cooked pasta were obtained and also for all the pasta formulations. Cooked samples were analyzed just after the cooking procedure, as described in Section 2.4. To evaluate changes in pasta texture while cooking, three parameters were taken into account: (i) force needed to cut tagliatelle (F) as a measure of firmness; (ii) the area compressed under the force-time curve (A) from the initial test time to the maximum cut force, which represents dough consistency. (iii) The initial slope of the force-time curve (S_i), which is related to the elasticity modulus, offers an idea of products' solid nature.

2.7. Statistical Analysis

Version 16.1.17 of the Statgraphics® Centurion XVI.I statistical software (StatPoint Technologies, Inc., 2011) was employed to fit the multiple regression models to the experimental data. This enabled the linear, quadratic, and interactive effects of hydrocolloids CMC, XG, and LBG on the selected dependent variables to be evaluated ($p < 0.05$). This statistical software was also used to produce surface response plots.

3. Results and Discussion

The experimental values for the cooking, optical and mechanical properties, and the chemical changes due to cooking for each experimental run are presented in Table 2. The results of the 15 runs were fitted to a second-order polynomial equation (Equation (1)). The removal of the non-significant terms ($p < 0.05$) was considered (stepwise regression).

The fitted model's goodness was assessed by an analysis of variance (ANOVA; Table 3), based mostly on probability (p-value) and the Fisher variation test (F-value), to gain a measurement of how much variability in the observed response values can be explained by the experimental factors and their interactions [43]. A p-value less than 0.0500 indicates that the model is statistically significant; therefore, only models where this value was greater than 0.0500 are shown in Table 3.

Table 3. Analysis of variance of response surface models.

Variables	Sources Of Variations	SS	DF	MS	F-Value	p-Value
F_o (N)	Model	211.398	3	70.466	18.85	0.0001
	Residual	41.131	11	3.739		
	Corrected total	252.529	14			
F_c (N)	Model	0.687	2	0.344	61.58	<0.0001
	Residual	0.067	12	0.006		
	Corrected total	0.754	14			
S_{ic} (N/s)	Model	0.006	3	0.002	14.63	0.0004
	Residual	0.002	11	0.0001		
	Corrected total	0.008	14			
A_c (N·s)	Model	9.012	2	4.506	26.12	<0.0000
	Residual	2.071	12	0.173		
	Corrected total	11.083	14			

Table 3. *Cont.*

Variables	Sources Of Variations	SS	DF	MS	F-Value	*p*-Value
WAI (g/g)	Model	0.057	1	0.057	20.82	0.0005
	Residual	0.036	13	0.003		
	Corrected total	0.092	14			
%CL	Model	22.383	4	5.596	31.57	0.0001
	Residual	1.418	8	0.177		
	Corrected total	23.802	12			
%SI	Model	0.361	2	0.180	12.79	0.0011
	Residual	0.169	12	0.014		
	Corrected total	0.530	14			
%FL	Model	0.031	4	0.008	15.14	0.0003
	Residual	0.005	10	0.001		
	Corrected total	0.037	14			

SS, sum of squares; DF, degree of freedom; MS, mean square; F-value, Fisher test; *p*-value, probability.

A Student's *t*-test was run to analyze the significance of the parameters' regression coefficients. Table 4 provides the results obtained for the *t*-values, the corresponding *p*-values, and the parameter estimates.

Table 4. Regression results from the data.

Variables		Coefficient Estimate	Standard Error	95% Confidence Interval Low	95% Confidence Interval High	*t*−Value	*p*−Value
F_o (N)	β_o	8.561	0.93	6.52	10.60	9.25	<0.0001
	CMC*CMC	12.015	3.04	5.33	18.70	3.95	0.0023
	XG*XG	16.790	3.01	10.17	23.41	5.58	0.0002
	CMC*XG	−36.539	5.03	−47.622	−25.46	−7.26	<0.0001
F_c (N)	β_o	1.449	0.04	1.36	1.54	35.19	<0.0001
	CMC	−0.226	0.07	−0.38	−0.07	−3.19	0.0078
	XG*XG	0.822	0.08	0.65	0.99	10.51	<0.0001
S_{ic} (N/s)	β_o	0.209	0.01	0.19	0.22	32.13	<0.0001
	XG	0.132	0.03	0.06	0.20	4.04	0.0019
	XG*XG	−0.118	0.04	−0.20	−0.03	−3.06	0.0109
	CMC*LBG	−0.084	0.02	−0.12	−0.05	−5.01	0.0004
A_c (N·s)	β_o	5.275	0.187	4.868	5.683	28.19	<0.0001
	XG*XG	2.544	0.435	1.597	3.491	5.85	0.0001
	CMC*LBG	−2.340	0.585	−3.614	−1.065	−4.00	0.0018
WAI (g/g)	β_o	1.154	0.019	1.113	1.194	61.97	<0.0001
	CMC*XG	0.343	0.075	0.180	0.505	4.56	0.0005
%CL	β_o	7.084	0.243	6.523	7.645	29.142	<0.0001
	CMC	6.614	1.435	3.306	9.923	4.611	0.0017
	CMC*CMC	−3.505	1.492	−6.945	−0.065	−2.350	0.0467
	CMC*XG	−10.003	1.135	−12.620	−7.387	−8.817	0.0000
	CMC*LBG	−2.542	0.789	−4.361	−0.723	−3.222	0.0122
%SI	β_o	1.144	0.071	0.990	1.299	16.162	<0.0001
	CMC	0.501	0.113	0.264	0.756	4.524	0.0007
	XG	0.238	0.105	0.009	0.466	2.263	0.0429
%FL	β_o	−0.358	0.020	−0.402	−0.314	−18.089	<0.0001
	CMC	−0.171	0.036	−0.252	−0.090	−4.694	0.0008
	XG	−0.076	0.020	−0.121	−0.031	−3.760	0.0037
	LBG	−0.179	0.045	−0.280	−0.078	−3.956	0.0027
	CMC*LBG	0.221	0.088	0.024	0.418	2.504	0.0312

The model's insights can be also obtained from determination coefficients (see Tables 5 and 6). R^2 quantitatively evaluates the correlation between the experimental data and the predicted responses, while R^2_{adj} defines the satisfactory fit of the polynomial model to experimental data. In practice, a model can be considered fairly good for describing the influence of the variable(s) when the coefficient of determination (R^2) is at least 80% [43] or the R^2_{adj} values exceed 70% [44].

Table 5. Constant values (β_o) and significant coefficients (β) at the 95% confidence interval of the stepwise multiple regression model for mechanical properties firmness (F), elasticity (Si), and consistency (A).

		F_o (N)	F_c (N)	S_{ic} (N/s)	A_c (N·s)
	Constant (β_o)	8.561	1.449	0.209	5.275
β	CMC	*ns*	−0.226	*ns*	*ns*
	XG	*ns*	*ns*	0.132	*ns*
	LBG	*ns*	*ns*	*ns*	*ns*
	CMC*CMC	12.015	*ns*	*ns*	*ns*
	XG*XG	16.790	0.822	−0.118	2.544
	LBG*LBG	*ns*	*ns*	*ns*	*ns*
	CMC*XG	−36.539	*ns*	*ns*	*ns*
	CMC*LBG	*ns*	*ns*	−0.084	−2.340
	XG*LBG	*ns*	*ns*	*ns*	*ns*
	Lack of fit	41.037	0.06682	0.001716	2.0096
	Pure error	0.094	0.00018	0.000284	0.0614
	Lack of fit *p*-value	0.999	0.999	0.823	0.997
	R^2	83.71	91.121	79.97	81.317
	R^2adj	79.27	89.641	74.50	78.204
	Standard error of est.	1.93	0.075	0.012	0.415
	Mean absolute error	1.40	0.048	0.007	0.324
	Durbin–Watson statistic (*p*-value)	1.417 (0.126)	2.214 (0.637)	2.727 (0.919)	2.047 (0.485)

Independent variables: CMC (carboximethylcelullose); XG (xanthan gum); LBG (locust bean gum). Subscripts o and c refer to the uncooked and cooked pasta samples, respectively. Only the significant relations are shown. Analysis of variance at the 95% confidence level. *ns*, no significant effect at level < 5%.

3.1. Mechanical Properties of the Uncooked and the Cooked Fresh Egg Pasta

The instrumental parameters of elasticity, firmness, and stickiness may be associated with consumer pasta acceptability. The expected high-quality cooked pasta should display good texture, resist stickiness and surface disintegration, and have a firm but consistent and elastic structure ("al dente"). Table 5 summarizes the estimated regression coefficients (Y_{Fo}, Y_{Fc}, Y_{Sic}, Y_{Ac}) of the second-order model, which were obtained for the mechanical properties of the uncooked/cooked tagliatelle and include the fitted parameters from the ANOVA. The predictive models developed for not only the firmness of the uncooked (F_o) and cooked (F_c) pasta but also for cooked pasta consistency (A_c) and elasticity (S_{ic}) were deemed suitable because the model significance and the R^2_{adj} values levels were satisfactory. The lack-of-fit parameter was always non-significant ($p > 0.05$), and the Durbin–Watson statistic *p*-value exceeded 0.05, which meant no indication of serial autocorrelation in the residuals at the 5% significance level.

Figure 1 depicts the response surface plots for the various mechanical parameters of the uncooked (a) and cooked pasta (b–f). The β and p values in Tables 4 and 5 show that the presence of XG significantly and positively influenced the mechanical fresh pasta properties at the tested concentration range. The firmness of the uncooked and cooked pasta quadratically rose with XG concentration. This impact was much stronger before cooking. Figure 1a,b show that the addition of 0.8% XG to the tiger-nut-based pasta brought about increases in the firmness of the uncooked (F_o) and cooked (F_c) tagliatelle pieces of 125.52% and 36.31% (values calculated from models), respectively. XG has been reported to enhance the firmness of bran-enriched spaghetti [45], composite semolina-flaxseed spaghetti [46], and GF tiger nut noodles [1]. Cooked pasta elasticity (Figure 1c), particularly consistency (Figure 1e), also improved (a maximum rise of 30.87% at the XG 0.8% concentration for consistency and one of 17.66% at the XG 0.56% concentration for elasticity). These results show the possibility of obtaining a better structure with continuous protein matrix entrapping starch granules, which absorb water and gelatinize with no major losses due to cooking. Soluble gums, such as CMC, LBG, and XG, have the potential to affect the internal pasta structure because of their interaction with starch and protein. The authors of ref [18] put forward the notion that forming a network by soluble fiber around starch granules could result in better cohesiveness in a pasta structure between protein and starch. In [5], the authors report that the rheological behavior of a tiger nut–wheat semolina composite dough was impacted by this XG at a 1% concentration because a more cohesive structure was obtained. Therefore, adding XG to formulations helps to enhance dough resistance to deformation. After hydrating this hydrocolloid, it can fill up any free space in the system, which makes dough structure stronger. As a previous study reports [47], adding up to 1% XG to corn-bean pasta results in a more compact internal structure with visible starch granule agglomerates embedded in the fibrous protein-gum matrix. In line with these results, it would be interesting to employ a XG concentration of approximately 0.6% to achieve improved textural characteristics to approach the sought "al dente" point.

Table 6. Constant values (Y_0) and significant coefficients (β) at 95% confidence interval of the stepwise multiple regression for cooking properties water absorption index (WAI), cooking loss (%CL), and swelling index (%SI).

		WAI (g/g)	%CL	%SI	%FL
	Constant (β_o)	1.154	7.084	1.144	−0.358
β	CMC	*ns*	6.614	0.510	−0.171
	XG	*ns*	*ns*	0.238	−0.076
	LBG	*ns*	*ns*	*ns*	−0.179
	CMC*CMC	*ns*	−3.505	*ns*	*ns*
	XG*XG	*ns*	*ns*	*ns*	*ns*
	LBG*LBG	*ns*	*ns*	*ns*	*ns*
	CMC*XG	0.343	−10.003	*ns*	*ns*
	CMC*LBG	*ns*	*ns*	*ns*	0.221
	XG*LBG	*ns*	−2.542	*ns*	*ns*
	Lack of fit	0.032733	1.3395	0.165749	0.003120
	Pure error	0.003267	0.0785	0.003369	0.002074
	Lack of fit *p*-value	0.244	0.135	0.096	0.870
	R^2	61.56	94.04	68.07	85.83
	R^2adj	58.61	91.06	62.75	80.16
	Standard error of est.	0.052	0.421	0.119	0.023
	Mean absolute error	0.040	0.289	0.09	0.014
	Durbin−Watson statistic (*p*-value)	1.676 (0.250)	1.737 (0.285)	1.552 (0.185)	1.553 (0.186)

Independent variables: CMC (carboximethylcelullose); XG (xanthan gum); LBG (locust bean gum). Subscripts o and c refer to the uncooked and cooked pasta samples, respectively. Only significant relations are shown. Analysis of variance at the 95% confidence level. *ns*, no significant effect at level < 5%.

Uncooked pasta firmness also improved (up to 47.32%) when only CMC was employed and had a quadratic positive effect (β value in Table 5; Figure 1c). This parameter slightly decreased after cooking (linear negative effect; β value in Table 5) after adding carboxymethylcellulose (with 12.48% at the 0.8% concentration). A significant and negative CMC and XG interaction was noted (Table 5; Figure 1a). Thus, uncooked pasta firmness significantly diminished when both hydrocolloids were combined. LBG affected only cooked pasta's consistency and elasticity when combined with CMC (Table 5; Figure 1d,f) but did not affect the mechanical response when employed alone.

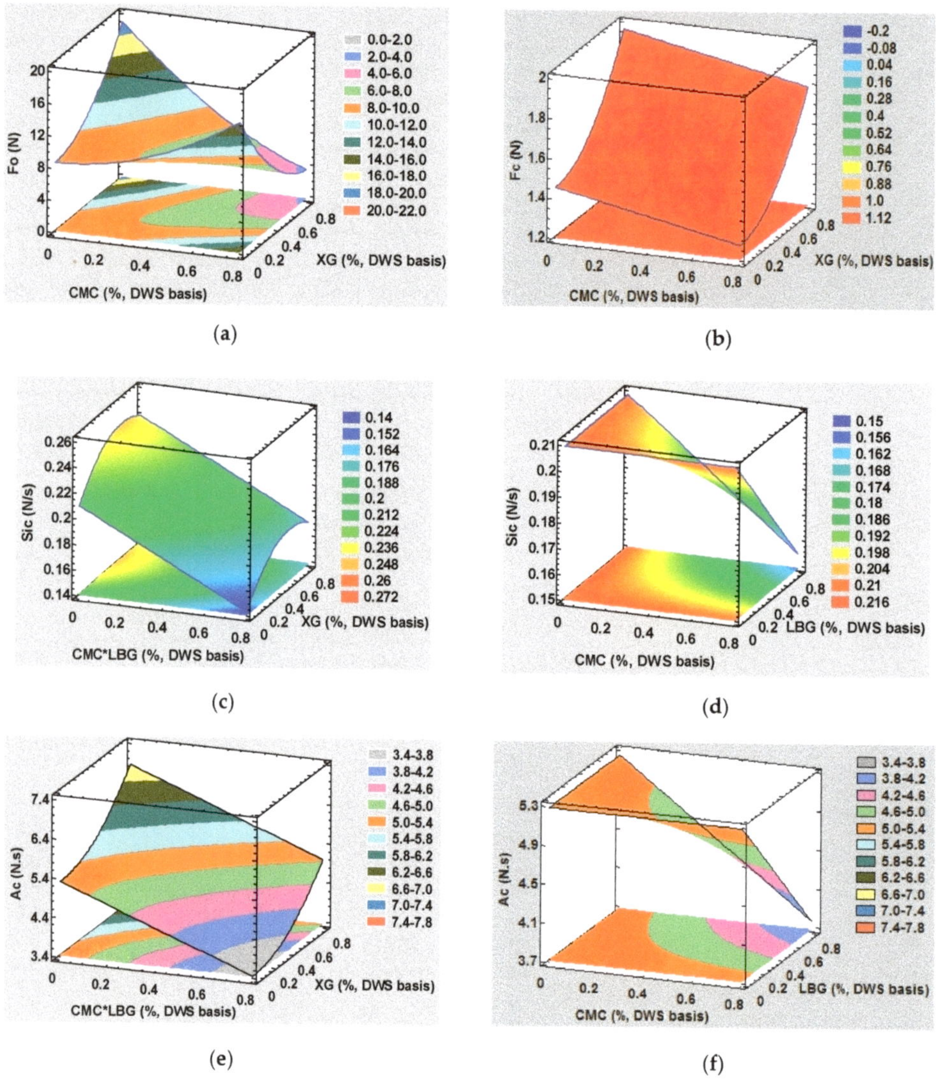

Figure 1. (**a**–**f**). Effect of the CMC, GX, and LBG levels on the uncooked and cooked firmness (F_o and F_c, respectively) and on cooked elasticity (S_{ic}) and consistency (A_c). CMC, carboximethylcelullose; XG, xanthan gum; LBG, locust bean gum.

We can conclude from these results that only XG implies a better, ready-to-eat tagliatelle texture within the test range, and the combination of CMC and LBG or XG ought to be avoided. No synergistic effect was observed between LBG and XG.

3.2. Cooking Quality and Color Attributes of Uncooked and Cooked Fresh Egg Pasta

Table 6 presents the regression summary and ANOVA for cooking quality (WAI, %CL, and %SI) and fiber loss during cooking (%FL). At the 95% confidence level, %CL and %FL proved to be significant variables for model construction (R^2_{adj} > 70%). The model was less suitable for the swelling index (%SI, R^2_{adj} = 62.75%) and the water absorption index (WAI, R^2_{adj} = 61.56%). However, an explanatory data analysis was run, which gave a reasonable initial solution for describing the tendency of these parameters.

The model developed for %SI and WAI was less predictive with an R^2_{adj} of 61.56% and 62.75%, respectively. This can be partly explained by the experimental response variables' narrow range (0.1–1.39 for WAI; 0.94–1.69 for %SI). Regression coefficients (Table 6) and surface plots (Figure 2a,b) were generated for the use models, as they give a reasonable initial solution for describing the quality response of both %SI and WAI. CMC (β = 0.510) and XG (β = 0.238) displayed a linear positive effect on swelling capacity, and a significant synergetic effect on WAI of both hydrocolloids was found (Figure 2a). This tendency is supported by greater swelling possibly being related to a large quantity of water bonded to proteins and starch because WAI displayed a similar pattern. CMC did not affect WAI, like other authors have found when substituting 0.25–1.5% wheat semolina for this hydrocolloid [22]. Cooking loss (%CL) is a pasta-quality measure that expresses resistance to disintegration when boiling. Figure 2c and 2d depict a drastic drop in %CL when combining XG and CMC (β = −10.003) or LBG (β = −2.542) in the pasta formula and, respectively, resulted in less cooking loss of 47.35% or 22.97% when these gums were employed at the 0.8% concentration. The obtained results also revealed that CMC increased the cooking loss by 43.03% when used at 0.8% concentration. However, it was not affected when XG or LBG was employed alone within the test range. From this viewpoint, the XG and CMC combination at 0.8% is recommendable. For wheat-based pasta, %CL is dependent on the degree of starch gelatinization and the strength of the retrograded starch network that surrounds gelatinized starch [48]. Solid loss while cooking is due mainly to the solubilization of loosely bound gelatinized starch from the product's surface [48]. In non-conventional pasta, starch polymers are entrapped less efficaciously in the matrix, which confers products a high CL, as expected given the concordance between the lower cooking loss and the better mechanical response obtained when XG was added. A significant difference in digestible carbohydrate losses while cooking is also expected. However, fiber loss during cooking was the only significant chemical component for model construction (of those assessed) (Table 6, Figure 2e,f). The ready-to-eat product had a final fiber content that went from 3.1 (0.6) for trial 13 to 3.6 (0.5) for trial 15. These results allow it to be labeled as a "source of fiber" (>3 g dietary fiber/100 g food) in line with Nutritional Requirements for Dietary Fiber Foods [42].

The color parameters were not statistically related ($p < 0.05$) to the hydrocolloids used within the test range.

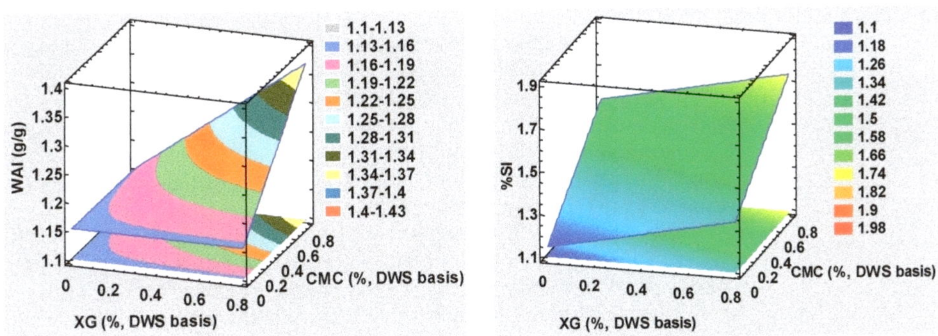

(a) (b)

Figure 2. *Cont.*

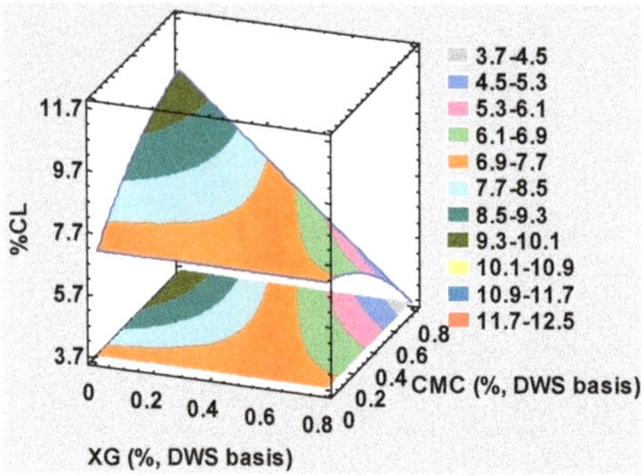

(c)

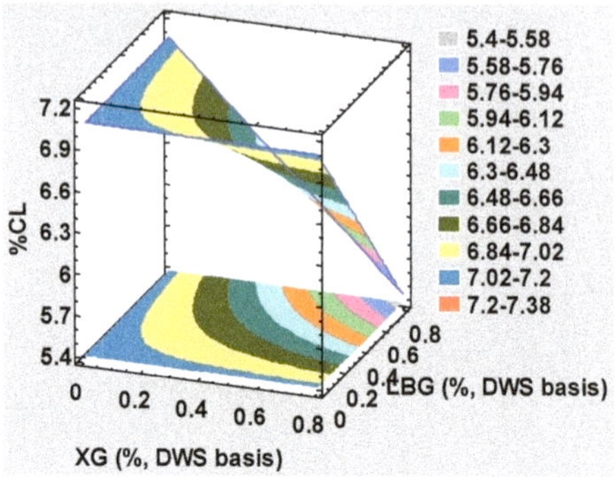

(d)

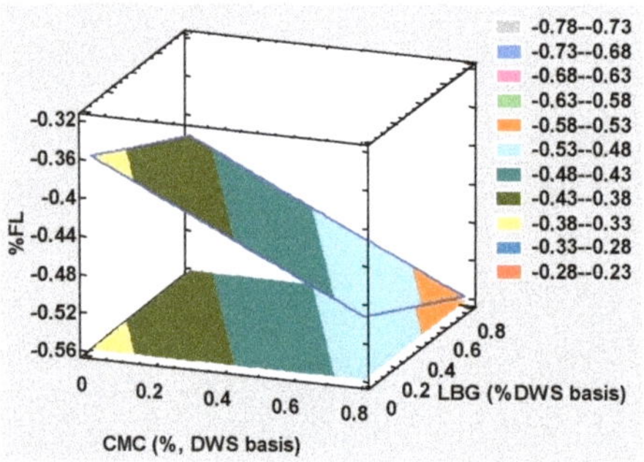

(e)

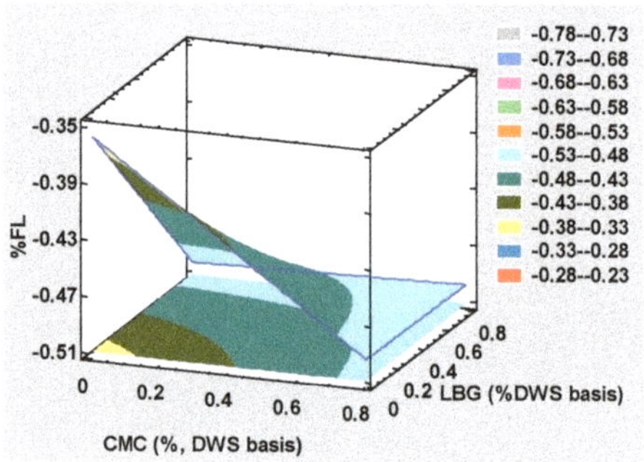

(f)

Figure 2. (**a–f**). Effect of the CMC, GX, and LBG levels on water absorption index (WAI), the swelling index (%SI), cooking loss (%CL), and fiber loss (%FL). CMC, carboximethylcelullose; XG, xanthan gum; LBG, locust bean gum.

4. Conclusions

The results from this manuscript address the improved nutritional value and fair techno-functional properties obtained with fresh tiger nut-based tagliatelle when XG was employed as a structural agent. Marked fiber and fat enhancement (rich in oleic and linoleic acids) contents, along with mineral enrichment, may be attained in tiger nut pasta. XG at 0.8% concentration considerably improved the textural characteristics and, accordingly, fresh pasta's cooking behavior. This means that a better structure with a continuous protein matrix to entrap starch granules is feasible.

It was not possible to accomplish an adequate hydrocolloid combination (CMC, XG, and LBG) within the test range (0–0.8%) with the RSM analysis. Nonetheless, the obtained results showed that employing XG at a concentration of about 0.6% would be interesting for obtaining better ready-to-eat, fresh pasta textural characteristics. Combining this gum with CMC at 0.8% can considerably reduce cooking losses while cooking. The cooked pasta can be labeled as a "source of fiber" (>3 g dietary fiber/100 g food) in all cases.

Author Contributions: Conceptualization, M.E.M.-E. and A.A.; funding acquisition, M.E.M.-E.; methodology, M.D.G.-M., M.D.R. and A.A.; supervision, M.E.M.-E.; writing—original draft, M.E.M.-E.; writing—review and editing, A.A. and M.E.M.-E. All authors have read and agreed to the published version of the manuscript.

Funding: This research was funded by the Conselleria de Empresa, Universidad y Ciencia, Generalitat Valenciana (Spain) throughout the project AICO/2016/056.

Institutional Review Board Statement: Not applicable.

Informed Consent Statement: Not applicable.

Acknowledgments: Also: authors express sincere thanks to Harinas Villamayor S.A.; Tigernuts Traders, S.L.; and Avícola Llombai S.A. for providing raw materials for conducting this study.

Conflicts of Interest: The authors declare no conflict of interest.

References

1. Gasparre, N.; Rosell, C.M. Role of hydrocolloids in gluten free noodles made with tiger nut flour as non-conventional powder. *Food Hydrocoll.* **2019**, *97*, 105194. [CrossRef]
2. Ukwuru, M.U.; Ibeneme, C.L.; Agbo, C.I. New Product Development from Tigernut (*Cyperus esculentus*) and Their Sensory, Proximate and Microbiological Evaluation. *Pak. J. Nutr.* **2011**, *10*, 101–105. [CrossRef]
3. Sánchez-Zapata, E.; Fernández-López, J.; Pérez-Alvarez, J.A. Tiger Nut (*Cyperus esculentus*) Commercialization: Health Aspects, Composition, Properties, and Food Applications. *Compr. Rev. Food Sci. F* **2012**, *11*, 366–377. [CrossRef]
4. Alegría-Torán, A.; Farré-Rovira, R. Horchata y Salud: Aspectos Nutricionales y Dietéticos. In *Jornada Chufa y Horchata: Tradición y Salud*, 1st ed.; Fundación Valenciana de Estudios Avanzados, Ed.; Consellería de Agricultura, Pesca y Alimentación: Valencia, Spain, 2003; pp. 55–70.
5. Martín-Esparza, M.E.; Raigón, M.D.; Raga, A.; Albors, A. Functional, thermal and rheological properties of high fibre fresh pasta: Effect of tiger nut flour and xanthan gum addition. *Food Bioprocess Tech.* **2018**, *11*, 2131–2141. [CrossRef]
6. Chinma, C.E.; Abu, J.O.; Abubakar, Y.A. Effect of tigernut (*Cyperus esculentus*) flour addition on the quality of wheat-based cake. *Int. J. Food Sci. Tech.* **2010**, *45*, 1746–1752. [CrossRef]
7. Albors, A.; Raigon, M.D.; Garcia-Martinez, M.D.; Martin-Esparza, M.E. Assessment of techno-functional and sensory attributes of tiger nut fresh egg tagliatelle. *Lebensm-Wiss. Technol.* **2016**, *74*, 183–190. [CrossRef]
8. Martín-Esparza, M.E.; Raigón, M.D.; Raga, A.; Albors, A. High fibre tiger nut pasta and xanthan gum: Cooking quality, microstructure, physico-chemical properties and consumer acceptance. *Food Sci. Biotechnol.* **2018**, *27*, 1075–1084. [CrossRef] [PubMed]
9. Aguilar, N.; Albanell, E.; Miñarro, B.; Guamis, B.; Capellas, M. Effect of tiger nut-derived products in gluten-free batter and bread. *Food Sci. Technol. Int.* **2014**, *21*, 323–331. [CrossRef]
10. Demirkesen, I.; Sumnu, G.; Sahin, S. Quality of Gluten-Free Bread Formulations Baked in Different Ovens. *Food Bioprocess Tech.* **2013**, *6*, 746–753. [CrossRef]
11. Foschia, M.; Peressini, D.; Sensidoni, A.; Brennan, C.S. The effects of dietary fibre addition on the quality of common cereal products. *J. Cereal Sci.* **2013**, *58*, 216–227. [CrossRef]
12. Mastromatteo, M.; Chillo, S.; Civica, V.; Iannetti, M.; Suriano, N.; Del Nobile, M.A. A multistep optimization approach for the production of healthful pasta based on nonconventional flours. *J. Food Process Eng.* **2012**, *35*, 601–621. [CrossRef]
13. RI. Fresh Pasta Market to 2025—Global Analysis, Industry Growth, Regional Share, Trends, Competitor Analysis, January 2020. Available online: https://www.reportsinsights.com/industry-forecast/fresh-pasta-market-analysis-1605 (accessed on 12 October 2020).
14. Arribas, C.; Cabellos, B.; Guillamón, E.; Pedrosa, M.M. Cooking and sensorial quality, nutritional composition and functional properties of cold-extruded rice/white bean gluten-free fettuccine fortified with whole carob fruit flour. *Food Funct.* **2020**, *11*, 7913. [CrossRef] [PubMed]
15. Borneo, R.; Aguirre, A. Chemical Composition, Cooking Quality, and Consumer Acceptance of Pasta made with Dried Amaranth Leaves Flour. *Lebensm-Wiss. Technol.* **2008**, *41*, 1748–1751. [CrossRef]
16. Chillo, S.; Laverse, J.; Falcone, P.M.; Protopapa, A.; Del Nobile, M.A. Influence of the addition of buckwheat flour and durum wheat bran on spaghetti quality. *J. Cereal Sci.* **2008**, *47*, 144–152. [CrossRef]
17. Nilusha, R.A.T.; Jayasinghe, J.M.J.K.; Perera, O.D.A.N.; Perera, P.I.P. Development of pasta products with nonconventional ingredients and their effect on selected quality characteristics: A brief overview. *Int. J. Food Sci.* **2019**, *2019*, 1–10. [CrossRef] [PubMed]
18. Larrosa, V.; Lorenzo, G.; Zaritzky, N.; Califano, A. Optimization of rheological properties of gluten-free pasta dough using mixture design. *J. Cereal Sci.* **2013**, *57*, 520–526. [CrossRef]
19. Mastromatteo, M.; Chillo, S.; Iannetti, M.; Civica, V.; del Nobile, M.A. Formulation optimisation of gluten-free functional spaghetti based on quinoa, maize and soy flours. *Int. J. Food Sci. Tech.* **2011**, *46*, 1201–1208. [CrossRef]
20. Mercier, S.; Moresoli, C.; Mondor, M.; Villeneuve, S.; Marcos, B. A Meta-Analysis of Enriched Pasta: What Are the Effects of Enrichment and Process Specifications on the Quality Attributes of Pasta? *Compr. Rev. Food Sci. Food Saf.* **2016**, *15*, 685–704. [CrossRef]
21. Wahanik, A.L.; Neri-Numa, I.A.; Pastore, G.M.; Chang, Y.K.; Pedrosa, M.T. Turmeric (*Curcuma longa* L.): New application as source of fiber and antioxidants in pasta with whole wheat flour. *Rev. Fac. Nac. Agron. Medellín* **2018**, *71*, 8423–8435. [CrossRef]
22. Aravind, N.; Sissons, M.; Fellows, C.M. Effect of soluble fibre (guar gum and carboxymethylcellulose) addition on technological, sensory and structural properties of durum wheat spaghetti. *Food Chem.* **2012**, *131*, 893–900. [CrossRef]
23. Aravind, N.; Sissons, M.; Egan, N.; Fellows, C. Effect of insoluble dietary fibre addition on technological, sensory, and structural properties of durum wheat spaghetti. *Food Chem.* **2012**, *130*, 299–309. [CrossRef]

24. Llavata, B.; Albors, A.; Martin-Esparza, M.E. High Fibre Gluten-Free Fresh Pasta with Tiger Nut, Chickpea and Fenugreek: Technofunctional, Sensory and Nutritional Properties. *Foods* **2020**, *9*, 11. [CrossRef]
25. Trevisan, S.; Pasini, G.; Simonato, B. An overview of expected glycaemic response of one ingredient commercial gluten free pasta. *Lebensm-Wiss Technol.* **2019**, *109*, 13–16. [CrossRef]
26. Mariotti, M.; Iametti, S.; Cappa, C.; Rasmussen, P.; Lucisano, M. Characterisation of gluten-free pasta through conventional and innovative methods: Evaluation of the uncooked products. *J. Cereal Sci.* **2011**, *53*, 319–327. [CrossRef]
27. Bustos, M.C.; Perez, G.T.; Leon, A.E. Combination of resistant starches types II and IV with minimal amounts of oat bran yields good quality, low glycaemic index pasta. *Int. J. Food Sci. Tech.* **2013**, *48*, 309–315. [CrossRef]
28. Lazaridou, A.; Duta, D.; Papageorgion, M.; Belc, N.; Biliaderis, C.G. Effect of hydrocolloids on dough rheology and bread quality parameters in gluten-free formulations. *J. Food Eng.* **2007**, *79*, 1033–1047. [CrossRef]
29. Alamprese, C.; Casiraghi, E.; Rossi, M. Modeling of fresh egg pasta characteristics for egg content and albumen to yolk ratio. *J. Food Eng.* **2009**, *93*, 302–307. [CrossRef]
30. Zhou, Y.; Zhao, D.; Foster, T.J.; Liu, Y.; Wang, Y.; Nirasawa, S.; Tatsumi, E.; Cheng, Y. Konjac glucomannan-induced changes in thiol/disulphide exchange and gluten conformation upon dough mixing. *Food Chem.* **2014**, *143*, 163–169. [CrossRef] [PubMed]
31. Li, J.; Yadav, M.P.; Li, J. Effect of different hydrocolloids on gluten proteins, starch and dough microstructure. *J. Cereal Sci.* **2019**, *87*, 85–90. [CrossRef]
32. Jia, F.; Ma, Z.; Hu, X. Controlling dough rheology and structural characteristics of chickpea-wheat composite flour-based noodles with different levels of *Artemisia sphaerocephala* Krasch. gum addition. *Int. J. Biol. Macromol.* **2020**, *150*, 605–616. [CrossRef]
33. Brennan, C.S.; Blake, D.E.; Ellis, P.R.; Schofield, J.D. Effects of guar galactomannan on wheat bread microstructure and on in vitro and vivo digestibility of starch from pasta. *J. Cereal Sci.* **1996**, *24*, 151–160. [CrossRef]
34. Giovanni, M. Response Surface Methodology and product optimization. *Food Technol.* **1983**, *37*, 41–45.
35. Abd Rahman, N.A.; Abdul Razak, S.Z.; Lokmanalhakim, L.A.; Taip, F.S.; Mustapa Kamal, S.M. Response surface optimization for hot air-frying technique and its effects on the quality of sweet potato snack. *J. Food Process Eng.* **2017**, *40*, 12507. [CrossRef]
36. Aliaa, A.N.; Mazlina, M.S.; Taip, F.S.; Abdullah, A.L. Response surface optimization for clarification of white pitaya juice using a commercial enzyme. *J. Food Process. Eng.* **2010**, *33*, 333–347. [CrossRef]
37. Shyu, S.; Hwang, L.S. Effects of processing conditions on the quality of vacuum fried apple chips. *Food Res. Int.* **2001**, *34*, 133–142. [CrossRef]
38. Shyu, S.; Hau, L.; Hwang, L.S. Effects of processing conditions on the quality of vacuum fried carrot chips. *J. Sci. Food Agric.* **2005**, *85*, 1903–1908. [CrossRef]
39. Sobukola, O.P.; Awonorin, S.O.; Sanni, L.O.; Bamiro, F.O. Optimization of blanching conditions prior to deep fat frying of yam slices. *Int. J. Food Prop.* **2008**, *11*, 379–391. [CrossRef]
40. Sobukola, O.P.; Awonorin, S.O.; Sanni, L.O.; Bamiro, F.O. Optimization of pre-fry drying of yam slices using response surface methodology. *J. Food Process Eng.* **2010**, *33*, 626–648. [CrossRef]
41. AACC. *Approved Methods of the American Association of Cereal Chemists*, 10th ed.; American Association of Cereal Chemists: St. Paul, MN, USA, 2005.
42. European Council. Regulation (EC) No 1924/2006 of the European Parliament and of the Council of 20 December 2006 on nutrition and health claims made on foods. *Off. J. Eur. Union* **2006**, *404*, 9–25.
43. Yaakob, H.; Ahmed, N.R.; Daud, S.K.; Malek, R.A.; Rahma, R.A. Optimization of ingredient and processing levels for the production of coconut yogurt using response surface methodology. *Food Sci. Biotechnol.* **2012**, *21*, 933–940. [CrossRef]
44. Cruz, A.G.; Faria, J.A.F.; Walter, E.H.M.; Andrade, R.R.; Cavalcanti, R.N.; Oliveira, C.A.F.; Granato, D. Processing optimization of probiotic yogurt containing glucose oxidasa using response surface methodology. *J. Dairy Sci.* **2010**, *93*, 5059–5068. [CrossRef]
45. Kaur, G.; Sharma, S.; Nagi, H.P.S.; Dar, B.N. Functional properties of pasta enriched with variable cereal brans. *J. Food Sci. Tech.* **2012**, *49*, 467–474. [CrossRef] [PubMed]
46. Manthey, F.A.; Sandhu, G.K. Effect of additives on the processing and cooking properties of pasta containing flax seed flour. In Proceedings of the 62nd Flax Institute of the USA, Fargo, ND, USA, 23–25 March 2000; Carter, J.F., Ed.; Flax Institute of the USA: Fargo, ND, USA, 2008; pp. 143–149.
47. Widelska, G.; Wójtowicz, A.; Kasprzak, K.; Dib, A.; Oniszczuk, T.; Olech, M.; Wojtunik-Kulesza, K.; Nowak, R.; Sujak, A.; Dobrzański, B.; et al. Impact of xanthan gum addition on phenolic acids composition and selected properties of new gluten free maize-field bean pasta. *Open Chem.* **2019**, *17*, 587–598. [CrossRef]
48. Resmini, P.; Pagani, M.A. Ultrastructure Studies of Pasta. A Review. *Food Microst.* **1983**, *2*, 1–12.

foods

MDPI

Article

Quality and Sensory Profile of Durum Wheat Pasta Enriched with Carrot Waste Encapsulates

Vanja Šeregelj [1], Dubravka Škrobot [2], Jovana Kojić [2], Lato Pezo [3], Olja Šovljanski [1,*], Vesna Tumbas Šaponjac [1], Jelena Vulić [1], Alyssa Hidalgo [4], Andrea Brandolini [5], Jasna Čanadanović-Brunet [1] and Gordana Ćetković [1]

1 Faculty of Technology Novi Sad, University of Novi Sad, Bulevar Cara Lazara 1, 21000 Novi Sad, Serbia; vanjaseregelj@tf.uns.ac.rs (V.Š.); vesnat@uns.ac.rs (V.T.Š.); jvulic@uns.ac.rs (J.V.); jasnab@uns.ac.rs (J.Č.-B.); gcetkovic@uns.ac.rs (G.Ć.)
2 Institute of Food Technology, University of Novi Sad, Bulevar Cara Lazara 1, 21000 Novi Sad, Serbia; dubravka.skrobot@fins.uns.ac.rs (D.Š.); jovana.kojic@fins.uns.ac.rs (J.K.)
3 Institute of General and Physical Chemistry, Studentski Trg 12, 11000 Belgrade, Serbia; latopezo@yahoo.co.uk
4 Department of Food, Environmental and Nutritional Sciences (DeFENS), Università Degli Studi di Milano, Via Celoria 2, 20133 Milan, Italy; alyssa.hidalgovidal@unimi.it
5 Council for Agricultural Research and Economics—Centre for Animal Production and Aquaculture (CREA-ZA), Viale Piacenza 29, 26900 Lodi, Italy; andrea.brandolini@crea.gov.it
* Correspondence: oljasovljanski@uns.ac.rs

Abstract: Consumer knowledge about pasta quality differs around the world. Modern consumers are more sophisticated compared to past times, due to the availability of information on pasta types and quality. Therefore, this study investigated the nutritional, physical, textural, and morphological quality of durum wheat pasta enriched with carrot waste encapsulates (10 and 20% freeze-dried encapsulate (FDE) and 10 and 20% spray-dried encapsulate (SDE)), as well as determining consumer preferences for this type of product. Replacement of semolina with FDE and SDE contributed to changes in the pasta nutritional quality, which was reflected in the increased protein, fat, and ash content. Additionally, changes in cooking quality, color, and texture were within satisfactory limits. The uncooked pasta enriched with 10 and 20% SDE was characterized by a lighter yellow intensity with color saturation, as well as an imperceptible waxy appearance compared to the control and enriched pasta with 10 and 20% FDE. After cooking, the yellow color was more intense in all the enriched pasta samples which can be linked to the raw cereal which was significantly greater in the control in comparison to the FDE and SDE containing samples. Overall, carrot waste can be a promising material for the food industry to produce high-quality pasta.

Keywords: carotenoids; cooking quality; nutritional quality; freeze-drying; spray-drying; durum wheat pasta; functional food

Citation: Šeregelj, V.; Škrobot, D.; Kojić, J.; Pezo, L.; Šovljanski, O.; Tumbas Šaponjac, V.; Vulić, J.; Hidalgo, A.; Brandolini, A.; Čanadanović-Brunet, J.; et al. Quality and Sensory Profile of Durum Wheat Pasta Enriched with Carrot Waste Encapsulates. *Foods* **2022**, *11*, 1130. https://doi.org/10.3390/foods11081130

Academic Editor: Mike Sissons

Received: 29 March 2022
Accepted: 13 April 2022
Published: 14 April 2022

1. Introduction

Pasta is a widely consumed cereal-based product all over the world. Pasta products are very popular due to their nutritional compositions, long shelf life, availability in the market, low cost, simplicity of preparation, and transportation [1]. Furthermore, about 15 million tons of pasta are produced annually with an expected increase in the range between 5 and 10% [2]. In summary, the role of pasta in the daily human diet can hardly be overestimated in any country [3]. Consumer knowledge about pasta quality differs around the world, but modern consumers are more sophisticated compared to the past, due to the availability of a large amount of information about the types and quality of pasta and all other diet-related characteristics [4]. It is worth remarking that consumer perception of pasta quality is very complex to understand, but the majority regard primary texture, color, health-improving effect, as well as price as the most important characteristics in making buying decisions [3]. Although all pasta manufacturers want to provide products based on consumers' preferences, they need to work within a good production practice, food safety principles, as well as functional food-management [5]. Targeting the mentioned requirements for pasta manufacturers, Hidalgo et al. [6] suggested fortification of pasta products, indicating that a partial replacement of flour with ingredients rich in nutritional and functional substances could be a good approach for the creation of higher pasta quality. One of those ingredients could be carrot waste rich in bioactive compounds because of the health-improving characteristics of its lipophilic compounds, primarily carotenoids [7–9]. Using carrot waste as a source rich in natural carotenoids provides an opportunity for fortification of food products, i.e., upgrading color properties and increasing antioxidant activity which is clinically related to several health benefits including inhibition of LDL oxidation, anti-inflammatory properties, alleviation of oxidative stress, and enhanced immune response [7]. On the other hand, using carotenoids in food products requires an encapsulation process due to the high sensitivity of these natural pigments to thermal degradation. The formation of a physical barrier for sensitive compounds provides longer shelf life under the variable storage conditions, preventing deleterious reactions and controlled release of targeted bioactives in food products [10,11]. Several scientific groups reported preliminary studies about pasta rich in carotenoids [12–14], but comprehensive research about this type of product and its quality and consumers' preferences have not yet been reported. Due to this fact, the objectives of this study were to determine the nutritional, physical, textural, and morphological quality of durum wheat pasta enriched with carrot waste encapsulates, as well as to investigate the sensory profile for this type of functional food product.

2. Materials and Methods

2.1. Materials

After food processing in the beverage industry ("Nectar", Bačka Palanka, Serbia), the obtained carrot waste was instantly sampled, freeze-dried, and kept at −20 °C until use. For carotenoid extraction, the sunflower oil ("Dijamant", Zrenjanin, Serbia) was selected, while the whey protein concentrate, inulin, and durum wheat semolina were obtained from Olimp Laboratories (Debica, Nagawczyna, Poland), Elephant Pharma (Belgrade, Serbia), and Molino Pagani (Borghetto Lodigiano, Italy), respectively.

2.2. Carrot Waste Extraction and Encapsulate Preparation

Freeze-dried carrot waste was mixed with sunflower oil (1:10 w/v) at 25 °C by stirring and using time shifts of 10 min blend and 5 min pause to avoid heating. After centrifugation at 4000 rpm for 10 min, the supernatant was recovered and kept at refrigerator temperature protected from light. The obtained carrot waste oil extract was encapsulated by freeze-drying and spray drying techniques, according to the optimal conditions reported by Šeregelj et al. [7]. The optimum wall materials imply 100% whey protein for freeze-drying as well as 71% whey protein and 29% inulin for spray drying [7]. The first formulation was kept at −80 °C during 24 h and then freeze-dried at −40 °C for 48 h to ensure complete drying. The second formulation was spray dried at an inlet temperature of 130 °C and an outlet temperature of 65 ± 2 °C. The spray-dried encapsulates (SDE) and the freeze-dried encapsulates (FDE) were kept at −20 °C.

2.3. Pasta Manufacturing

Preparation of the pasta was carried out in a small-scale pilot plant (Mac30, Italpast, Parma, Italy) using the procedure described in detail in the author's previous work [9]. Briefly, the control pasta dough was produced from durum wheat semolina (32% final humidity), while the carrot waste enriched pastas were created by replacing semolina with 10% or 20% FDE or SDE (Figure 1).

Figure 1. Manufactured pasta. Control: Durum wheat pasta without carrot waste encapsulates. 10% FDE: Durum wheat pasta with 10% freeze-dried carrot waste encapsulate. 10% SDE: Durum wheat pasta with 10% spray-dried carrot waste encapsulate. 20% FDE: Durum wheat pasta with 20% freeze-dried carrot waste encapsulate. 20% SDE: Durum wheat pasta with 20% spray-dried carrot waste encapsulate.

2.4. Nutritional Quality

The pasta samples were examined for moisture (M), crude protein (CP), crude fat (CF), and ash according to the method described in AACC (2000), while total carbohydrate content (TC) was calculated by subtracting the sum of M, CP, CF, and ash from 100.

2.5. Cooking Quality

To assess pasta cooking quality, short-cut macaroni samples (100 g) were cooked in boiling water with salt addition (5 g/L). All samples were cooked for the optimum cooking time which was defined by squashing a cut-open macaroni between two glass plates at different cooking times. The pasta was considered cooked when the white, opaque core had disappeared.

2.5.1. Cooked Weight

The cooked weight (CW) was measured as the weight of 100 g dry pasta after cooking.

2.5.2. Swelling Index

The swelling weight index (SI) was determined based on the weight of cooked pasta (Wcp) dried to a constant mass (Wdp) at 105 °C [1]. It was calculated by using the Equation (1).

$$SI = \frac{Wcp - Wdp}{Wdp} \tag{1}$$

2.5.3. Protein Loss

Protein loss (PL) was determined as the amount of proteins measured in the cooking water and expressed as the percentage of total protein in the pasta.

2.6. Color Properties of Pasta

The Yellow Index (YIAE; YICE), the Brown Index (BIAE; BICE), and total color difference (ΔE—DEAE; DECE) between uncooked and cooked pasta were used to evaluate pasta color properties. Color characteristics were measured by using the Minolta

Chromameter (Model CR-400, Minolta Co., Osaka, Japan) equipped with attachment CR-A33b. All the samples were illuminated with D65-artificial daylight (10° standard angle). Each measure was performed in triplicate. YI [15], BI, and ΔE [16] were calculated following Equations (2)–(4):

$$YIAE/YICE = \frac{142.86 \cdot b^*}{L^*} \tag{2}$$

$$BIAE/BICE = 100 - L^* \tag{3}$$

$$\Delta E^* = \sqrt{\Delta a^{*2} + \Delta b^{*2} + \Delta L^{*2}} \tag{4}$$

where L* is lightness, a* is red/green, and b* is yellow/blue. For evaluation of encapsulate addition (DEAE) the effect on ΔE, ΔL*, Δa*, and Δb* was calculated as the difference in L*, a*, and b* values (respectively) between control and enriched pasta samples. For evaluation of pasta cooking the effect on ΔE, ΔL*, Δa*, and Δb* was calculated as the difference in L*, a* and b* values (respectively) between uncooked and cooked pasta samples.

2.7. Textural Properties of Pasta

Optimally cooked pasta was washed with 500 mL of distilled water, drained, and allowed to balance at room temperature for 10 min in plates with lids before analysis. The texture analyzer (TA.XT Plus, Exponent Stable Micro System, Godalming, Surry, UK) was equipped with a 30 kg load cell, a P/36R probe was attached to the load cell, and the heavy duty platform was positioned centrally below the probe. Samples were centrally aligned under the retaining plate in as flat a position as possible. Pasta hardness (Hard), cohesiveness (Coh), springiness (Spr), chewiness (Chew), gumminess (Gum), and resilience (Res) were determined from the recorded force–time curve. The experimental procedure was as follows: 1 mm/s pre-test speed, 5 mm/s test speed and post-test speed, 75% strain, trigger type 5 g—auto. All texture measurements were carried out in six replicates.

2.8. Structural Morphology

Uncooked and cooked pastas were cut transversely without damaging the structure. The inner parts of the uncooked and cooked samples were used for analysis on a SEM Hitachi TM3030 scanning electron microscope (acceleration voltage 15 kV, beam current 20 nA, spot size 1 mm). For covering samples with gold the LeyboldHeraus L560Q putter coating device was used. The cooked pasta samples were dehydrated and prepared for SEM analysis using the preparation protocol reported by Ribotta et al. [17].

2.9. Sensory Analysis

The sensory profiles of uncooked and cooked pasta were determined by a trained sensory panel (2 males and 8 females, 23 to 45 years old) that consisted of members of the scientific team of the Institute of Food Technology, University of Novi Sad. A descriptive analysis was performed to obtain the complete description of a product's sensory properties by using the checklist method for the selection of sensory descriptors [18]. The final list of sensory descriptors, reached after discussion and training sessions, is presented in Table 1. The perceived intensity of evaluated sensory properties was expressed on a 100 mm linear scale. The sensory analysis was performed in individual booths at 22 °C. The samples were presented in plastic closed boxes coded with three-digit random numbers and were evaluated in two consecutive sessions, within ten minutes after pasta cooking. The sample presentation was in a completely balanced order. Before sensory testing, all participants were asked about possible food allergies and were required to sign written consent to participate in the study.

Table 1. Sensory attributes, descriptors, and definitions with end anchors.

Sensory Attributes	Descriptors	Definition with End Anchors
Appearance	Yellow color intensity—YIUP (uncooked pasta)	The intensity of yellow color (light—dark)
	Color saturation—CSUP (uncooked pasta)	The degree of color pureness (relative to pure grey) (washed out/pale—pure/vivid)
	Waxy appearance—WAUP (uncooked pasta)	Resembling wax in appearance (imperceptible—very pronounced)
	Yellow color intensity—YICP (cooked pasta)	The intensity of yellow color (light—dark)
Odour	Cereal odour intensity—COCP (cooked pasta)	The intensity of odour associated with raw cereals topped with boiling water (none—intensive)
	Fat odour intensity—FACP (cooked pasta)	The intensity of odour associated with fat or oil (none—intensive)
	Boiled eggs odour intensity—BECP (cooked pasta)	The intensity of odour associated with boiled eggs (none—intensive)
Flavour	Fat flavour intensity—FFI (cooked pasta)	The intensity of flavour associated with fat or oil (none—intensive)
Taste	Saltiness—Salt (cooked pasta)	The intensity of salty taste associated with sodium chloride solution (none—intensive)
Texture	Firmness—FCP (cooked pasta)	Force required biting down on pasta strands between the molars (not at all firm—very firm)
	Surface stickiness—SSCP (cooked pasta)	The degree to which pasta strands adhering to each other (not at all sticky—very sticky)
	Elasticity tactile—ETCP (cooked pasta)	Ability of the sample to return to the starting position after compression (not at all elastic—very elastic)
	Brittleness—Brit	The tendency of pasta to break without being significantly exposed to a high level of stress (not at all brittle—very brittle)
Residual	Oiliness—Oil	The degree to which the oily sensation in the mouth lags behind after pasta swallowing (none—intensive)

2.10. Principal Component Analysis (PCA)

Principal component analysis (PCA) enables insight into the presence of patterns in available data by providing information of defined variables, which behave similarly to each other. The results of PCA analysis of the five samples according to the investigated variable nutritional characteristics (M, CP, CF, Ash, TC), cooking quality and color parameters (CW, SI, PL, DEAE, YIAE, BIAE, DECE, YICE, BICE), textural parameters (Hard, Spr, Coh, Gumm, Chew, and Res), and sensor profile (YIUP, CSUP, WAUP, YICP, COCP, FACP, BECP, FCP, SSCP, ETCP, Brit, Salt, FFI, Oil) were presented in the form of five biplot plots.

2.11. Standard Scores Analysis

For a more complex ranking investigation of durum wheat pasta enriched with carrot waste encapsulates, standard scores (SS) were evaluated by integrating the obtained values of different nutritional and cooking quality parameter evaluation methods. The min–max normalization was used to compare nutritional and cooking quality parameters of samples obtained using experiments, in which samples were ranked according to extreme values of experimental data. Normal standard scores of all variables, for each sample, were derived by the Equations (5) and (6).

$$\bar{x}_i = \frac{x_i - \min_i x_i}{\max_i x_i - \min_i x_i}, \ \forall i, \text{ in case of ``the higher, the better'' criteria, or} \quad (5)$$

$$\overline{x}_i = 1 - \frac{x_i - \min_i x_i}{\max_i x_i - \min_i x_i}, \ \forall i, \text{ in case of "the lower, the better" criteria,} \quad (6)$$

where x_i are the experimental data. The averaged normalized scores sum applied for each sample gives a unitless value, which is termed as "standard score" (SS).

2.12. Statistical Analysis

To assess differences for the traits analyzed among samples, one-way analyses of variance (ANOVA) were calculated. In the case where significant differences were discovered, Fisher's least significant differences (LSD) at $p \leq 0.05$ were determined. The data were enumerated statistically applying the results obtained in the software package XLSTAT July 2018.

3. Results

3.1. Nutritional Quality

Table 2 shows the nutritional composition of control and enriched durum wheat pasta with carrot waste encapsulates. The carrot waste encapsulate enrichment decreased moisture content and total carbohydrates, whereas it increased crude protein, crude fat, and ash. According to Gupta et al. [1], a greater protein–polysaccharides interaction in enriched samples compared to control leads to a reduction in moisture content. Significantly superior protein contents were detected in pasta enriched with freeze-dried carrot waste encapsulates (FDE) which was expected due to the content of whey protein in the wall material, while spray-dried encapsulates (SDE) included inulin (29%) as well in wall material. The increases in protein content were 3.12 and 5.37 g/100 g for the replacement of semolina with 10% and 20% of FDE. When semolina was replaced with 10% and 20% of SDE, the increases in protein content were 1.69 and 3.85 g/100 g. Crude fat contributed to approximately 0.8% of the control pasta weight. Enriched pasta with FDE and SDE showed a high crude fat content, ranging from 4.23 to 7.20 g/100 g. This increase could be due to the inclusion of nutrients such as fatty acids present in the encapsulated carrot waste oil extract [7]. The significantly lower carbohydrate content in enriched pasta samples could be attributed to the decrease in semolina level in the blend.

Table 2. Nutritional quality of pasta enriched with carrot waste encapsulates.

Composition (g/100 g)	Polarity	Control	10% FDE	10% SDE	20% FDE	20% SDE
Moisture (M)	–	9.63 ± 0.09^{a}	9.61 ± 0.08^{a}	9.60 ± 0.08^{a}	8.87 ± 0.04^{b}	8.73 ± 0.06^{c}
Crude Protein (CP)	+	13.84 ± 0.04^{e}	16.96 ± 0.09^{c}	15.53 ± 0.06^{d}	19.21 ± 0.12^{a}	17.69 ± 0.08^{b}
Crude Fat (CF)	+	0.79 ± 0.05^{d}	4.23 ± 0.08^{c}	4.32 ± 0.11^{c}	5.51 ± 0.12^{b}	7.20 ± 0.10^{a}
Ash	–	0.84 ± 0.02^{c}	0.91 ± 0.02^{b}	0.87 ± 0.00^{bc}	0.97 ± 0.04^{a}	0.88 ± 0.03^{bc}
Total carbohydrates (TC)	–	74.90 ± 0.04^{a}	68.29 ± 0.07^{c}	69.68 ± 0.06^{b}	65.40 ± 0.08^{d}	65.50 ± 0.16^{d}

Results are presented in form of mean ± standard deviation (n = 3). Different letters in superscripts within the same row are significantly different at $p \leq 0.05$ according to Fisher's least significant differences (LSD) test; Polarity: '+' = the higher the better criteria, '−' = the lower the better criteria.

The PCA of the nutritional composition of durum wheat pasta samples showed that the first two principal components summarized 93.67% of the total variance in the five nutritional parameters (M, CP, CF, Ash, TC). According to the biplot of the PCA analysis of the collected data, moisture content (which provided 15.7% of the whole variance, established on correlations) and total carbohydrates (23.8%) showed a positive influence on the first principal component (PC1), while crude protein content (24.0%), crude fat content (20.6%) and ash content (16.0%) exerted a negative score in line with the PC1 component (Figure 2).

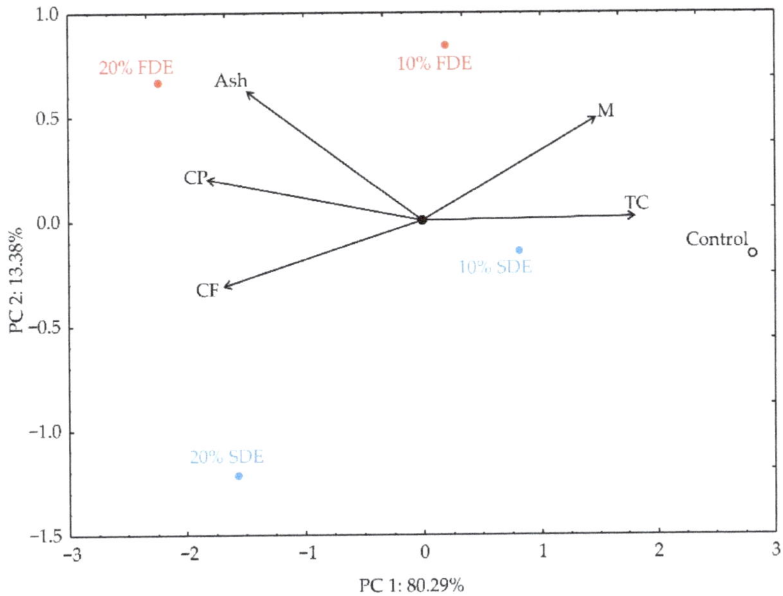

Figure 2. PCA ordination of variables based on the nutritional composition (M—moisture; CP—crude protein; CF—crude fat; TC—total carbohydrate content). Control: Durum wheat pasta without carrot waste encapsulates. 10% FDE: Durum wheat pasta with 10% freeze-dried carrot waste encapsulate. 10% SDE: Durum wheat pasta with 10% spray-dried carrot waste encapsulate. 20% FDE: Durum wheat pasta with 20% freeze-dried carrot waste encapsulate. 20% SDE: Durum wheat pasta with 20% spray-dried carrot waste encapsulate.

A positive leverage on the second principal component (PC2) was observed for moisture content (31.4% of the whole variance, identified on correlations) and ash content (50.3%), whereas a negative influence on PC2 was obtained for crude fat (13.1%). PC1 explained the differences in samples according to the nutritional composition of durum wheat pasta enriched with carrot waste encapsulates. Samples 20% FDE and 20% SDE achieved the required ash, crude protein, and crude fat content, while the control sample achieved the moisture and total carbohydrate content. The contents of ash and crude protein were higher in samples 10% FDE and 20% FDE.

3.2. Cooking Quality

Cooking pasta quality could be estimated based on cooked weight, swelling index, and protein loss. The results obtained for the cooking quality of pasta enriched with carrot waste encapsulates are shown in Table 3. Several authors have reported that the good cooking quality of pasta is related to the quality and content of protein, and the possibility to form optimum carbohydrates—protein network [2,19,20]. Table 3 illustrates that the replacement of semolina with carrot waste encapsulates pointedly affected the cooking weight of pasta; the values increased from 229.8 to 236.3 g and 237.5 g for pasta containing 10% and 20% of FDE respectively, while higher weights of 242.7 and 248.9 g were noted for pasta containing 10% and 20% of SDE respectively. The whey protein is rich in polar amino acids, so the increase in the cooked weight of pasta could be attributed to the water-binding capacity of the protein. On the other hand, SDE contains inulin which is highly hydrophilic and by this means the pasta cooking weight values increased more. Gupta et al. [1] reported cooking weight increase in quinoa protein isolate supplemented pasta because of the presence of polar amino acids. Reddy Surasani et al. [21] also found that the increased cooking weight of pasta with pangas protein isolate was a consequence of the protein being rich in polar amino acids.

Table 3. Cooking quality, color, and textural properties.

	Polarity	Control	10% FDE	10% SDE	20% FDE	20% SDE
			Cooking quality			
Cooked weight—CW (g)	+	229.8 ± 0.0 [e]	236.3 ± 0.1 [d]	242.7 ± 0.1 [b]	237.5 ± 0.1 [c]	248.9 ± 0.1 [a]
Swelling Index—SI (g/g)	+	0.77 ± 0.0 [e]	0.94 ± 0.1 [d]	1.02 ± 0.0 [c]	1.31 ± 0.0 [a]	1.13 ± 0.1 [b]
Protein loss—PL (%)	−	0.14 ± 0.01 [c]	0.15 ± 0.01 [b]	0.14 ± 0.00 [c]	0.16 ± 0.01 [a]	0.15 ± 0.00 [b]
			Color properties			
ΔE encapsulate addition effect—DEAE	+		3.96 ± 1.2 [b]	6.89 ± 3.3 [a]	6.58 ± 2.0 [a]	8.16 ± 2.3 [a]
Uncooked pasta						
YIAE		92.5 ± 3.1 [c]	96.6 ± 2.3 [b]	75.6 ± 4.2 [d]	107.8 ± 1.5 [a]	78.4 ± 4.4 [d]
BIAE		43.0 ± 0.8 [a]	41.6 ± 0.8 [b]	34.8 ± 1.4 [c]	40.5 ± 2.5 [b]	32.8 ± 1.9 [d]
Cooked pasta						
ΔE cooking effect—DECE	−	22.4 ± 2.1 [b]	23.9 ± 1.4 [ab]	11.9 ± 2.0 [c]	24.9 ± 2.1 [a]	12.0 ± 2.6 [c]
YICE		45.9 ± 4.1 [c]	46.1 ± 3.5 [c]	57.5 ± 4.4 [a]	51.7 ± 2.8 [b]	57.2 ± 3.9 [a]
BICE		25.1 ± 1.5 [a]	23.2 ± 1.3 [b]	24.7 ± 1.3 [a]	23.2 ± 1.5 [b]	23.2 ± 1.5 [b]
			Textural properties			
Hardness—Har (N)	−	16.0 ± 2.1 [bc]	21.8 ± 2.1 [b]	16.3 ± 3.5 [c]	30.75 ± 7.6 [a]	20.1 ± 1.6 [b]
Springiness—Spr	+	0.99 ± 0.0 [a]	0.96 ± 0.0 [ab]	0.96 ± 0.1 [ab]	0.90 ± 0.1 [b]	0.77 ± 0.1 [c]
Cohesiveness—Coh	−	0.66 ± 0.0 [a]	0.54 ± 0.1 [ab]	0.54 ± 0.0 [ab]	0.48 ± 0.2 [b]	0.42 ± 0.1 [b]
Gumminess—Gum (N)	−	10.5 ± 1.7 [b]	11.8 ± 2.5 [ab]	7.24 ± 2.4 [b]	14.9 ± 6.5 [a]	8.81 ± 2.3 [b]
Chewiness—Chew	+	10.4 ± 1.6 [ab]	11.4 ± 2.6 [ab]	6.94 ± 2.2 [b]	13.3 ± 6.0 [a]	6.80 ± 1.8 [b]
Resilience—Res	−	16.0 ± 2.1 [bc]	21.8 ± 2.1 [b]	16.3 ± 3.5 [c]	30.75 ± 7.6 [a]	20.1 ± 1.6 [b]

Results are presented in form of mean ± standard deviation (n = 3). Different letters in superscripts within the same row are significantly different at $p \leq 0.05$ according to Fisher's least significant differences (LSD) test. ΔE encapsulate addition effect: color difference between control and pasta with encapsulates. ΔE cooking effect: color difference between uncooked and cooked pasta. YIAE/YICE: yellow index. BIAE/BICE: brown index. Polarity: '+' = the higher the better criteria, '−' = the lower the better criteria.

Pasta enriched with carrot waste encapsulates showed a significantly higher swelling index than the control pasta (Table 3). The swelling index of the control was 0.77 g/g which increased in the range from 0.94 to 1.31 g/g when FDE or SDE was added and showed a directly proportional relation with increased concentration. Desai et al. [2] and El-Sohaimy et al. [22] obtained results which are in agreement with the present study. This property could be interpreted as referring to the water-binding and gelling ability of the proteins. Protein loss was not significantly affected by the substitution of semolina with 10% SDE, and thereafter it significantly increased for other enriched pasta. Higher values of protein loss were noted for pasta enriched with FDE, which could be due to 100% of whey protein in encapsulates and the high solubility of this protein. Mahmoud et al. [23] and Gupta et al. [1] also noticed an increase in protein loss for pasta with quinoa protein isolate and noodles fortified with protein products from lupine.

The PCA of the cooking quality of durum wheat pasta samples (Figure 3) showed that the first two principal components summarized 97.00% of the whole variance in the three parameters (CW, SI and PL). The cooked weight (which provided 19.3% of the whole variance, calculated based on correlations), swelling index (45.9%), and protein loss (34.8%) showed negative influence on PC1 (Figure 3). CW (70.0%) exerted a negative score according to PC2 component, while PL (30.0%) showed a positive influence on PC2 component. PL, SI, and CW parameters were the highest in samples 10% FDE and 10% FDE.

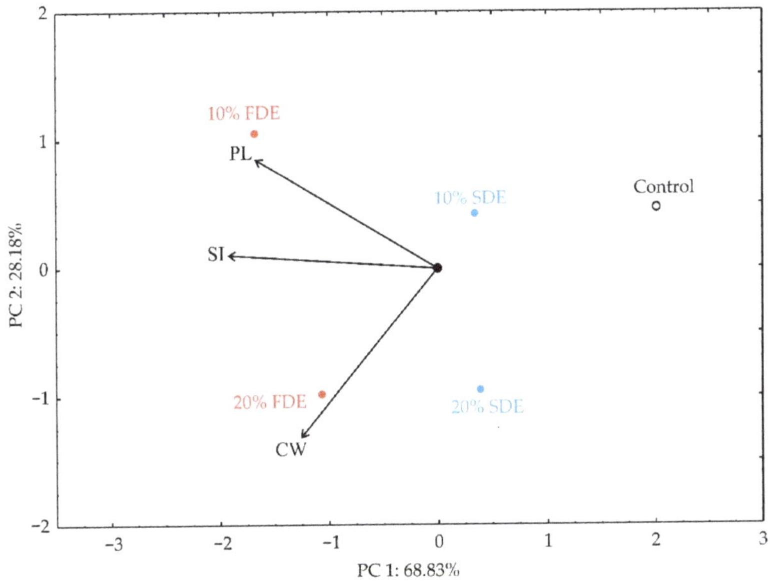

Figure 3. PCA ordination of variables based on cooking quality of durum wheat pasta enriched with carrot waste encapsulates (CW—Cooked weight (g); SI—Swelling Index (g/g); PL—Protein loss (%)). Control: Durum wheat pasta without carrot waste encapsulates. 10% FDE: Durum wheat pasta with 10% freeze-dried carrot waste encapsulate. 10% SDE: Durum wheat pasta with 10% spray-dried carrot waste encapsulate. 20% FDE: Durum wheat pasta with 20% freeze-dried carrot waste encapsulate. 20% SDE: Durum wheat pasta with 20% spray-dried carrot waste encapsulate.

3.3. Color Properties

Color is the single most important food-intrinsic sensory cue and hence contributes to consumers differing expectations regarding the likely taste and flavor. In order to evaluate the pasta color change due to carrot waste encapsulate addition, the color differential index (ΔE) was defined. Table 3 illustrates that the ΔE (DEAE) values of carrot waste encapsulate enriched pasta increased with increasing levels of FDE and SDE, whereas significantly higher ΔE was exhibited by the SDE-containing pasta. For all samples, the DEAE values were more than 3.0, which means that color changes are perceptible by visual observation [24].

In durum wheat, lutein and zeaxanthin as representatives of carotenoids, are the main color components [9] and contribute substantially to the YIAE of semolina and pasta. The durum wheat pasta enriched with FDE is characterized by a significantly higher YIAE, which is a consequence of the presence of α-carotene, β-carotene, and cis β-carotene in encapsulated carrot waste extract [8]. On the other hand, the YIAE of durum wheat pasta enriched with SDE is lower than the control, and this could be explained by the presence of inulin in the composition of these samples. Additionally, the BIAE was significantly lower than in enriched pasta samples compared to the control. Giannone et al. [25] reported that a yellowish color and high YIAE are appreciated by consumers of durum wheat pasta, while BIAE should be low, allowing a perception of brilliant and luminous color in the final product. Cooked samples exhibited higher ΔE (DECE) values, indicative of the pigments released after cooking the pasta. This is also confirmed with higher YICE values and lower BICE values of cooked pasta samples. Gull et al. [26] also observed an increase in yellow color in cooked pasta from millet flour and carrot pomace and explained this change as the consequence of swelling and conversion of pigments during cooking.

The PCA of the color properties of durum wheat pasta samples showed that the first two principal components totaled 95.53% of the whole variance in the six parameters (DEAE, YIAE, BIAE, DECE, YICE, and BICE). BIAE (which contributed 24.0% of the total variance, calculated according to correlations), YIAE (16.0%), and DECE (21.9%) exhibited a positive influence on PC1, while YICE (22.3%) and DEAE (15.5%) exerted a negative score according to the PC1 component. The positive influence on PC2 was exerted by BICE (56.5% of the total variance, based on correlations), while the negative influence on PC2 was obtained for DEAE (20.5%) and YIAE (16.6%) (Figure 4).

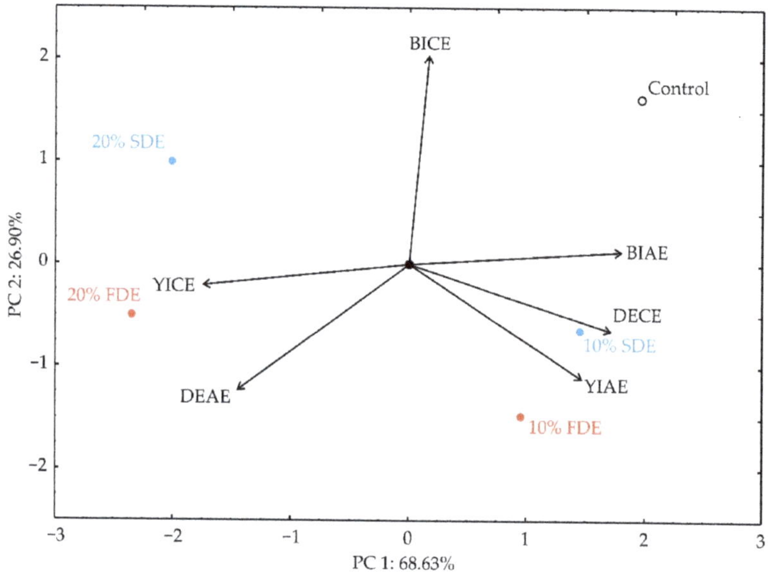

Figure 4. PCA ordination of variables based on color properties of durum wheat pasta enriched with carrot waste encapsulates (DEAE—ΔE encapsulate addition effect; YIAE—yellow index addition effect; BIAE—brown index addition effect; DECE—ΔE cooking effect; YICE—yellow index cooking effect; BICE—brown index cooking effect). Control: Durum wheat pasta without carrot waste encapsulates. 10% FDE: Durum wheat pasta with 10% freeze-dried carrot waste encapsulate. 10% SDE: Durum wheat pasta with 10% spray-dried carrot waste encapsulate. 20% FDE: Durum wheat pasta with 20% freeze-dried carrot waste encapsulate. 20% SDE: Durum wheat pasta with 20% spray-dried carrot waste encapsulate.

3.4. Textural Properties

The texture of pasta is one of the most significant indicators of quality that influences sensory attributes and final consumer acceptance. Table 3 illustrates the textural properties of cooked pasta represented by hardness, springiness, cohesiveness, gumminess, chewiness, and resilience parameters. An increase in hardness was recorded when comparing control and enriched pasta with carrot waste encapsulates. Ogawa and Adachi [27] ascribed the strength of the gluten network as the main factor which governed the hardness of enriched pasta. The samples enriched with 20% FDE and SDE showed a significant decrease in springiness and cohesiveness. This meant that for the 20% enriched pasta, it was more difficult to hold the structure together as time proceeded [28]. Gumminess and chewiness showed increasing values for the increasing concentration of FDE in pasta, while in terms of pasta enriched with SDE these values were lower. It was also observed that the addition of carrot waste encapsulates up to 20% did not have a significant influence on pasta resilience.

The PCA of the textural properties of durum wheat pasta samples (Figure 5) showed that the first two principal components totaled 54.96% of the whole variance in the six textural properties (hardness, springiness, cohesiveness, gumminess, chewiness, and resilience). Springiness (which contributed 16.4% of the total variance, based on correlations), cohesiveness (10.1%), gumminess (17.8), chewiness (26.4%), and resilience (24.4%) exhibited negative influence on PC1 (Figure 2). The positive influence on the second principal component (PC2) was noticed for cohesiveness (23.4% of the total variance, based on correlations), and springiness (16.2%), whereas a negative influence on PC2 was obtained for gumminess (16.6%) and hardness (32.3%). The textural properties were augmented in the samples of control, 10% FDE, and 10% SDE.

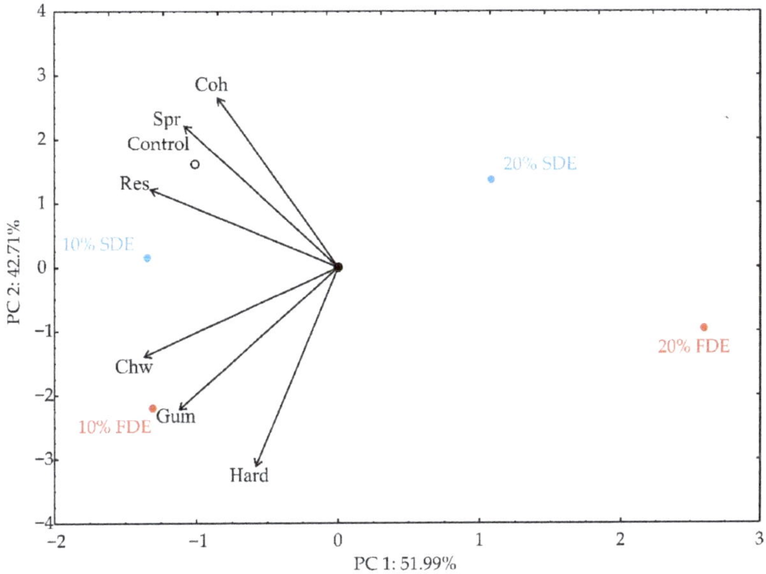

Figure 5. PCA biplot of variables based on textural properties of durum wheat pasta enriched with carrot waste encapsulates (Hard—Hardness (N); Spr—Springiness; Coh—Cohesiveness; Gum—Gumminess; Chw—Chewiness; Res—Resilience). Control: Durum wheat pasta without carrot waste encapsulates. 10% FDE: Durum wheat pasta with 10% freeze-dried carrot waste encapsulate. 10% SDE: Durum wheat pasta with 10% spray-dried carrot waste encapsulate. 20% FDE: Durum wheat pasta with 20% freeze-dried carrot waste encapsulate. 20% SDE: Durum wheat pasta with 20% spray-dried carrot waste encapsulate.

3.5. Standard Score

The "higher the better" or the "lower the better" criteria were used according to the sign in the "Polarity" column in Tables 2 and 3 for nutritional quality and cooking quality, color, and textural properties of durum wheat pasta enriched with carrot waste encapsulates. The standard score (SS) was calculated using Equation (7) by summing the normal scores for all variables which were multiplied by their weight.

$$SS = \frac{\overline{X_1} + \overline{X_2} + \ldots + \overline{X_n}}{n} \quad (7)$$

where $\overline{X_i}$ were the nutritional quality and cooking quality, color, and textural properties defined in Tables 2 and 3. The maximum of SS represents the optimal nutritional quality and cooking quality, color, and textural parameters. SS evaluation results are presented in Figure 6. The optimal parameters were for the sample 20% SDE, which exhibited the highest SS value of 0.662.

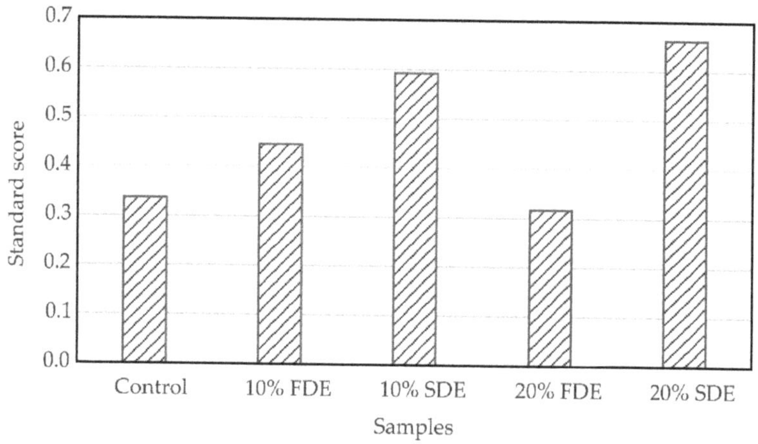

Figure 6. Standard score for samples. Control: Durum wheat pasta without carrot waste encapsulates. 10% FDE: Durum wheat pasta with 10% freeze-dried carrot waste encapsulate. 10% SDE: Durum wheat pasta with 10% spray-dried carrot waste encapsulate. 20% FDE: Durum wheat pasta with 20% freeze-dried carrot waste encapsulate. 20% SDE: Durum wheat pasta with 20% spray-dried carrot waste encapsulate.

3.6. SEM Observations

On the micrographs (Figure 7), the surfaces of the pasta are present in dry form (uncooked) and after contact with water during the whole cooking process (cooked). Differences in structure can be observed depending on the different stage of gelatinization, the fibrillar structure of the protein network, or some of its fragments. Regardless of the type of pasta sample, uncooked pasta samples have an irregular outer surface in which starch granules are entirely entrenched in the gluten matrix. It is well-known that gluten exists with irregular edges and starch granules are distributed in the direction of the force applied during the initial phase of extrusion [29]. Considering that all the tested pasta samples were made in the same extruder, the pattern in the gluten network was similar, which was to be expected. The micrographs of the samples showed a loose fibrillar protein network which was subject to big changes after the cooking process. This is in agreement with reported remarks by Gull et al. [26], who explained that the cooking process expands the pasta in volume resulting in the enveloping protein film and pasta surface becoming smoother. These complex changes are reflected in the appearance of cavities of different sizes and shapes (origin from gluten), as well as starch granules in different phases of gelatinization. Comparing the micrographs obtained for 10 and 20% SDE pasta, as well as 10 and 20% FDE pasta, differences in the structure of the protein network and the degree of deformation of the starch granules can be observed. Briefly, for these samples only the fibrillar structure of the protein network is visible, and the starch granules are completely gelatinized or deformed and decomposed. The difference in rupture surface is more evident compared to the surface micrographs of enriched pasta with control samples. Observed morphological differences between control and enriched pasta samples can be related to differences in the quality characteristics of the examined pasta (texture, behavior during cooking, sensory quality, etc.).

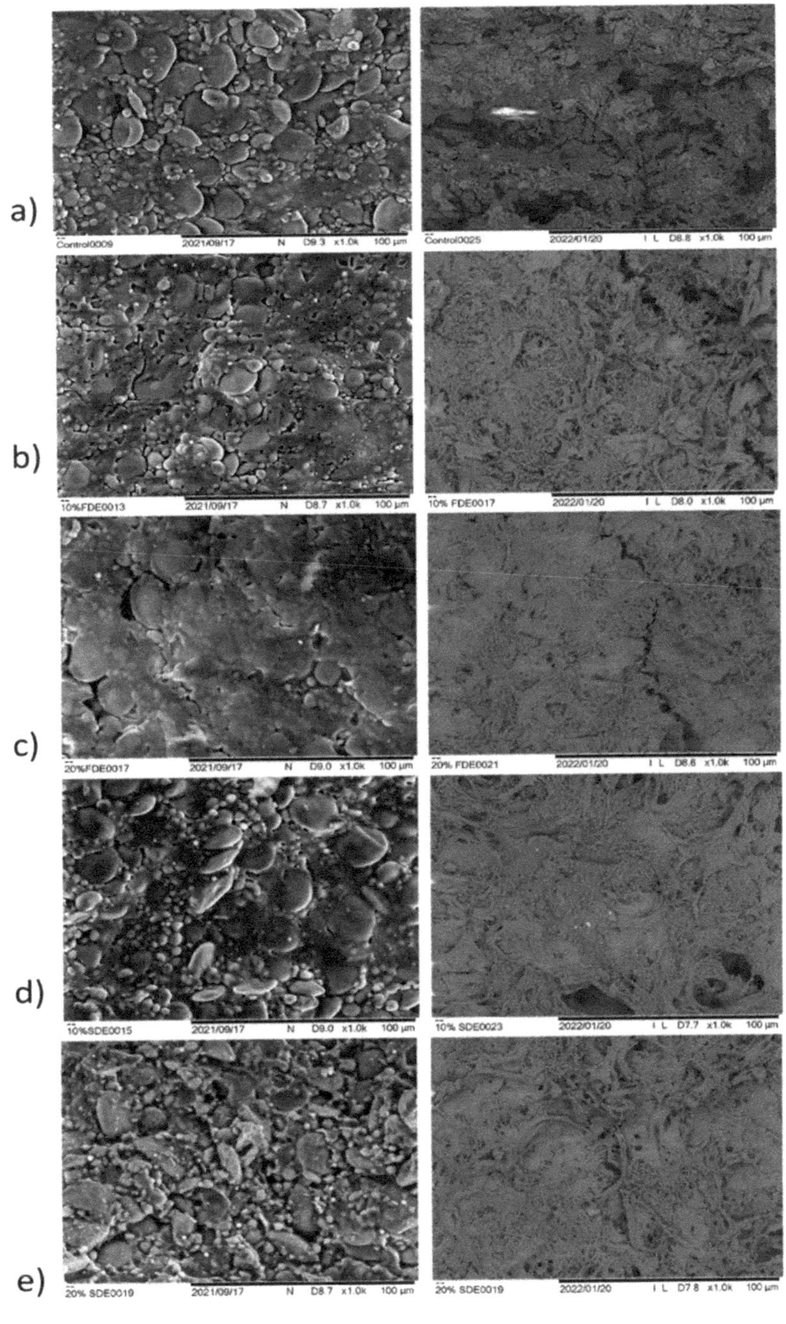

Figure 7. SEM images of uncooked and cooked pasta: (**a**) Control: Durum wheat pasta without carrot waste encapsulates. (**b**) 10% FDE: Durum wheat pasta with 10% freeze-dried carrot waste encapsulate. (**c**) 20% FDE: Durum wheat pasta with 20% freeze-dried carrot waste encapsulate. (**d**) 10% SDE: Durum wheat pasta with 10% spray-dried carrot waste encapsulate. (**e**) 20% SDE: Durum wheat pasta with 20% spray-dried carrot waste encapsulate.

3.7. Sensory Analysis

Descriptive sensory values are summarized in Figure 8. The uncooked pasta samples enriched with 10 and 20% SDE were characterized by lighter yellow intensity and color saturation, as well as imperceptible waxy appearance compared with the control and enriched pasta with 10 and 20% FDE. Furthermore, uncooked 20% FDE pasta was marked with the highest values of yellow color intensity and saturation, followed by 10% FDE and the control pasta. These results are comparable with the previously presented comments for the color properties in Section 3.3. The very pronounced waxy appearance of 20% FDA pasta is probably because of the high carrot waste extracted in oil delivered by encapsulates [7,9]. On the other hand, after cooking, the yellow color was more intensive in all the enriched pasta samples, which is in good agreement with the findings reported by Gull et al. [26]. The intensity of odour linked with raw cereal was significantly more intense in the control in comparison to the FDE and SDE containing samples. Namely, the addition of FDE and SDE in pasta contributed to a more intense fat and boiled egg odour. Therefore, the more intensive flavour associated with fat or oil was thought to be in these samples, due to the presence of sunflower oil in the encapsulate composition [9]. The saltiness, i.e., the salty taste associated with sodium chloride solution was barely noticeable in enriched pasta samples, while in the control this taste was more intensive. Carrot waste encapsulate addition meaningfully contributed to changes in the textural properties of the pasta. The firmness of cooked pasta reflects the force required to bite down on pasta strands between the molars. Enriched pasta with FDE and SDE was characterized as firmer than control.

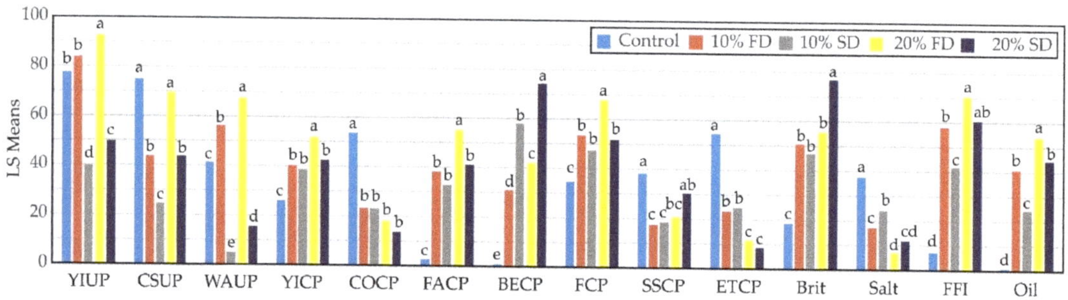

Figure 8. Descriptive sensory analysis of of durum wheat pasta enriched with carrot waste encapsulates. YIUP—Yellow color intensity (uncooked pasta); CSUP—Color saturation (uncooked pasta); WAUP—Waxy appearance (uncooked pasta); YICP—Yellow color intensity (cooked pasta); COCP—Cereal odour intensity (cooked pasta); FACP—Fat odour intensity (cooked pasta); BECP—Boiled egg odour intensity (cooked pasta); FCP—Firmness (cooked pasta); SSCP—Surface stickiness (cooked pasta); ETCP—Elasticity tactile (cooked pasta); Brit—Brittleness; Salt—Saltiness (cooked pasta); FFI—Fat flavour intensity (cooked pasta); Oil—Oiliness. Control: Durum wheat pasta without carrot waste encapsulates. Pasta samples with: 10% FD: 10% freeze-dried carrot waste encapsulate. 20% FD: 20% freeze-dried carrot waste encapsulate. 10% SD: 10% spray-dried carrot waste encapsulate. 20% SD: 20% spray-dried carrot waste encapsulate. Different letters in superscripts within the same row are significantly different at $p \leq 0.05$ according to Fisher's least significant differences (LSD) test.

The obtained results are in agreement with the report by Gupta et al. [1] which explained that the thermal protein denaturation during cooking enhances the firmness. Laleg et al. [30] also reported that protein additives increase the firmness of pasta. Since the protein content is higher in FDE pasta samples, higher values of firmness were recorded for these samples. The surface stickiness of FDE and SDE enriched pasta was lower than control, which may be attributed to the addition of whey protein and sunflower oil. The protein network entraps the starch granules which then prevent starch leaching [31], while sunflower oil acts as a lubricant. Elasticity was inversely dependent on the protein content. Pasta enriched with encapsulated carrot waste extract with pure whey protein was characterized as barely elastic and very brittle. The control pasta was almost found to have no oiliness, i.e., none of the oily sensations in the mouth remained after pasta swallowing. Samples enriched with carrot waste encapsulates were described with a noticeable more intense oil sensation after swelling, which is to be expected, given the composition of these samples.

The PCA of the sensory analysis of durum wheat pasta samples showed that the first two principal components totaled 92.11% of the total variance in the fourteen sensory analysis parameters. COCP (which contributed 10.0% of whole variance, calculated, according to correlations), ETCP (10.3%) and Salt (10.0%) exhibited a positive influence on PC1, while FCP (8.4%), YICP (9.5%), Oil (9.9%), FACP (10.2%), FFI (10.3%), and Brit (8.3%) showed a negative score according to PC1 component (Figure 9). The positive influence on PC2 was noticed for WAUP (28.2%), YIUP (28.5%), and CSUP (19.0%), while a negative influence on PC2 was obtained for BECP (10.3%).

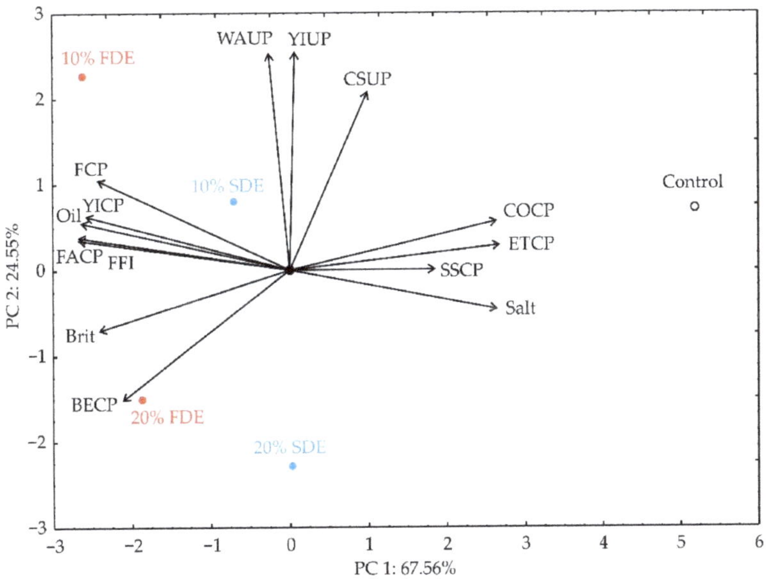

Figure 9. PCA ordination of variables based on sensory analysis of durum wheat pasta enriched with carrot waste encapsulates. YIUP—Yellow color intensity (uncooked pasta); CSUP—Color saturation (uncooked pasta); WAUP—Waxy appearance (uncooked pasta); YICP—Yellow color intensity (cooked pasta); COCP—Cereal odour intensity (cooked pasta); FACP—Fat odour intensity (cooked pasta); BECP—Boiled egg odour intensity (cooked pasta); FCP—Firmness (cooked pasta); SSCP—Surface stickiness (cooked pasta); ETCP—Elasticity tactile (cooked pasta); Brit—Brittleness; Salt—Saltiness (cooked pasta); FFI—Fat flavour intensity (cooked pasta); Oil—Oiliness.

PC1 explained the differences in the cooked pasta samples according to the sensory analysis parameters, while PC2 explained the differences in the uncooked durum wheat pasta samples enriched with carrot waste encapsulates (Figure 8). Samples 10% FDE and 20% FDE were characterized by augmented Firmness, Yellow color intensity, Oiliness, Fat odour intensity, Fat flavour intensity, Brittleness and Boiled egg odour intensity in cooked pasta, while the control sample was characterized by increased Cereal odour intensity, Elasticity tactile, Saltiness, and Surface stickiness in cooked pasta. The increased values of Waxy appearance, Yellow color intensity, and Color saturation were noticed in the 10% FDE uncooked pasta sample.

4. Conclusions

Pasta enrichment with carrot waste encapsulates significantly improved the protein, fat and ash contents. In addition, cooking quality, color, and textural properties were affected within acceptable limits. Sensory descriptive analysis revealed that enriched durum wheat pasta with 10% FDE exhibited more intensive yellow color intensity, color saturation, and waxy appearance. Enriched samples were also characterized by lower surface stickiness. Overall, the supplementation of durum wheat semolina with encapsulated carrot waste extract could be a good approach for producing pasta with better nutritional ingredients and satisfactory technical and sensory qualities.

Author Contributions: Conceptualization, V.Š. and O.Š.; methodology, G.Ć.; validation, D.Š., L.P. and J.V.; formal analysis, V.Š. and V.T.Š.; investigation, V.Š., J.K., A.H. and A.B.; resources, A.H. and A.B.; data curation, L.P.; writing—original draft preparation, V.Š. and O.Š.; writing—review and editing, V.Š., L.P. and O.Š.; visualization, V.Š., L.P. and O.Š.; supervision, J.Č.-B., G.Ć. and A.H. All authors have read and agreed to the published version of the manuscript.

Funding: This research received no external funding.

Institutional Review Board Statement: The study was approved by the Ethics Committee of Institute of Food Technology in Novi Sad, University of Novi Sad, Serbia (Ref. No. 175/I/14-3).

Informed Consent Statement: Informed consent was obtained from all subjects involved in the study.

Data Availability Statement: The data presented in this study are available in article.

Acknowledgments: This research study was supported by the Ministry of Education, Science, and Technological Development of the Republic of Serbia (Grant No. 451-03-9/2022-14/200134). The authors would like to thank the BioSense Institute, the University of Novi Sad, Serbia, for the use of measurement equipment.

Conflicts of Interest: The authors declare no conflict of interest.

References

1. Gupta, A.; Sharma, S.; Reddy Surasani, V.K. Quinoa protein isolate supplemented pasta: Nutritional, physical, textural and morphological characterization. *LWT* **2021**, *135*, 110045. [CrossRef]
2. Desai, A.S.; Brennan, M.A.; Brennan, C.S. Influence of semolina replacement with salmon (*Oncorhynchus tschawytscha*) powder on the physicochemical attributes of fresh pasta. *Int. J. Food Sci. Technol.* **2018**, *54*, 1497–1505. [CrossRef]
3. Wang, J.; Brennan, M.A.; Brennan, C.S.; Serventi, L. Effect of Vegetable Juice, Puree, and Pomace on Chemical and Technological Quality of Fresh Pasta. *Foods* **2021**, *10*, 1931. [CrossRef] [PubMed]
4. Miskelly, D. *Cereal Grains*; Woodhead Publishing: Cambridge, UK, 2017; pp. 653–688.
5. Nilusha, R.A.T.; Jayasinghe, J.M.J.K.; Perera, O.D.A.N.; Perera, P.I.P. Development of Pasta Products with Nonconventional Ingredients and Their Effect on Selected Quality Characteristics: A Brief Overview. *Int. J. Food Sci.* **2019**, *2019*, 27. [CrossRef]
6. Hidalgo, A.; Ferraretto, A.; De Noni, I.; Bottani, M.; Cattaneo, S.; Galli, S.; Brandolini, A. Bioactive compounds and antioxidant properties of pseudocereals-enriched water biscuits and their in vitro digestates. *Food Chem.* **2018**, *240*, 799–807. [CrossRef]
7. Šeregelj, V.; Cetkovic, G.; Canadanovic-Brunet, J.; Tumbas Saponjac, V.; Vulic, J.; Levic, S.; Nedovic, V.; Brandolini, A.; Hidalgo, A. Encapsulation of carrot waste extract by freeze and spray drying techniques: An optimization study. *LWT* **2021**, *138*, 110696. [CrossRef]
8. Šeregelj, V.; Vulic, J.; Cetkovic, G.; Canadanovic-Brunet, J.; Tumbas Saponjac, V.; Stajcic, S. Natural bioactive compounds in carrot waste for food applications and health benefits. In *Bioactive Natural Products, Atta-ur-Rahman, Ed.*; Elsevier Science: Amsterdam, The Netherlands, 2021; pp. 307–344.
9. Ćetković, G.; Šeregelj, V.; Brandolini, A.; Čanadanović-Brunet, J.; Tumbas Šaponjac, V.; Vulić, J.; Šovljanski, O.; Četojević-Simin, D.; Škrobot, D.; Mandić, A.; et al. Composition, texture, sensorial quality, and biological activity after in vitro digestion of durum wheat pasta enriched with carrot waste extract encapsulates. *Int. J. Food Sci. Nutr.* **2022**, *in press*.
10. Nedovic, V.; Kalusevic, A.; Manojlovic, V.; Petrovic, T.; Bugarski, B. Encapsulation systems in the food industry. In *Advances in Food Process Engineering Research and Applications*; Yanniotis, S., Taoukis, P., Stoforos, N.G., Karathanos, V.T., Eds.; Springer: Boston, MA, USA, 2013; pp. 229–253.
11. Cvetković, D.; Ranitović, A.; Šeregelj, V.; Šovljanski, O.; Vulić, J.; Jović, B.; Pavlović, V. Encapsulation of peach waste extract in *Saccharomyces cerevisiae* cells. *J. Serb. Soc. Chem.* **2021**, *86*, 367–380. [CrossRef]

12. Di Marco, F.; Trevisani, F.; Vignolini, P.; Urciuoli, S.; Salonia, A.; Montorsi, F.; Romani, A.; Vago, R.; Bettiga, A. Preliminary Study on Pasta Samples Characterized in Antioxidant Compounds and Their Biological Activity on Kidney Cells. *Nutrients* **2021**, *13*, 1131. [CrossRef]
13. Oduro-Obeng, H.; Fu, B.X.; Beta, T. Influence of cooking duration on carotenoids, physical properties and in vitro antioxidant capacity of pasta prepared from three Canadian durum wheat cultivars. *Food Chem.* **2021**, *363*, 130016. [CrossRef]
14. Scarton, M.; Ribeiro, T.; Godoy, H.T.; Behrens, J.H.; Campelo, P.H.; Clerici, M.T.P.S. Gluten free pasta with natural ingredient of color and carotene source. *Res. Soc. Dev.* **2021**, *10*, 21310413959. [CrossRef]
15. Francis, F.J.; Clydesdale, F.M. *Food Colorimetry: Theory and Applications*; AVI Publishing Co.: Westport, CT, USA, 1975.
16. Rhim, J.; Wu, Y.; Weller, C.; Schnepf, M. Physical characteristics of a composite film of soy protein isolate and propyleneglycol alginate. *J. Food Sci.* **1999**, *64*, 149–152. [CrossRef]
17. Ribotta, P.D.; Pérez, G.T.; León, A.E.; Añón, M.C. Effect of emulsifier and guar gum on microstructural, rheological and baking performance of frozen bread dough. *Food Hydrocoll.* **2004**, *18*, 305–313. [CrossRef]
18. Lawless, H.; Heymann, H. Qualitative Consumer Research Methods. In *Sensory Evaluation of Food*; Springer: New York, NY, USA, 2010.
19. Troccoli, A.; Borrelli, G.M.; De Vita, P.; Fares, C.; Di Fonzo, N. Mini Review: Durum Wheat Quality: A Multidisciplinary Concept. *J. Cereal Sci.* **2000**, *32*, 99–113. [CrossRef]
20. De Noni, I.; Pagani, M.A. Cooking Properties and Heat Damage of Dried Pasta as Influenced by Raw Material Characteristics and Processing Conditions. *Crit. Rev. Food Sci. Nutr.* **2010**, *50*, 465–472. [CrossRef] [PubMed]
21. Reddy Surasani, V.K.; Singh, A.; Gupta, A.; Sharma, S. Functionality and cooking characteristics of pasta supplemented with protein isolate from pangas processing waste. *LWT* **2019**, *111*, 443–448. [CrossRef]
22. El-Sohaimy, S.A.; Brennan, M.; Darwish, A.M.G.; Brennan, C. Physicochemical, texture and sensorial evaluation of pasta enriched with chickpea flour and protein isolate. *Ann. Agric. Sci.* **2020**, *65*, 28–34. [CrossRef]
23. Mahmoud, E.A.M.; Nassef, S.L.; Basuny, A.M.M. Production of high protein quality noodles using wheat flour fortified with different protein products from lupine. *Ann. Agric. Sci.* **2012**, *57*, 105–112. [CrossRef]
24. Cacak-Pietrzak, G.; Dziki, D.; Gawlik-Dziki, U.; Sułek, A.; Kalisz, S.; Sujka, K. Effect of the Addition of Dried Dandelion Roots (*Taraxacum officinale* F. H. Wigg.) on Wheat Dough and Bread Properties. *Molecules* **2021**, *26*, 7564. [CrossRef]
25. Giannone, V.; Giarnetti, M.; Spina, A.; Todaro, A.; Pecorino, B.; Summo, C.; Pasqualone, A. Physico-chemical properties and sensory profile of durum wheat Dittaino PDO (Protected Designation of Origin) bread and quality of re-milled semolina used for its production. *Food Chem.* **2018**, *241*, 242–249. [CrossRef]
26. Gull, A.; Prasad, K.; Kumar, P. Effect of millet flours and carrot pomace on cooking qualities, color and texture of developed pasta. *LWT* **2015**, *63*, 470–474. [CrossRef]
27. Ogawa, T.; Adachi, S. Drying and rehydration of pasta. *Dry. Technol.* **2017**, *35*, 1919–1949. [CrossRef]
28. Kosović, I.; Jukić, M.; Jozinović, A.; Ačkar, Đ.; Koceva Komlenić, D. Influence of chestnut flour addition on quality characteristics of pasta made on extruder and minipress. *Czech J. Food Sci.* **2016**, *34*, 166–172. [CrossRef]
29. Bustos, M.C.; Perez, G.T.; Leon, A.E. Structure and quality of pasta enriched with functional ingredients. *RSC Adv.* **2015**, *5*, 30780–30792. [CrossRef]
30. Laleg, K.; Cassan, D.; Barron, C.; Prabhasankar, P.; Micard, V. Structural, Culinary, Nutritional and Anti-Nutritional Properties of High Protein, Gluten Free, 100% Legume Pasta. *PLoS ONE* **2016**, *11*, e0160721. [CrossRef] [PubMed]
31. Lamacchia, C.; Di Luccia, A.; Baiano, A.; Gambacorta, G.; la Gatta, B.; Pati, S.; La Notte, E. Changes in pasta proteins induced by drying cycles and their relationship to cooking behaviour. *J. Cereal Sci.* **2007**, *46*, 58–63. [CrossRef]

foods

MDPI

Review

Increasing the Versatility of Durum Wheat through Modifications of Protein and Starch Composition and Grain Hardness

Domenico Lafiandra [1,*], Francesco Sestili [1], Mike Sissons [2,*], Alecia Kiszonas [3] and Craig F. Morris [3,†]

[1] Department of Agriculture and Forest Sciences, University of Tuscia, 01100 Viterbo, Italy; francescosestili@unitus.it
[2] NSW Department of Primary Industries, Tamworth 2340, Australia
[3] United States Department of Agriculture, Agriculture Research Service, Western Wheat Quality Lab, Pullman, WA 99164, USA; alecia.kiszonas@usda.gov (A.K.); craig.morris@usda.gov (C.F.M.)
* Correspondence: lafiandr@unitus.it (D.L.); mike.sissons@dpi.nsw.gov.au (M.S.)
† Deceased.

Abstract: Although durum wheat (*Triticum durum* L. ssp. *durum* Desf.) has traditionally been used to make a range of food products, its use has been restricted due to the absence of the D-genome glutenin proteins, the relatively low variability in starch composition, and its very hard grain texture. This review focuses on the manipulation of the starch and protein composition and modification of the hardness of durum wheat in order to improve its technological and nutritional value and expand its utilization for application to a wider number of end products. Starch is composed of amylopectin and amylose in a 3:1 ratio, and their manipulation has been explored for achieving starch with modified composition. In particular, silencing of the genes involved in amylose and amylopectin synthesis has made it possible to isolate durum wheat lines with amylose content varying from 2–3% up to 75%. This has created opportunities for new products with different properties and enhanced nutritional value. Durum-made bread has generally inferior quality to bread made from common wheat. Attempts to introduce the *Glu-D1* subunits 1Dx5 + 1Dy10 and 1Dx2 + 1Dy12 produced stronger dough, but the former produced excessively strong, inelastic doughs, and loaf volume was either inferior or not affected. In contrast, the 1Dx2 + 1Dy12 sometimes improved bread loaf volume (LV) depending on the glutenin subunit background of the genotype receiving these genes. Further breeding and selection are needed to improve the dough extensibility to allow higher LV and better texture. The versatility of durum wheat has been greatly expanded with the creation of soft-textured durum via non-GMO introgression means. This soft durum mills like soft hexaploid wheat and has similar baking properties. The pasta quality is also not diminished by the soft-textured kernels. The *Glu-D1* locus containing the subunits 1Dx2 + 1Dy12 has also been introgressed to create higher quality soft durum bread.

Keywords: durum wheat; grain hardness; D-genome glutenin subunits; dough strength; waxy; amylose

Citation: Lafiandra, D.; Sestili, F.; Sissons, M.; Kiszonas, A.; Morris, C.F. Increasing the Versatility of Durum Wheat through Modifications of Protein and Starch Composition and Grain Hardness. *Foods* **2022**, *11*, 1532. https://doi.org/10.3390/foods11111532

Academic Editor: Ken'ichi Ohtsubo

Received: 8 April 2022
Accepted: 19 May 2022
Published: 24 May 2022

1. Introduction

Durum wheat (*Triticum turgidum* subsp. *durum*) is widely grown in the Mediterranean area representing a major staple crop used for the preparation of different end products whose history and utilization have accompanied the journey of man for thousands of years. Though commonly known for its use in pasta, durum wheat, also thanks to its glassy texture, is used for the preparation of different types of bread, biscuits, pastries, and other kinds of foods, and the processing can include whole or crushed kernels [1]. About 25% of the durum wheat produced in the world is used for breadmaking and up to 70–90% in some Middle East countries [2,3]. This review covers approaches used to manipulate the

(1) protein composition to improve durum breadmaking quality, (2) starch composition to modify the amylose/amylopectin ratio to alter the nutritional value, and (3) durum wheat kernel hardness to a soft texture for expanded use of durum by making milling accessible to those with traditional roller mills without sacrificing pasta quality.

It is generally known that durum makes pan breads that are inferior in their bread-baking performance to that made from hexaploid wheat varieties. Specifically, durum bread has inferior loaf volume, structure, and texture compared with common wheat bread [4–8], although durum bread is preferred in some markets for its peculiar and distinctive sensory properties, with a better shelf life than conventional bread [2]. However, it is possible to produce acceptable bread made from blends of durum with bread wheat flour [9].

Although increasing protein content can improve loaf volume, the typically inelastic and poorly extensible gluten in durum prevents full gas expansion, as dough extensibility is an important trait for obtaining good loaf volume [4]. Guzman et al. [7] noted that several durum varieties produced loaf volume similar to bread wheat in some environments, particularly under drought stress, indicating the potential for durum in breadmaking.

The inelastic nature of durum gluten is related to the glutenin subunit composition. Two factors are important: the total number of high molecular weight glutenin subunits (HMW-GS) and the absence of the HMW-GS associated to the D genome. The maximum possible number of HMW-GS in durum wheat is three, compared with the hexaploid wheat maximum of five [10]. The HMW-GS enables the strong doughs that are critical in baking for trapping small bubbles of carbon dioxide gas formed by yeast during proofing, thereby enabling the dough to rise and ensure a good loaf volume and structure to leavened bread. In bread wheat, the most prominent locus that contributes to dough elasticity and extensibility is the *Glu-D1* locus, with two allelic variants, 1Dx2 + 1Dy12 (2 + 12) and 1Dx5 + 1Dy10 (5 + 10). Functionality studies of specific HMW subunits have found that subunit 1Dx5 + 1Dy10 is considered the "stronger" allele and is more desirable for bread quality prepared from hexaploid wheat [11,12]. Durum wheat lacks the D genome, and thus, it also lacks the *Glu-D1* locus. Consequently, the elasticity and extensibility of durum doughs are often viewed as inferior to bread wheat [4]. Dough strength results from a complex interplay between the HMW glutenin subunits, low molecular weight glutenin subunits (LMW-GS), gliadins, and non-protein endosperm constituents. Transferring the storage protein genes that are present on chromosome 1D into durum has been attempted to try to improve durum breadmaking quality.

A technology that has been very effective in introducing D-genome-related proteins into durum wheat is chromosome engineering. In wheat normally, recombination is restricted to homologous chromosomes from the same genome due to the presence of the *Ph1* locus present on the long arm of chromosome 5B. The availability of mutants (*ph1*) with a deletion at this locus has resulted in the possibility of inducing homoeologous pairing (similar chromosomes from different genomes, e.g., 1A vs. 1B or 1A vs. 1D) and realize the transfer of chromosomal segments, carrying genes of interest from wild or cultivated wheat into bread and durum wheat [13].

However, for good pasta quality, it has been shown that the most important genes are those associated at the *Glu-A3*, *Glu-B3*, and *Glu-B2* loci, which encode the B-type LMW-GS [14–16]. Broadening the uses of durum wheat to improve breadmaking quality while maintaining pasta-making quality is a desirable goal.

Starch makes up more than 75% of the wheat kernel and represents an important source of energy in the human diet, contributing to >50% of caloric intake in the Western world and up to 90% in developing countries [17]. According to Hardy et al. [18], starch has played an essential role in human evolution by providing accessible carbohydrates to promote a significant increase in brain size. Starch contributes greatly to the textural properties of many foods and is widely used in industrial and food applications as a thickener, colloidal stabilizer, gelling agent, bulking agent, and water retention agent [19]. Starch is composed of two different types of polymers, amylopectin and amylose, in a 3:1 ratio, differing in degree of polymerization and number of side branches. Amylose is essentially formed by a linear chain of D-glucose molecules with a low degree of polymerization, whereas amylopectin shows a higher degree of polymerization. Together, the two polymers are assembled to form insoluble semi-crystalline starch granules.

The application of mutagenesis in association with reverse genetics has provided a new powerful advance in both functional genomics and crop-breeding applications: A combination of chemical mutagens with PCR detection of point mutations in a gene of interest for which the sequence is known has resulted in the development of the TILLING (Targeting induced local lesions in genomes) strategy [20]. Chemical mutagens such as ethyl methane sulfonate (EMS) are very effective in inducing randomly distributed single-nucleotide changes in genomes of many different crop species. Its use generates allelic series of mutations, including missense changes, with amino acid substitutions that can have a range of effects on protein function and non-sense or splice site changes that can cause truncation of the gene product and, depending on the location, probable loss of function. Such an approach has been used to generate and detect mutations in different types of genes including those involved in the biosynthesis of amylose and amylopectin in durum and bread wheat [21–25].

Durum wheat traditionally has a very hard and glassy-textured kernel. This hard texture necessitates specialized durum mills, as opposed to mills for hexaploid wheat, which can be changed easily for hard and soft bread wheat. Not only are durum mills specialized, but they produce semolina. Semolina is a much coarser product with a greater average particle size (~250–300 µm). Durum flour can be made with a particle size distribution more similar to that of bread wheat (~80–100 µm), but the level of damaged starch is so great that the flour is challenging to work with as a base flour [26,27]. The particle size and properties of semolina make durum best for pasta and couscous but limit its further uses. In bread, wheat grain hardness is an important parameter that influences its mechanical properties, milling behavior, flour yield, starch damage, water absorption, dough rheological properties, volume, and crumb structure [28]. Grain hardness is primarily controlled by the genes present at the Hardness locus, *Ha*, present on the distal part of the short arm of chromosome 5D, which encodes the so-called Puroindolines a and b [28,29]. Both genes, when present in their wild-type (*Pina-D1a* and *Pinb-D1a*), create a soft-textured kernel. Any changes in either gene create a harder texture [30]. Traditionally, soft-textured wheat is used for cookies, cakes, pastries, etc. whereas hard wheat is used for bread. More recently, to increase the versatility of durum, both of the wild-type alleles of the puroindoline genes were transferred into durum using the *Ph1* gene as described above. The result was that a small section of the distal tip of chromosome 5DS was transferred to chromosome 5BS of the durum wheat. This generated drastic changes in the milling characteristics of the grain and end-product properties [31,32]. Gazza et al. [31] introduced the two *Pina-D1a* and *Pinb-D1a* alleles in the Italian durum wheat variety Colosseo and showed that the hardness of the grain was strongly reduced as demonstrated by the low SKCS-HI values. In addition, the reduction of the kernel hardness produced several changes in the flour, such as higher flour extraction rates (24%) compared with their hard durum sister line, decreased farinograph water absorption, decreased dough tenacity (P), and, accordingly, alveograph P/L ratio, but increased farinograph stability, mixing tolerance and dough extensibility (L). Moreover, spaghetti cooking quality, as determined by firmness, stickiness, and bulkiness, was unaffected by the kernel hardness, whereas the loaf volume exhibited a 10% increase associated with kernel softening.

2. Introduction of D Genome Gluten Proteins in Durum Wheat

There are two approaches that can be used to introduce the *Glu-D1* subunits. Transgenic lines expressing additional HMW-GS genes have been reported to mostly involve adding subunits 5 + 10 to improve dough strength with the hope of improving breadmaking quality in bread and durum wheat [33–37]. These works showed increases in polymeric protein, mixing times, and mixing tolerance but often produced overly strong doughs. Gadaleta et al. [38] transformed four durum cultivars with 1Dx5 and 1Dy10. Higher SDS sedimentation volume and overly long mixograph peak time and very stable but inextensible dough resulted. Overexpression of subunit 1Dx5 in the durum wheat cultivar Ofanto led to the production of doughs that were too strong for conventional mixograph analysis, resulting in erroneously low mixing time and peak resistance [37]. Given that GMOs still lack widespread consumer acceptance, this approach has not proved effective.

Using chromosome engineering, segments carrying the *Glu-D1* loci containing genes encoding the pairs 1Dx5 + 1Dy10 or 1Dx2 + 1Dy12 have been transferred in durum wheat, replacing the null allele present at the *Glu-A1* locus on the long arm of the 1A chromosome using translocation lines 1AS.1AL-1DL, in an attempt to improve durum breadmaking quality [39–43] (for a recent review on the effects of *Glu-D1* gene introgressions in durum wheat see Morris [44]) (Figure 1). Klindworth et al. [45] reported that dough from the durum 1AS.1AL-1DL translocation lines produced by Joppa et al. [39] had exceptionally strong mixing characteristics but did not improve loaf volume over their parental cultivars. Klindworth et al. [40] noted that RugbyT genotypes, carrying the low molecular weight glutenin subunits LMW-1 banding pattern, which conditions weak gluten, had a lower glutenin-to-gliadin ratio and better loaf volume than genotypes carrying LMW-2, associated with superior quality, suggesting that higher gliadin content and improved extensibility are needed to improve breadmaking quality of 1AS.1AL-1DL translocation genotypes.

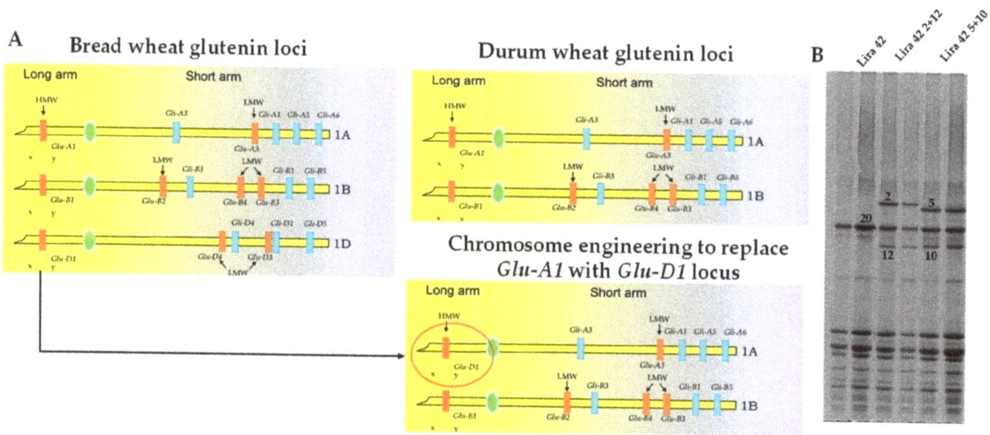

Figure 1. Chromosome localization of glutenin loci in bread and durum wheat. The red circle highlights the *Glu-D1* locus introgressed in durum wheat through chromosome engineering (**A**). SDS-PAGE of glutenin subunits of three durum wheats: Lira 42 and two chromosome engineered lines (Lira 42 2 + 12 and Lira 42 5 + 10) (**B**).

Ammar et al. [46] backcrossed lines carrying 1Dx5 + 1Dy10, produced by Lukaszewski [41] into three durum cultivars and observed a greatly increased SDS sedimentation volume. One of these lines was crossed into the Italian cultivar Svevo, and this increased mixograph peak development time (one indicator of dough strength) from 5.1 to 15.0 min [47]. Sissons et al. [42] used these Lukaszewski [41] lines to develop BC6 NILs possessing 1Dx2 + 1Dy12 or 1Dx5 + 1Dy10 in the durum cv. Svevo, which has the *Glu-B1* 7 + 8 subunit pair present. Mixograph peak development times followed this order, Svevo < Svevo 1Dx2 + 1Dy12 < Svevo 1Dx5 + 1Dy10, the same order as that for Farinograms run at 180 rpm. The addition of 5 + 10 made the dough overstrong and inextensible, and both subunit pairs failed to improve loaf volume over Svevo. The inclusion of these subunits led to a greater amount of the larger polymeric glutenin (higher UPP%), probably due to additional disulfide bond formation, leading to dough that shows more resistance to mixing with higher dough strength. An increased glutenin-to-gliadin ratio has also been reported in lines having a 5 + 10 inclusion [40,46]. Further work by Sissons et al. [48] explored the impact of

Glu-D1 subunits on bread quality in different glutenin subunit backgrounds to see if this provided the dough with more extensibility to improve loaf volume. Durum wheat variety Svevo missing *Glu-B1* subunits 7 + 8 and Lira biotypes with low molecular weight glutenin subunit types 1 (Lira 42) and 2 (Lira 45) and HMW-GS 20, which is known for its negative effect on dough characteristics, were evaluated for their dough strength and breadmaking potential with and without high molecular weight glutenin subunits 2 + 12 or 5 + 10. The absence of subunit pair 7 + 8 in Svevo reduced the overly strong dough-strengthening effect from 5 + 10 but also 2 + 12 compared to what was found previously when 7 + 8 is present. The weak gluten variety Lira 42 genotype had stronger dough from the addition of 2 + 12 and 5 + 10, and both Lira biotypes showed much larger effects on dough strength from the *Glu-D1* pairs than with Svevo, which has moderate strength. There were minor impacts on pasta quality with 2 + 12 or 5 + 10 additions, which should allow flexibility to develop durum with a better balance of glutenin subunits more suited to bread making [49].

Bread prepared using blends of durum flour with commercial wheat flour indicated that at any blend level, durum flour reduced loaf volume in a dose-dependent manner, showing that none of the genotypes could match 100% hexaploid commercial flour for LV and texture. In weaker gluten genotypes (Lira 42, Lira 45) with added 2 + 12 or 5 + 10, LV improved beyond its control genotype but was inferior to 100% bread wheat loaves. However, in the stronger genotype, Svevo, there was no improvement in LV with 2 + 12 and indeed a decline in LV with 5 + 10 present [48]. This indicates that there is an interaction between the added *Glu-D1* subunits and the background glutenin composition. Key *Glu-D1* subunits critical for good breadmaking in hexaploid wheat appear to have limited value in improving loaf structure and volume in durum bread, especially when the proportion of durum to bread wheat flour increases above 25%. The balance of glutenin to gliadin subunits is still not ideal because what is needed is more extensibility in the dough. Klindworth et al. [40] determined that a translocation line carrying the LMW-1 banding pattern, which conditions weak gluten, had better loaf volume and mixing characteristics than lines carrying LMW-2, which conditions strong gluten, supporting this hypothesis. However, Ammar et al. [4] noted that durum carrying HMW-GS 6 + 8 produced bread loaves with higher LV than those produced by genotypes having 7 + 8 or 20, probably due to their higher dough extensibility. These researchers concluded that in order to produce durum wheat with baking performance equivalent to bread wheat, greater dough strength but, more importantly, extensibility is needed. The 1AS.1AL-1DL translocation lines have been reported to produce dough with very low extensibility [50]. The effect of the 2 + 12 addition does depend on the genetic background because when present in a soft durum, dough strength and bread LV were greatly improved [51] but not in lines where the 5 + 10 subunit pair was added because the dough was too strong and inelastic.

3. Manipulation of Starch Composition

The synthesis of amylose and amylopectin is carried out by different classes of enzymes. In particular, a granule-bound starch synthase (GBSSI) is involved in amylose synthesis, whereas amylopectin is produced by the concerted action of different starch synthases (SSI, SSII, SSIII), starch-branching enzymes (SBEI, SBEIIa and SBEIIb) and starch-debranching enzymes of isoamylase- and limit dextrinase-type (ISA and LD) [52–54]. The manipulation of starch composition has been the target of many researchers, thanks to the identification of mutants involved in the synthesis of the amylose and amylopectin and the availability of genomic resources and of new high-throughput technologies (Figure 2). This has made it possible to generate durum and bread wheat with large variations in amylose content and starches with unique functionalities.

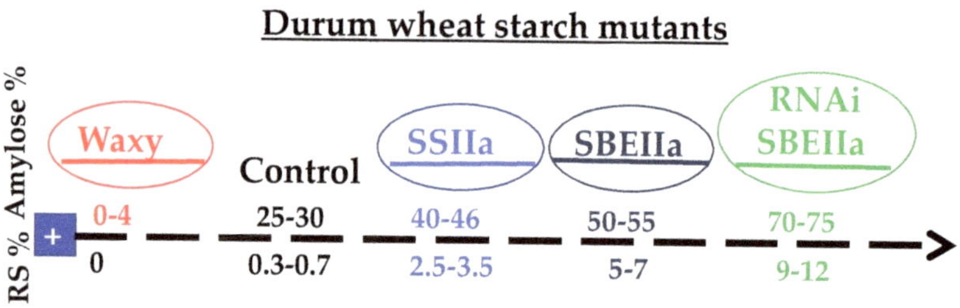

Figure 2. Schematic representation of amylose content and resistant starch (RS) in a panel of durum wheat starch mutants along with a transgenic line obtained through RNA interference (RNAi). SSIIa and SBEIIa indicate mutant lines not expressing the starch synthase of class IIa (SSIIa) and the starch branching enzyme of class IIa (SBEIIa).

3.1. Low Amylose Durum Wheat

The gene-encoding GBSSI enzymes, also termed waxy due to the appearance of the kernel, are located at chromosomes 7A (*Wx-A1*), 4A (*Wx-B1*) and 7D (*Wx-D1*) of bread wheat, while in durum wheat, only the *Wx-A1* and *Wx-B1* genes are present. When only one gene is silenced, partial waxy wheat is obtained, whereas when all the genes are silenced a full waxy is realised.

The different studies of these proteins have identified polymorphism at the three loci, their molecular characterisation and evaluation of their effects on starch composition [55]. The presence of one or two *GBSSI* null alleles results in the production of starch with reduced amylose content in bread wheat. Complete waxy genotypes have been produced in different durum wheat cultivars through two main strategies, the introgression of natural mutations in cultivated varieties and the use of chemical mutagenesis to suppress the activity of their encoding genes [21,56]. The first waxy wheat genotype was produced by Nakamura and colleagues [57] by crossing the bread wheat cv. Kanto 107 (a natural mutant lacking *Wx-A1* and *Wx-B1*) and the wheat landrace Baihuomai. The complete waxy genotype showed a drastic reduction of amylose (close to zero) and an endosperm with a waxy phenotype. A similar approach was followed by Urbano and colleagues [56] that produced a waxy durum wheat line by crossing the cv. Kanto 107 with the durum wheat cv. Svevo. The mutant line had low amylose content (less than 3%) [58].

The complete durum wheat waxy mutants showed drastic changes in starch composition and properties compared with the nonwaxy, suggesting possible new applications in food and non-food industries. Grant et al. [59] highlighted significant differences in the chemical and functional properties of waxy durum starch compared with the starch in the wild type. The full waxy starches had different pasting properties (by RVA) characterised by higher peak viscosities, earlier peak times, lower stabilities and final viscosities [55]. Further differences were also detected in the gelatinization properties: starches gelatinized at higher temperatures and needed more energy to be melted compared with the control durum wheat. Vignaux et al. [60] investigated the quality characteristics of the durum wheat null waxy lines, finding a slight worsening of gluten quality and semolina yields. However, the overall quality characteristics of waxy durum grain were still considered satisfactory. Complete and partial waxy durum wheat genotypes were used to evaluate the effect of waxy mutations on spaghetti quality [61]. A significant worsening was observed for all the main parameters associated with pasta quality (firmness, cooking loss, stickiness). In detail, the cooked waxy pasta was softer and had more cooking loss compared with the control pasta. The higher cooking losses were associated with the absence of the amylose-proteins interaction that is essential for forming a strong network that can trap the exudates [61].

Similarly, Gianibelli and colleagues [62] found that the incorporation of waxy starch isolated from bread wheats into semolina had negative effects on pasta quality due to the increase of stickiness and the reduction of firmness in cooked pasta. Sharma et al. [63], using partial waxy durum wheat genotypes, observed a minor impact on the worsening of pasta quality. From the above cited literature, it is evident that waxy semolina is not suitable for making pasta because of its lower firmness and increased stickiness; however, the softening properties of waxy dough may open new opportunities in other industrial applications.

More recently, Sissons et al. [48] prepared bread made from blends of a set of durum waxy flours and a commercial baker's flour and found that the presence of the 5 + 10 improved pasta quality by increasing the firmness and reducing stickiness and cooking loss of pasta. Though durum wheat is mostly used for the production of pasta, its use for different kind of breads and pastries is very popular in the Mediterranean area, and waxy durum wheat can be used as an antistaling agent [64].

White pan bread was baked from 10, 20, and 30% waxy durum wheat flour composites and evaluated for loaf volume and crumb firmness over a period of 0, 3, and 5 days. Loaf volumes were unaffected by waxy flour blends, but as staling progressed over 3–5 days, significant firming of crumb was observed in the control sample compared with the loaves containing waxy flour. Bread firmness was inversely proportional to the level of waxy flour used in the blend, with a 20% waxy wheat flour blend being optimal in retarding staling while producing bread quality comparable with the control. It was further established that bread made with 20% waxy flour gave lower firmness values after 5 days of storage in comparison with bread made with 3% shortening [65]. According to Shevkani et al., [17] the incorporation of waxy wheat flour positively influences bread making due to the retardation of staling and the extension of shelf life. The breads prepared with waxy wheat (5–30% incorporation levels) had soft and tasty crumb and showed improved shelf life [66]. The waxy common and durum wheat flour incorporated (25%) into common flour also improved loaf expansion during baking and reduced loaf firmness of breads, attributed to the higher gelatinization temperatures and swelling ability of waxy starches that reduced overall water availability [17].

More recently, Sissons et al. [48] used the durum wheat variety Svevo (Sv), a low-amylose line (SvLA, 14,9% amylose) and SvLA 5 + 10, to make blends at 10%, 25%, 50% durum and baker's flour and prepare 100 g loaves. Low-amylose Svevo showed similar loaf volume to Svevo, while the presence of the 5 + 10 HMW-GS had minimal impact on LV, except at 50% with a small improvement in loaf quality. Bread stored up to 7 days became firmer partly due to increased starch retrogradation, and loaves were similar to bread made from baker's flour. Low-amylose Svevo kept the loaf fresher but only up to 3 days of storage. Subunit pair 5 + 10 made the loaf firmer after 7 days compared with control.

3.2. *High Amylose Durum Wheat*

There is strong evidence that numerous chronic health conditions could be prevented or moderated by correct dietary behaviours. In this regard, an important factor is starch digestibility. Based on the digestibility rate, the starches can be divided into rapidly digestible starches, slowly digestible starches, and resistant starches. Foods containing rapidly digested starches may contribute to the onset of chronic diseases because they are easily accessible to digestive enzymes and rapidly digested to glucose molecules with detrimental effects on human health [67]. On the contrary, several studies have demonstrated the existence of a positive correlation between the amylose content in flour or semolina and the resistant starch (RS) in foods, attracting the interest of consumers and the food industry for its role in the prevention of several diet-related diseases [68]. Indeed, RS escapes the digestion in the small intestine and reaches the large bowel, where it is fermented by the bacterial microflora, playing a role as dietary fibre. Short-chain fatty acids (SCFAs), especially butyrate, contribute to the well-being of colonocytes with a possible role in the prevention of colon cancer [69–71]. In addition, increases in SCFAs result in lowering of the bowel pH, and this contributes to hindering the proliferation of pathogenic bacteria and reducing inflammation in irritable bowel disease [72]. In addition, the foods with high resistant starch have low glycaemic index and can help in preventing the onset of type II diabetes and obesity [69,73]. The mechanisms responsible for the reported effects of dietary fibre on metabolic health are still being investigated, but it is speculated to be a result of changes in intestinal viscosity, nutrient absorption, rate of passage, production of short chain fatty acids and production of gut hormones [74]. The beneficial effects for human health associated with the consumption of RS have promoted a growing interest in increasing the amount of amylose in cereal grains [73].

Two main strategies have been followed to raise the amount of amylose in wheat. The first one has focused on the silencing of a key gene of amylopectin biosynthesis involved in the formation of α-1,4 linkages, corresponding to the starch synthases of class IIa (SSIIa) [58,75]. The second strategy targeted another key gene of amylopectin synthesis involved in the formation of α-1,6 linkages, corresponding to the starch branching enzyme of class IIa (SBEIIa) [21,23,24,76]. Lafiandra et al. [58] produced a durum wheat mutant line with an increased amylose content by about 89% compared with the control (37.3% vs. 28.9% amylose) by introgressing, in the cultivar Svevo, null mutations on *SSIIa-A* and *SSIIa-B* homeoalleles, previously identified in bread wheat [77]. A set of 14 SSIIa mutant lines, derived from the backcross with Svevo, was subsequently characterized by Botticella et al. [78], who investigated some major traits such as amylose, resistant starch, starch content, arabinoxylans, α-glucans, seed colour and thousand grain weight. These analyses highlighted large variability for all the parameters among the fourteen lines; the amylose content ranged from 37% to 46% and was positively correlated to the content of resistant starch, which reached up to 3.2% vs. 0.4% in Svevo, confirming their elevated nutritional value. However, the major issue of SSIIa mutant lines is the drastic reduction of some parameters associated with yield (total starch, thousand grain weight) [77].

Hogg et al. [75] produced a new set of partial and complete SSIIa mutants in durum wheat by combining natural and EMS-induced mutations. The authors introgressed a natural knockout mutation on the *SSIIa-A1* gene in the variety Montrail and treated the obtained lines with a chemical mutagen with the aim of knocking out the other homoeoallele (*SSIIa-B1*). Double null SSIIa mutant lines showed a drastic reduction of total starch (up to ~33%); the loss in seed weight was significant but less important (−6%) compared with the reduction reported by Botticella et al. [78], which ranged from −32% to −46%. Martin et al. [79] investigated the influence of the *null* allele at the locus *SSIIa-A* on rheological properties and noodle quality in two segregating durum wheat populations derived from the cross of two SSIIa-A1 null mutants (PI330546 and IG86304) with the cultivar Montrail [75]. Although the swelling power was lower in both the crosses, the amylose amount and noodle firmness were higher in the IG86304 cross compared with the control (cv Montrail), whereas no significant differences were observed in the other cross. In a more recent study, the same research group evaluated pasta quality and nutritional value of the SSIIa null durum wheat genotype compared to the wild-type control line, reporting an increase in pasta firmness and resistance to overcooking, parameters that have a positive influence from a quality point of view [80]. Nevertheless, negative quality effects were described about pasta cooking time and colour, which were diminished, and cooking loss, which increased. From a nutritional view, high amylose pasta had several improved attributes, such as an increased amount of resistant starch and dietary fibre and fewer free carbohydrates. The lower release of carbohydrates, in particular glucose, in the stomach has been associated with beneficial effects for human health in the prevention of diet-related diseases, such as type II diabetes and obesity [81].

The second strategy, based on the suppression of the activity of the two paralogs *SBEIIa* and *SBEIIb*, present in cereals, proved to be more efficient in increasing the amount of amylose in wheat. In bread wheat, Regina et al. [82] used RNA interference and showed that the silencing of the *SBEIIa* isoform was associated with a highly increased proportion of amylose in the transgenic lines (up to 70% of total starch).

Sestili et al. [24] used the RNA interference strategy to down-regulate *SBEIIa* genes in order to increase the amylose content in two durum wheat cultivars, Svevo and Ofanto. Genetic transformation was carried out with the biolistic approach for the cultivar Svevo and with Agrobacterium for Ofanto. The silencing of the *SBEIIa* gene caused marked modifications to amylose content, starch composition and granule morphology. Amylose values ranged from 30.8% up to 75% for Svevo and from 27.7% to 56.4 for Ofanto. Both grain weight and total starch were significantly decreased in all the SBEIIa null lines compared with the controls, but the reduction was lower than that observed in the SSIIa null genotypes.

Following the first attempts to modify starch composition by transgenic approaches, the introduction of the TILLING approach, a non-GM technology, was soon explored in bread and durum wheat [21–23,76]. The silencing of these *SBEIIa* genes by a TILLING approach increased the amylose and resistant starch up to 47.4% and 6.21% in the cultivar Kronos [21] and up to 52.7% and 6.79% in the cultivar Svevo [76], respectively. A modest increase of amylose (+22%) and resistant starch (+115%) was reported by Hazard et al. [23] in the cultivar Kronos, where *SBEIIa* genes were targeted by TILLING.

Subsequently, Hazard et al. [83] pyramiding *SBEIIa* and *SBEIIb* genes found that the mutations in the four starch-branching enzyme II genes of durum wheat resulted in larger increases of amylose and resistant starch content. The presence of the mutations was also associated with an average 5.2% reduction in kernel weight (p = 0.0007) and 15% reduction in grain yield (p = 0.06) compared with the wild type. Technological quality analysis showed that the mutant lines have acceptable quality, with positive effects on pasta firmness but negative effects on semolina extraction and pasta colour. Positive fermentation responses were detected in rats (*Rattus* spp.) fed with diets incorporating SBEIIa/b-AB pasta compared with controls. The differences included significant increases in cecal contents, decreases in cecal pH and increases in cecal SCFAs. Sissons et al. [84] used semolina obtained from the SSIIa and SBEIIa mutants with high amylose content (43.5% and 57.8%, respectively), to prepare spaghetti, with the objective of reducing the glycaemic index of pasta while maintaining acceptable technological properties. The appearance of the pasta showed that both high-amylose (HA) pastas were darker than Svevo, with the SBEIIa pasta being the darkest but having a more desirable appearance than commercial wholemeal pasta. Both HA pastas had reduced fully cooked times, with SSIIa having the lowest, compared with Svevo pasta, which is very likely due to the reduced amount of starch that can gelatinize in the HA pasta. In addition, the cooked HA pasta was softer, with higher cooking loss but lower stickiness compared with Svevo. The in vitro starch digestion extent decreased in both mutants, but much more in SBEIIa, while the human in vivo GI was only significantly reduced in SBEIIa pasta (50 to 38). Overall pasta quality was acceptable in both mutants, but the SBEIIa mutation provides a clear glycaemic benefit and would be much more appealing than wholemeal spaghetti. In addition, the authors suggested that a minimum RS content in spaghetti of ~7% is needed to lower GI, which corresponded to an amylose content of ~58. High-amylose starches do not gelatinize fully at the typical temperatures used in RVA profiles, resulting in very low pasting viscosities [85].

4. Characterization of Soft Durum Wheat

The overexpression of the *Pina* gene was obtained by Li et al. (2014) [86], through a transgenic approach, in order to reduce the hardness of a durum wheat line and observe the effects of the PINA protein on the kernel texture and other characteristics. The analysis of grain hardness showed that the PINA overexpression reduced grain hardness with medium–hard durum wheat grain. Li et al. (2012) [87] also analysed lines coexpressing the HMW-GS 1Ax1 and lines separately expressing 1Ax1 subunit or the PINA protein. Dough mixing analysis of these lines showed that expression of subunit 1Ax1 positively influenced dough strength and overmixing tolerance, while expression of PINA negatively influenced dough resistance to extension. Lines coexpressing 1Ax1 and PINA showed faster hydration of flour during mixing, very likely due to the lower water absorption and damaged starch associated with PINA expression. Moreover, the presence of the 1Ax1 HMW-GS seemed to compensate the negative effect of PINA on dough resistance to extension.

The first commercial durum variety carrying the translocation with the puroindoline genes was Soft Svevo, the soft-textured kernel version of the durum variety Svevo [32]. In a direct comparison of Svevo and Soft Svevo grown at the same location, the single-kernel characterization (SKCS) hardness index of Soft Svevo was ~31, whereas the hardness index of Svevo was ~78.6 [88]. Soft wheat typically averages around 25 for hardness index, hard wheat approximately 75.

When Soft Svevo and Svevo were milled on a traditional hexaploid-type mill along with check varieties of soft white winter and hard red spring wheats, Soft Svevo had a milling profile similar to, and in some cases better than, that of the soft white winter wheat. Svevo had poor break flour and total flour yields (~17 and 52%, respectively), which was to be expected on a hexaploid-type mill [88]. The average flour ash for Soft Svevo was 0.50, compared with 0.62 for Svevo. Furthermore, the starch damage for Soft Svevo was 1.7%, whereas that for Svevo was 6.4% [88]. Although this level of starch damage for Svevo was on a hexaploid-type mill, the levels of starch damage seen in hard-textured durum when milled are prohibitively high for products other than pasta and couscous.

The milling energy of Svevo vs. Soft Svevo was examined for paired, triplicate samples of each from 12 growing locations [89]. Several key parameters were studied, including the energy required to grind wheat (kJ/kg) as well as the energy required to produce 1 kg of flour (kJ/kg), the total flour produced, and the starch damage. At the first break rolls, Soft Svevo required 13.9 kJ/kg, compared with 17.4 kJ/kg in Svevo. Similarly, in the second break rolls, Soft Svevo and Svevo required 19.4 and 29.8 kJ/kg, respectively. The difference in energy requirements was more pronounced when comparing the energy required to produce 1 kg of flour. At the first break rolls, Soft Svevo and Svevo required 153 and 789 kJ/kg flour, respectively. At the second break rolls, Soft Svevo and Svevo required 103 and 288 kJ/kg flour, respectively. These differences amount to between three and five times more energy needed for Svevo compared to Soft Svevo to produce 1 kg of flour. Furthermore, the total flour yield from this milling was 17.7% in Soft Svevo, compared with 8.7% in Svevo. The resultant flour from Soft Svevo had 2.4% starch damage compared with 11.0% in Svevo [89]. Not only was the milling much more efficient for Soft Svevo but produced a higher-quality flour.

Soft Svevo, along with a second soft durum, Soft Alzada, have been examined for baking properties compared to Svevo [90]. Both soft- and hard-wheat products were tested, although continued emphasis was more on bread products. Soft Svevo made sugar snap cookies with a similar diameter to those of the soft white winter wheat Xerpha. However, the soft durum line Soft Alzada made cookies of a similar diameter to the elite soft white wheat on the market [90]. Cookies made by Svevo were quite small, as had been expected for durum. In a similar study using populations made from a cross between Soft Svevo and a number of CIMMYT durum lines, the soft durum populations had excellent soft wheat quality. The sugar snap cookies were as large as or larger than the elite released varieties from the Pacific Northwest [91].

5. Introduction of the HMW-GS 1Dx2 + 1Dy12, 1Dx5 + 1Dy10 and (*Gpc-B1*) Allele in Soft Durum

Although soft durum made bread with better quality than the soft white winter wheat Xerpha, the gluten strength and loaf volume were not on par with those of hard wheat [90,92]. One challenge with durum is the lack of the D genome and the HMW glutenin subunits found there. The 5 + 10 and 2 + 12 alleles have a strong influence on gluten properties and strength. Soft Svevo and Soft Alzada showed moderate gluten strength, although comparatively much weaker than that of a hard red spring variety [90]. Three approaches were taken to improve gluten strength of soft durums: introgression into soft durum of the *Glu-D1* alleles of 2 + 12 and 5 + 10 and crossing the functional *Gpc-B1* allele into soft durum, the hypothesis being that introducing the *Gpc-B1* allele would increase protein content and potentially the dough strength.

Two experimental lines, UCRD01-05 and UCRD01-01, which were two 1D-1B chromosomal translocation lines with the 2 + 12 and 5 + 10 alleles, respectively, were crossed with Soft Svevo and carried out to the F5 and F7 generations, respectively [51]. The gluten strength measurements of flour sodium dodecyl sulfate (SDS) and lactic acid solvent retention capacity increased dramatically with the introgression of the 2 + 12 alleles. Furthermore, bread dough mixing qualities improved substantially from a very weak, poor mixing dough without the 2 + 12 alleles to a moderately strong mixing dough resembling that of a hard winter wheat. The introgression of the 2 + 12 alleles also improved the loaf volume by 131 cm^3, from 762 cm^3 without 2 + 12, 893 cm^3 with 2 + 12. Although the 2 + 12 allele, known to be generally weaker, did improve the dough and bread quality substantially, the introgression of the 5 + 10 alleles, known to be stronger, did not have the same effect. The introgression of the 5 + 10 alleles did improve some of the mixing parameters measured with the Mixograph. However, in both the Mixograph and the bread baking, the dough showed too much elasticity and not enough extensibility to create sufficient oven spring. In combination with the other alleles in Soft Svevo, adding the 5 + 10 alleles created a dough that was too strong and not extensible enough [51].

The third approach to improving gluten strength of Soft Svevo was gaining a functional Grain protein content-B1 (*Gpc-B1*) allele in Soft Svevo by crossing with Desert King-High Protein. The protein content of the Soft Svevo with *Gpc-B1* increased by 1.7% along with an increase in the flour SDS sedimentation volume. However, the Mixograph dough parameters were not markedly improved with *Gpc-B1*. Bread loaf volume was similarly improved only marginally [92].

The traditional idea that pasta needed to be made from durum semolina to result in a high-quality product was challenged in a pasta study comparing commercial durum semolina, durum flour and three varieties of soft durum [93]. The pasta weight increase during cooking was similar across the soft durum flour and commercial semolina samples. However, the soft durum samples had lower cooking loss than durum semolina (average 3.93% vs. 5.12%, respectively). Optimum cooking time was also slightly lower in soft durum samples compared with durum semolina. Pasta firmness is an important sensory parameter, contributing greatly to the pleasant pasta-eating experience. Soft durum flour had pasta firmness comparable with that of durum semolina. Additionally, stickiness is a negative attribute in pasta. Soft durum pasta had similar stickiness to, and for some samples, less than that of durum semolina pasta [93].

6. Conclusions

Durum wheat has limited uses compared with bread wheat, with pasta being the major end product, and this can be associated with its high hardness and lack of D genome. The use of classical and innovative biotechnological tools has made it possible to modify the processing and nutritional characteristics of durum wheat. In particular, modification of kernel texture and introduction of D-genome HMW-GS have both been achieved via *ph1*-mediated homoeologous recombination and the transfer of genetic material from bread wheat to durum wheat, whereas mutagenesis via TILLING has proved very effective in modifying starch composition.

Durum wheat for pan bread use is limited due to its weak and/or inextensible gluten. The introduction of *Glu-D1* alleles from bread wheat improved dough strength in a range of genetic backgrounds. Although the 1Dx5 + 1Dy10 produced stronger dough than 1Dx2 + 1Dy12, the dough was excessively strong and inelastic, and loaf volume was either inferior or not affected. In contrast, the 1Dx2 + 1Dy12 sometimes improved bread LV particularly when the background genotype had weak gluten strength, as in the Lira biotypes. Generally, greater dough strength did not result in better loaf volume, so the modified durum was still unable to match the bread quality of the hexaploid wheat. It is suggested that dough extensibility needs to be improved to allow higher LV. This could be achieved by exploiting better the genetic variation that exists in durum and/or using the *Glu-D1* gene introgressions.

The use of mutagenesis of the genes involved in starch biosynthesis has permitted selecting durum wheat lines with large variation in amylose content, capable to satisfy the demand of foods with high nutritional value and counteract the onset of important diseases.

Durum wheat kernel hardness has been modified, making milling accessible to those with traditional roller mills without sacrificing pasta quality. Soft durum also has some unique properties that make it a good candidate for niche products like extruded snack foods. The current and future work concerning soft durum and glutenins is focused on lines carrying the 2 + 12 alleles and continuing to make crosses with high-quality soft durum lines for both bread quality and agronomic properties. Soft durum has also shown promise with extrusion for snack foods and breakfast cereals [94]. Currently, there are soft durum lines crossed with waxy (low- to zero-amylose) and high-amylose lines in an effort to create even higher-quality end products with improved characteristics related to the transformation processes and to the consumer demand for healthy food.

Author Contributions: Conceptualization, D.L., F.S., M.S., A.K. and C.F.M.; methodology, D.L., F.S., M.S., A.K. and C.F.M.; software, D.L., F.S., M.S., A.K. and C.F.M.; validation, D.L., F.S., M.S., A.K. and C.F.M.; formal analysis, D.L., F.S., M.S., A.K. and C.F.M.; investigation, D.L., F.S., M.S., A.K. and C.F.M.; resources, D.L., F.S., M.S., A.K. and C.F.M.; data curation, D.L., F.S., M.S., A.K. and C.F.M.; writing—original draft preparation, D.L., F.S., M.S., A.K. and C.F.M.; writing—review and editing, D.L., F.S., M.S., A.K. and C.F.M.; visualization, not applicable; supervision, not applicable; project administration, D.L., F.S., M.S., A.K. and C.F.M.; funding acquisition, D.L., F.S., M.S., A.K. and C.F.M. All authors have read and agreed to the published version of the manuscript.

Funding: Work on high-amylose and resistant starch in the laboratory of Genetics and Biochemistry of Plant Proteins (University of Tuscia) was supported by the AGER (Agroalimentare e Ricerca) "From Seed to Pasta" project, by the Italian Ministry of Education, University and Research (MIUR) in the frame of the MIUR initiative "Departments of excellence", Law 232/2016 and by PRIMA Foundation, PRIMA SECTION 1 2020 AGROFOOD VALUE CHAIN IA TOPIC: 1.3.1-2020 (IA), MEDWHEALTH project grant no. 2034.

Institutional Review Board Statement: Not applicable.

Informed Consent Statement: Not applicable.

Data Availability Statement: No new data were created or analyzed in this study. Data sharing is not applicable to this article.

Conflicts of Interest: The authors declare no conflict of interest.

References

1. Abecassis, J.; Cuq, B.; Boggini, G.; Namoune, H. Other traditional durum-derived products, Chapter 10. In *Durum Wheat Chemistry and Technology*, 2nd ed.; Sissons, M., Abecassis, J., Marchylo, B., Carcea, M., Eds.; AACC International Inc.: St. Paul, MN, USA, 2012; pp. 177–199.
2. Pasqualone, A. Italian durum wheat breads. In *Bread Consumption and Health*; Silva Clerici, M.T.P., Ed.; Nova Biomedical: New York, NY, USA, 2012; pp. 57–80.
3. Quaglia, G.B. Other durum wheat products. In *Durum Wheat: Chemistry and Technology*; Fabriani, G., Lintas, C., Eds.; American Association of Cereal Chemistry: St. Paul, MN, USA, 1988; pp. 263–282.
4. Ammar, K.; Kronstad, W.E.; Morris, C.F. Breadmaking quality of selected durum wheat genotypes and its relationship with high molecular weight glutenin subunits allelic variation and gluten protein polymeric composition. *Cereal Chem.* **2000**, *77*, 230–236. [CrossRef]
5. Boggini, G.; Tusa, P.; Pogna, N.E. Bread making quality of durum genotypes with some novel glutenin compositions. *J. Cereal Sci.* **1995**, *22*, 105–113. [CrossRef]
6. Dexter, J.E.; Matsuo, R.R.; Preston, K.R.; Kilborn, R.H. Comparison of gluten strength, mixing properties, baking quality and spaghetti quality of some Canadian durum and common wheats. *Can. Inst. Food Technol. J.* **1981**, *14*, 108–111. [CrossRef]
7. Guzman, C.; Autrique, J.E.; Mondal, S.; Singh, R.P.; Govindan, V.; Morales-Dorantes, A.; Pena, R.J. Response to drought and heat stress on wheat quality, with special emphasis on breadmaking quality, in durum wheat. *Field Crops Res.* **2016**, *186*, 157–165. [CrossRef]
8. Hareland, G.A.; Puhr, D.P. Baking performance of durum and soft wheat flour in a sponge-dough breadmaking procedure. *Cereal Chem.* **1999**, *75*, 830–835. [CrossRef]

9. Sissons, M.J. Role of durum wheat composition on the quality of pasta and bread. *Food* **2008**, *2*, 75–90.
10. Shewry, P.R.; Halford, N.G. Cereal seed storage proteins: Structures, properties and role in grain utilization. *J. Exp. Bot.* **2002**, *53*, 947–958. [CrossRef]
11. Kiszonas, A.M.; Morris, C.F. Wheat breeding for quality: A historical review. *Cereal Chem.* **2018**, *95*, 17–34. [CrossRef]
12. Payne, P.L. Genetics of wheat storage proteins and the effect of allelic variation on bread making quality. *Plant Physiol.* **1987**, *38*, 141–153. [CrossRef]
13. Ceoloni, C.; Forte, P.; Gennaro, A.; Micali, S.; Carozza, R.; Bitti, A. Recent developments in durum wheat chromosome engineering. *Cytogenet. Genome Res.* **2005**, *109*, 328–334. [CrossRef]
14. Payne, P.I.; Jackson, E.A.; Holt, L.M. The association between -gliadin 45 and gluten strength in durum wheat varieties: A direct causal effect or the result of g netic linkage? *J. Cereal Sci.* **1984**, *2*, 73–81. [CrossRef]
15. Pogna, N.E.; Autran, J.-C.; Mellini, F.; Lafiandra, D.; Feillet, P. Chromosome 1B encoded gliadins and glutenin subunits in durum wheat: Genetics and relationship to gluten strength. *J. Cereal Sci.* **1990**, *11*, 15–34. [CrossRef]
16. Ruiz, M.; Giraldo, P. The influence of allelic variability of prolamins on gluten quality in durum wheat: An overview. *J. Cereal Sci.* **2021**, *101*, 103304. [CrossRef]
17. Shevkani, K.; Singh, N.; Bajaj, R.; Kaur, A. Wheat starch production, structure, functionality and applications—A review. *Int. J. Food Sci. Technol.* **2017**, *52*, 38–58. [CrossRef]
18. Hardy, K.; Brand-Miller, J.; Brown, K.D.; Thomas, M.G.; Copeland, L. The importance of dietary carbohydrate in human evolution. *Q. Rev. Biol.* **2015**, *90*, 251–268. [CrossRef]
19. Wang, P.; Xu, X. Modified starches and the stability of frozen foods. In *Starch in Food*; Woodhead Publishing: Sawston, UK, 2018; pp. 581–593.
20. McCallum, C.M.; Comai, L.; Greene, E.A.; Henikoff, S. Targeting induced local lesions in genomes (TILLING) for plant functional genomics. *Plant Physiol.* **2000**, *123*, 439–442. [CrossRef]
21. Slade, A.J.; McGuire, C.; Loeffler, D.; Mullenberg, J.; Skinner, W.; Fazio, G.; Holm, A.; Brandt, K.M.; Steine, M.N.; Goodstal, J.F.; et al. Development of high amylose wheat through TILLING. *BMC Plant Biol.* **2012**, *12*, 69. [CrossRef]
22. Dong, C.; Vincent, K.; Sharp, P. Simultaneous mutation detection of three homoeologous genes in wheat by High Resolution Melting analysis and Mutation Surveyor. *BMC Plant Biol.* **2009**, *9*, 143. [CrossRef]
23. Hazard, B.; Zhang, X.; Colasuonno, P.; Uauy, C.; Beckles, D.M.; Dubcovsky, J. Induced mutations in the Starch Branching Enzyme II (SBEII) genes increase amylose and resistant starch content in durum wheat. *Crop Sci.* **2012**, *52*, 1754–1766.
24. Sestili, F.; Janni, M.; Doherty, A.; Botticella, E.; D'Ovidio, R.; Masci, S.; Jones, H.D.; Lafiandra, D. Increasing the amylose content of durum wheat through silencing of the SBEIIa genes. *BMC Plant Biol.* **2010**, *10*, 144. [CrossRef]
25. Botticella, E.; Sestili, F.; Hernandez-Lopez, A.; Phillips, A.; Lafiandra, D. High resolution melting analysis for the detection of EMS induced mutations in wheat *Sbella* genes. *BMC Plant Biol.* **2011**, *11*, 156. [CrossRef] [PubMed]
26. Fabriani, G.; Lintas, C. Durum wheat. In *Chemistry and Technology*; American Association of Cereal Chemists: St. Paul, MN, USA, 1988.
27. Posner, E.S.; Hibbs, A.N. Milling of soft and durum wheats. In *Wheat Flour Milling*; AACC International: St. Paul, MN, USA, 1997; pp. 257–283.
28. Bhave, M.; Morris, C.F. Molecular genetics of puroindolines and related genes: Regulation of expression, membrane binding properties and applications. *Plant Mol. Biol.* **2008**, *66*, 221–231. [CrossRef] [PubMed]
29. Bhave, M.; Morris, C.F. Molecular genetics of puroindolines and related genes: Allelic diversity in wheat and other grasses. *Plant Mol. Biol.* **2008**, *66*, 205–219. [CrossRef] [PubMed]
30. Morris, C.F. 2016 Cereals—Overview of uses: Accent on wheat grain. In *Encyclopedia of Food Grains*, 2nd ed.; Wrigley, C., Corke, H., Seetharaman, K., Faubion, J., Eds.; Academic Press: Oxford, UK, 2016; pp. 1–7.
31. Gazza, L.; Sgrulletta, D.; Cammerata, A.; Gazzelloni, G.; Perenzin, M.; Pogna, N.E. Pastamaking and breadmaking quality of soft-textured durum wheat lines. *J. Cereal Sci.* **2011**, *54*, 481–487. [CrossRef]
32. Morris, C.F.; Simeone, M.C.; King, G.E.; Lafiandra, D. Transfer of soft kernel texture from *Triticum aestivum* to durum wheat, *Triticum turgidum* ssp. *durum*. *Crop Sci.* **2011**, *51*, 114–122. [CrossRef]
33. Barro, F.; Rooke, L.; Bekes, F.; Gras, P.; Tatham, A.S.; Fido, R.; Barcelo, P. Transformation of wheat with high molecular weight subunit genes results in improved functional properties. *Nat. Biotechnol.* **1997**, *15*, 1295–1299. [CrossRef]
34. Blechl, A.; Lin, J.; Nguyen, S.; Chan, R.; Anderson, O.D.; Dupont, F.M. Transgenic wheats with elevated levels of Dx5 and/or Dy10 high-molecular-weight glutenin subunits yield doughs with increased mixing strength and tolerance. *J. Cereal Sci.* **2007**, *45*, 172–183. [CrossRef]
35. Butow, B.J.; Tatham, A.S.; Savage, A.W.J.; Gilbert, S.M.; Shewry, F.M.; Solomon, R.G.; Bekes, F. Creating a balance—The incorporation of a HMW glutenin subunit into transgenic wheat lines. *J. Cereal Sci.* **2003**, *38*, 181–187. [CrossRef]
36. Graybosch, R.A.; Seabourn, B.; Chen, Y.R.; Blechl, A.E. Quality and agronomic effects of three high-molecular-weight glutenin subunit transgenic events in winter wheat. *Cereal Chem.* **2011**, *88*, 95–102. [CrossRef]
37. He, G.Y.; Rooke, L.; Steele, S.; Bekes, F.; Gras, P.; Tatham, A.S.; Lazzeri, P.A. Transformation of pasta wheat (*Triticum turgidum* L. var. *durum*) with high-molecular weight glutenin subunit genes and modification of dough functionality. *Mol. Breed.* **1999**, *5*, 377–386. [CrossRef]

38. Gadaleta, A.; Blechl, A.E.; Nguyen, S.; Cardone, M.F.; Ventura, M.; Quick, J.S.; Blanco, A. Stably expressed D-genome derived HMW glutenin subunit genes transformed into different durum wheat genotypes change dough mixing properties. *Mol. Breed.* **2008**, *22*, 267–279. [CrossRef]
39. Joppa, L.R.; Klindworth, D.L.; Hareland, G.A. Transfer of high molecular weight glutenins from spring wheat to durum wheat. In Proceedings of the 9th International Wheat Genetics Symposium, Saskatoon, SK, Canada, 2–8 August 1998; Slinkard, A.E., Ed.; University of Saskatchewan Extension Press: Saskatoon, SK, Canada, 1998; pp. 257–260.
40. Klindworth, D.L.; Hareland, G.A.; Elias, E.M.; Ohm, J.-B.; Puhr, D.; Xu, S.S. Interactions of genotype and glutenin subunit composition on breadmaking quality of durum 1AS.1AL.-1DL translocation lines. *Cereal Chem.* **2014**, *91*, 211–217. [CrossRef]
41. Lukaszewski, A.J. Registration of six germplasms of durum wheat with introgressions of the *Glu-D1* locus. *Crop Sci.* **2003**, *43*, 1138–1139. [CrossRef]
42. Sissons, M.; Pleming, D.; Margiotta, B.; D'Egidio, M.G.; Lafiandra, D. Effect of the introduction of D-genome related gluten proteins on durum wheat pasta and bread making quality. *Crop Pasture Sci.* **2014**, *65*, 27–37. [CrossRef]
43. Vitellozzi, F.; Ciaffi, M.; Dominici, L.; Ceoloni, C. Isolation of a chromosomally engineered durum wheat line carrying the common wheat *Glu-D1d* allele. *Agronomie* **1997**, *17*, 413–419. [CrossRef]
44. Morris, C.F. Bread-baking quality and the effects of Glu-D1 Gene introgressions in durum wheat (*Triticum turgidum* ssp. *durum*). *Cereal Chem.* **2021**, *98*, 1151–1158. [CrossRef]
45. Klindworth, D.L.; Hareland, G.A.; Elias, E.M.; Xu, S.S. Agronomic and quality characteristics of 1as. *1al-1dl translocation lines of durum wheat carrying glutenin allele Glu-D1d. Crop Sci.* **2005**, *45*, 77–84.
46. Ammar, K.; Lukaszewsky, A.J.; Banowetz, G.M. Effect of *Glu-D1* (5+10) on gluten strength and polymeric protein composition in durum wheat. *Cereal Foods World* **1997**, *42*, 610.
47. Lafiandra, D.; Margiotta, B.; Colaprico, G.; Masci, S.I.; Roth, M.R.; MacRitchie, F. Introduction of the D-genome related high-and low-Mr glutenin subunits into durum wheat and their effect on technological properties. In Proceedings of the 7th International Gluten Workshop, Bristol, UK, 2–6 April 2000; Shewry, P.R., Tatham, A.S., Eds.; Royal Society of Chemistry: London, UK, 2000; pp. 51–54.
48. Sissons, M.; Pleming, D.; Sestili, F.; Lafiandra, D. Effect of *Glu-D1* introgression and amylose content on breadmaking potential of blends of durum and hexaploid wheat. *Cereal Chem.* **2019**, *96*, 193–206. [CrossRef]
49. Sissons, M.; Sestili, F.; Lafiandra, D. Effect of Glu-D1 introgression on dough and pasta making quality of durum wheat lines with different glutenin composition and amylose content. *Cereal Chem.* **2019**, *96*, 207–219. [CrossRef]
50. Rao, B.N. A Study of the Rheological Properties and Gluten Protein Components Associated with Enhanced Baking Quality in Durum Wheat (*Triticum turgidum* L. var. *durum*). Master's Thesis, University of Saskatchewan, Saskatoon, SK, Canada, 2008. Available online: https://harvest.usask.ca/handle/10388/etd-09142008-143422 (accessed on 20 May 2022).
51. Kiszonas, A.M.; Ibba, M.I.; Boehm, J.D., Jr.; Morris, C.F. Effects of *Glu-D1* gene introgressions on soft white spring durum wheat (*Triticum turgidum* ssp. durum) quality. *Cereal Chem.* **2021**, *98*, 1112–1122. [CrossRef]
52. Botticella, E.; Savatin, D.V.; Sestili, F. The triple jags of Dietary Fibers in cereals: How biotechnology is longing for High Fiber grains. *Front. Plant Sci.* **2021**, *12*, 745579. [CrossRef] [PubMed]
53. Ball, S.G.; Morell, M.K. From bacterial glycogen to starch: Understanding the biogenesis of the plant starch granule. *Ann. Rev. Plant Biol.* **2003**, *54*, 207–233. [CrossRef] [PubMed]
54. Sestili, F.; Sparla, F.; Botticella, E.; Janni, M.; D'Ovidio, R.; Giuseppe, F.; Marri, L.; Cuesta-Seijo, J.A.; Trost, P.; Lafiandra, D. The down-regulation of the genes encoding Isoamylase 1 alters the starch composition of the durum wheat grain. *Plant Sci.* **2016**, *252*, 230–238. [CrossRef] [PubMed]
55. Guzmán, C.; Alvarez, J.B. Wheat waxy proteins: Polymorphism, molecular characterization and effects on starch properties. *Theor. Appl. Genet.* **2016**, *129*, 1–16. [CrossRef]
56. Urbano, M.; Margiotta, B.; Colaprico, G.; Lafiandra, D. Waxy proteins in diploid, tetraploid and hexaploid wheats. *Plant Breed.* **2002**, *121*, 465–469. [CrossRef]
57. Nakamura, T.; Kuroda, A. Variations in the content of starch-granule bound protein among several Japanese cultivars of common wheat (*Triticum aestivum* L.). *Euphytica* **1992**, *64*, 215–219.
58. Lafiandra, D.; Sestili, F.; D'Ovidio, R.; Janni, M.; Botticella, E.; Ferrazzano, G.; DeAmbrogio, E. Approaches for modification of starch composition in durum wheat. *Cereal Chem.* **2010**, *87*, 28–34. [CrossRef]
59. Grant, L.A.; Vignaux, N.; Doehlert, D.C.; McMullen, M.S.; Elias, E.M.; Kianian, S. Starch characteristics of waxy and nonwaxy tetraploid (*Triticum turgidum* L. var. durum) wheats. *Cereal Chem.* **2001**, *78*, 590–595. [CrossRef]
60. Vignaux, N.; Doehlert, D.C.; Hegstad, J.; Elias, E.M.; McMullen, M.S.; Grant, L.A.; Kianian, S.F. Grain quality characteristics and milling performance of full and partial waxy durum lines. *Cereal Chem.* **2004**, *81*, 377–383. [CrossRef]
61. Vignaux, N.; Doehlert, D.C.; Elias, E.M.; McMullen, M.S.; Grant, L.A.; Kianian, S.F. Quality of spaghetti made from full and partial waxy durum wheat. *Cereal Chem.* **2005**, *82*, 93–100. [CrossRef]
62. Gianibelli, M.C.; Sissons, M.J.; Batey, I.L. Effect of source and proportion of waxy starches on pasta cooking quality. *Cereal Chem.* **2005**, *82*, 321–327. [CrossRef]
63. Sharma, R.; Sissons, M.J.; Rathjen, A.J.; Jenner, C.F. The Null-4A allele at the waxy locus in durum wheat affects pasta cooking quality. *J. Cereal Sci.* **2002**, *35*, 287–297. [CrossRef]
64. Šárka, E.; Dvořáček, V. New processing and applications of waxy starch (a review). *J. Food Eng.* **2017**, *206*, 77–87. [CrossRef]

65. Bhattacharya, M.; Erazo-Castrejón, S.V.; Doehlert, D.C.; McMullen, M.S. Staling of bread as affected by waxy wheat flour blends. *Cereal Chem.* **2002**, *79*, 178–182. [CrossRef]
66. Hayakawa, K.; Tanaka, K.; Nakamura, T.; Endo, S.; Hoshino, T. End use quality of waxy wheat flour in various grain-based foods. *Cereal Chem.* **2004**, *81*, 666–672. [CrossRef]
67. Birt, D.F.; Boylston, T.; Hendrich, S.; Jane, J.L.; Hollis, J.; Li, L.; McClelland, J.; Moore, S.; Phillips, G.J.; Rowling, M.; et al. Resistant starch: Promise for improving human health. *Adv. Nutr.* **2013**, *4*, 587–601. [CrossRef]
68. Sestili, F.; Botticella, E.; Lafiandra, D. TILLING for improved starch composition in wheat. In *Genomics of Plant Genetic Resources*; Tuberosa, R., Graner, A., Frison, E., Eds.; Springer: Berlin/Heidelberg, Germany, 2014; pp. 467–487.
69. Topping, D. Cereal complex carbohydrates and their contribution to human health. *J. Cereal Sci.* **2007**, *46*, 220–229. [CrossRef]
70. Lockyer, S.; Nugent, A.P. Health effects of resistant starch. *Nutr. Bull.* **2017**, *42*, 10–41. [CrossRef]
71. Carretta, M.D.; Quiroga, J.; López, R.; Hidalgo, M.A.; Burgos, R.A. Participation of short-chain fatty acids and their receptors in gut inflammation and colon cancer. *Front. Physiol.* **2021**, *12*, 662739. [CrossRef]
72. Montroy, J.; Berjawi, R.; Lalu, M.M.; Podolsky, E.; Peixoto, C.; Sahin, L.; Stintzi, A.; Mack, D.; Fergusson, D.A. The effects of resistant starches on inflammatory bowel disease in preclinical and clinical settings: A systematic review and meta-analysis. *BMC Gastroenterol.* **2020**, *20*, 372. [CrossRef] [PubMed]
73. Lafiandra, D.; Riccardi, G.; Shewry, P.R. Improving cereal grain carbohydrates for diet and health. *J. Cereal Sci.* **2014**, *59*, 312–326. [CrossRef]
74. Lattimer, J.M.; Haub, M.D. Effects of dietary fiber and its components on metabolic health. *Nutrients* **2010**, *2*, 1266–1289. [CrossRef]
75. Hogg, A.C.; Gause, K.; Hofer, P.; Martin, J.M.; Graybosch, R.A.; Hansen, L.E.; Giroux, M.J. Creation of a high-amylose durum wheat through mutagenesis of starch synthase II (SSIIa). *J. Cereal Sci.* **2013**, *57*, 377–383. [CrossRef]
76. Sestili, F.; Palombieri, S.; Botticella, E.; Mantovani, P.; Bovina, R.; Lafiandra, D. TILLING mutants of durum wheat result in a high amylose phenotype and provide information on alternative splicing mechanisms. *Plant Sci.* **2015**, *233*, 127–133. [CrossRef] [PubMed]
77. Yamamori, M.; Endo, T.R. Variation of starch granule proteins and chromosome mapping of their coding genes in common wheat. *Theor. Appl. Genet.* **1996**, *93*, 275–281. [CrossRef]
78. Botticella, E.; Sestili, F.; Ferrazzano, G.; Mantovani, P.; Cammerata, A.; D'Egidio, M.G.; Lafiandra, D. The impact of the SSIIa null mutations on grain traits and composition in durum wheat. *Breed. Sci.* **2016**, 16025. [CrossRef]
79. Martin, J.M.; Hogg, A.C.; Hofer, P.; Manthey, F.A.; Giroux, M.J. Impacts of *SSIIa-A* null allele on durum wheat noodle quality. *Cereal Chem.* **2014**, *91*, 176–182. [CrossRef]
80. Hogg, A.C.; Martin, J.M.; Manthey, F.A.; Giroux, M.J. Nutritional and quality traits of pasta made from SSIIa null high-amylose durum wheat. *Cereal Chem.* **2015**, *92*, 395–400. [CrossRef]
81. Nugent, A.P. Health properties of resistant starch. *Nutr. Bull.* **2005**, *30*, 27–54. [CrossRef]
82. Regina, A.; Bird, A.; Topping, D.; Bowden, S.; Freeman, J.; Barsby, T.; Kosar-Hashemi, B.; Li, Z.; Rahman, S.; Morell, M. High-amylose wheat generated by RNA interference improves indices of large-bowel health in rats. *Proc. Natl Acad. Sci. USA* **2006**, *103*, 3546–3551. [CrossRef] [PubMed]
83. Hazard, B.; Zhang, X.; Naemeh, M.; Hamilton, M.K.; Rust, B.; Raybould, H.E.; Newman, J.W.; Martin, R.; Dubcovsky, J. Mutations in durum wheat *SBEII* genes affect grain yield components, quality, and fermentation responses in rats. *Crop Sci.* **2015**, *55*, 2813–2825. [CrossRef] [PubMed]
84. Sissons, M.; Sestili, F.; Botticella, E.; Masci, S.; Lafiandra, D. Can manipulation of durum wheat amylose content reduce the glycaemic index of spaghetti? *Foods* **2020**, *9*, 693. [CrossRef] [PubMed]
85. Soh, H.N.; Sissons, M.J.; Turner, M.A. Effect of starch granule size distribution and elevated amylose content on durum dough rheology and spaghetti cooking quality. *Cereal Chem.* **2006**, *83*, 513–519. [CrossRef]
86. Li, Y.; Wang, Q.; Li, X.Y.; Xiao, X.; Sun, F.S.; Wang, C.; Hu, W.; Feng, Z.; Chang, J.; Chen, M.; et al. Coexpression of the high molecular weight glutenin subunit 1Ax1 and puroindoline improves dough mixing properties in durum wheat (*Triticum turgidum* L. ssp. *durum*). *PLoS ONE* **2012**, *7*, e50057. [CrossRef] [PubMed]
87. Li, Y.; Mao, X.; Wang, Q.; Zhang, J.R.; Li, X.Y.; Ma, F.Y.; Sun, F.; Chang, J.; Chen, M.; Wang, Y.; et al. Overexpression of *Puroindoline a* gene in transgenic durum wheat (*Triticum turgidum* ssp. *durum*) leads to a medium-hard kernel texture. *Mol. Breed.* **2014**, *33*, 545–554. [CrossRef]
88. Murray, J.C.; Kiszonas, A.M.; Wilson, J.; Morris, C.F. Effect of soft kernel texture on the milling properties of soft durum wheat. *Cereal Chem.* **2016**, *93*, 513–517. [CrossRef]
89. Heinze, K.; Kiszonas, A.M.; Murray, J.C.; Morris, C.F.; Lullien-Pellerin, V. Puroindoline genes introduced into durum wheat reduce milling energy and change milling behavior similar to soft common wheats. *J. Cereal Sci.* **2016**, *71*, 183–189. [CrossRef]
90. Murray, J.C.; Kiszonas, A.M.; Morris, C.F. Influence of soft kernel texture on the flour, water absorption, rheology, and baking quality of durum wheat. *Cereal Chem.* **2017**, *94*, 215–222. [CrossRef]
91. Boehm, J.D., Jr.; Ibba, M.I.; Kiszonas, A.M.; Morris, C.F. End-use quality of CIMMYT-derived soft-kernel durum wheat germplasm: I. Grain, milling, and soft wheat quality. *Crop Sci.* **2017**, *57*, 1475–1484. [CrossRef]
92. Kiszonas, A.M.; Ibba, M.I.; Boehm, J.D., Jr.; Morris, C.F. Effects of the functional *Gpc-B1* allele on soft durum wheat grain, milling, flour, dough, and breadmaking quality. *Cereal Chem.* **2021**, *98*, 1250–1258. [CrossRef]

93. Murray, J.C.; Kiszonas, A.M.; Morris, C.F. Influence of soft kernel texture on fresh durum pasta. *J. Food Sci.* **2018**, *83*, 2812–2818. [CrossRef] [PubMed]
94. Gu, B.-J.; Kerr, C.J.; Morris, C.F.; Ganjyal, G.M. Soft durum wheat as a potential ingredient for direct expanded extruded products. *J. Cereal Sci.* **2021**, *98*, 103184. [CrossRef]

MDPI
St. Alban-Anlage 66
4052 Basel
Switzerland
Tel. +41 61 683 77 34
Fax +41 61 302 89 18
www.mdpi.com

Foods Editorial Office
E-mail: foods@mdpi.com
www.mdpi.com/journal/foods

www.ingramcontent.com/pod-product-compliance
Lightning Source LLC
Chambersburg PA
CBHW061827210926
51850CB00006B/826